Geology of the Fortrose and eastern Inverness district

An appreciation of the geology of any area undergoing rapid development is essential. Not only is it important to know the general disposition of potential resources such as sand, gravel, building stone, limestone and water, but some knowledge of the history of change must be considered by planners and civil engineers in dealing with future developments.

In the case of eastern Inverness and its environs, the recent 1:10 000-scale geological resurvey has substantially updated the previous surveys of the area. This memoir provides an interpretive explanation of the published 1:50 000 Geological Sheet 84W (Fortrose) covering this district and largely satisfies the need for an up-to-date account of the local geology. It will act as a guide to the archive of more detailed data held by the British Geological Survey.

The district lies astride Inverness Firth at the head of the Inner Moray Firth. It incorporates the south-eastern part of the Black Isle from just north of Rosemarkie to Kilmuir and an area east of Inverness as far as Cawdor in the east and Moy in the south.

The upland areas are mainly underlain by Precambrian–Lower Palaeozoic metamorphic and granitic rocks that form an irregular crystalline basement to Devonian sedimentary rocks deposited at the south-western margin of the freshwater Orcadian Old Red Sandstone Basin. The district is geologically important for its position at the north-eastern onshore end of the Great Glen Fault Zone, one of the most conspicuous geological features in Britain.

The whole region was variously glaciated during Pleistocene ice ages that ended about 10 000 years ago. The present topography and surface deposits are a legacy of such times and are of sufficient variety to establish this district as one of the most important in documenting British Pleistocene events, especially those relating to changes in sea level.

Cover photograph

View looking south-westwards over the Devensian and Flandrian deposits of Chanonry Peninsula to Inverness and the Great Glen. Rocks of the Precambrian Rosemarkie Metamorphic Complex underlie the immediate foreground and, along the Black Isle cliffline, Quaternary deposits patchily cover both conglomerates and sandstones of the Devonian Old Red Sandstone. The coastline of the Black Isle parallels the NE-trending Great Glen Fault Zone that lies just offshore. Across the Moray and Inverness firths, the low ground is underlain by a different Old Red Sandstone sequence. These rocks are covered by extensive Quaternary deposits responsible for the varied topography of eastern Inverness, notable features being the headlands of Fort George and the more distant Alturlie (D4886A).

The south-western side of Ardersier Peninsula. View looking north and west across Fort George to Rosemarkie on the Black Isle.

The slope in the foreground is part of the Main Flandrian Cliffline above the shelving coastal flat of Flandrian marine sand and gravel. To the right of Rosemarkie, the Rosemarkie Metamorphic Complex is exposed in cliff sections; to the left, the Devonian Old Red Sandstone sedimentary rocks are largely masked by Quaternary deposits (D4302).

BRITISH GEOLOGICAL SURVEY

T P FLETCHER
C A AUTON
A J HIGHTON
J W MERRITT
S ROBERTSON and
K E ROLLIN

CONTRIBUTORS

Offshore geology
C C Graham

Hydrogeology
N S Robins

Geology of Fortrose and eastern Inverness district

Memoir for 1:50 000 Geological Sheet 84W (Scotland)

LONDON: HMSO 1996

ISBN 0 11 884511 X

Bibliographical reference

FLETCHER, T P, AUTON, C A, HIGHTON, A J, MERRITT, J W, ROBERTSON, S, and ROLLIN, K E. 1996. Geology of the Fortrose and eastern Inverness district. *Memoir of the British Geological Survey*, Sheet 84W (Scotland).

Authors

T P Fletcher, MSc, PhD, FGS
C A Auton, BSc, CGeol, FGS
A J Highton, BSc, PhD, CGeol, FGS
J W Merritt, BSc
S Robertson, BSc, PhD
British Geological Survey, Edinburgh

K E Rollin, BSc
British Geological Survey, Keyworth

Contributors

C C Graham, BSc
British Geological Survey, Edinburgh

N S Robins, BSc
British Geological Survey, Wallingford

Printed in the UK for HMSO

Dd 301321 2/96 C8

Other publications of the Survey dealing with this and adjacent districts

BOOKS

Memoirs
The geology of the Lower Findhorn and Lower Strath Nairn including part of the Black Isle near Fortrose, Sheet 84 and part of 94, 1923.
The geology of mid-Strathspey and Strathdearn including the country between Kingussie and Grantown, Sheet 74, 1915.
The geology of the country around Beauly and Inverness including a part of the Black Isle, Sheet 83, 1914.
The geology of Ben Wyvis, Carn Chuinneag, Inchbae and the surrounding country, Sheet 93, 1912.
The geology of the Elgin district, Sheet 95, 1968.

British regional geology
The Grampian Highlands, 4th edition, 1995
The Northern Highlands, 4th edition, 1989

United Kingdom offshore regional report
The geology of the Moray Firth, 1990

Regional geochemical atlas
Great Glen, 1987

Reports
The Old Red Sandstone east of Loch Ness, Inverness-shire. IGS Report No. 82/13, 1983
Sand and gravel resources of the Highland Region. IGS Report No. 78/8, 1978.

MAPS

1:2 500 000
Sub-Pleistocene geology of the British Isles and the adjacent Continental Shelf, 1979

1:1 000 000
Sea-bed sediments around the United Kingdom (North Sheet), 1987
Geology of the United Kingdom, Ireland and the adjacent Continental Shelf (North Sheet), 1991

1:625 000
Geological map of the United Kingdom North, Solid 3rd edition, 1979
Quaternary map of the United Kingdom North, 1st edition, 1977
Aeromagnetic map of Great Britain, Sheet 1, 1st edition, 1972

1:250 000
Moray-Buchan (57°N-04°W), Solid geology, 1977
Moray-Buchan (57°N-04°W), Aeromagnetic Anomaly Map, 1978
Moray-Buchan (57°N-04°W), Bouguer Gravity Anomaly Map 1977
Moray-Buchan (57°N-04°W), Sea-bed Sediments and Quaternary Geology, 1984
Great Glen (57°N-06°W), Solid geology, 1989
Great Glen (57°N-06°W), Aeromagnetic Anomaly Map, 1978
Great Glen (57°N-06°W), Bouguer Gravity Anomaly Map, 1976

1:63 360
Grantown-on-Spey (Sheet 74) Solid and Drift, 1917
Inverness (Sheet 83) Solid and Drift, 1918, reprinted 1964
Nairn (Sheet 84) Solid and Drift, 1923, reprinted 1954 and 1958
Alness (Sheet 93) Solid and Drift, 1912, reprinted 1962
Cromarty (Sheet 94) Solid and Drift, 1889, reprinted 1973

1:50 000
Fortrose (Sheet 84W) Drift, 1978
Nairn (Sheet 84E) Drift, 1978
Fortrose (Sheet 84W) Solid and Drift, 1994
Sheet 84W (Fortrose)

CONTENTS

FIGURES

LIST OF SIX-INCH AND 1:10 000 GEOLOGICAL MAPS

The following is a list of six-inch and 1:10 000 geological maps included, wholly or in part, in 1:50 000 Sheet 84W (Fortrose) with the initials of the surveyors and dates of survey. The surveyors on the original six-inch County sheets were L W Hinxman, J Horne and B N Peach, those on the 1:10 000 National Grid sheets were C A Auton, D Ball, T P Fletcher, A Grout, A L Harris, A J Highton, J W Merritt, S Robertson and C W Thomas. The 1:10 000 surveys were supervised by Drs D J Fettes and D I J Mallick as Regional Geologists.

Manuscript copies of the County maps can be examined in the British Geological Survey library in Edinburgh. Manuscript copies of all complete National Grid maps have been deposited for public reference in the libraries of the British Geological Survey in Keyworth and Edinburgh and can be purchased from the British Geological Survey, Murchison House, West Mains Road, Edinburgh EH9 3LA. Incomplete maps are marked with an asterisk.

Six-inch County maps

INVERNESS-SHIRE

I	Ardersier	JH	1877
IV	Alturlie	JH	1878–84
V	Croy	JH	1878–79
XII	Inverness	JH	1875–1900
XIII	Easter Town	JH	1879
XX	Faillie	JH	1885–1900
XXI	Loch Moy	LWH, JH, BNP	1885–1900
XXII	Gleann Seileach	LWH	1900
XXXI	Beinn Bhreac	JH	1900
XXXII	Tomatin	LWH, JH	1900–05
XIII	Croft	LWH	1904

NAIRNSHIRE

I	Nairn	JH	1876
IV	Cawdor	JH	1878
VI	Clava	JH	1877
VII	Dalcharn	JH	1878–79
IX	Beinn Bhuide Mhor	JH	1879–89
X	Carn nan Tri-tighearnan	JH, BNP	1879–1904

ROSS-SHIRE AND CROMARTYSHIRE

LXXVII	Millbuie	JH	1885
LXXVIII	Flowerburn	JH	1885
LXXXIX	Munlochy	JH	1886
XC	Avoch	JH	1886
C	Kilmuir	JH	1885
CI	Craigiehow	JH	1885

1:10 000 National Grid maps

* NH 63 NE	Leys	TPF, JWM, SR	1988–89
* NH 63 SE	Farr	JWM, SR	1988–89
* NH 64 NE	Inverness north	TPF, JWM	1987–89
* NH 64 SE	Inverness south	TPF, JWM	1987–89
* NH 65 NE	Avoch	TPF	1989
* NH 65 SE	Craigiehowe	TPF	1989
* NH 66 SE	Millbuie	TPF	1989
NH 73 NW	Daviot	TPF, JWM, SR	1988–89
NH 73 NE	Moy	JWM, SR	1988–89
* NH 73 SW	Carn na h-Easgainn	JWM, SR	1988–89
* NH 73 SE	Loch Moy	JWM, SR	1988–89
NH 74 NW	Smithton	TPF, JWM	1987–89
NH 74 NE	Cantray	TPF, JWM, SR	1987–89
NH 74 SW	Muckovie	TPF, JWM, SR	1987–89
NH 74 SE	Clava	TPF, JWM, CWT, SR	1987–89
NH 75 NW	Rosemarkie	TPF, ALH, AJH	1969–89
NH 75 NE	Fort George	DB, TPF, ALH, AJH, JWM	1969–89
NH 75 SW	Fisherton	DB, TPF, JWM	1986–90
NH 75 SE	Ardersier	DB, TPF, JWM	1986–89
* NH 76 SW	Whitebog	TPF	1989
* NH 76 SE	Hillockhead	ALH, AJH	1969–89
NH 83 NW	Ballachrochin	CAA, JWM, SR	1987–90
* NH 83 NE	Drynachan	CAA, SR	1987–89
* NH 83 SW	Balvraid	CAA, JWM, SR	1988–89
* NH 83 SE	Carn an t-Sean Liathanach	CAA, JWM, SR	1988–89
NH 84 NW	Dalcharn	CAA, AG, JWM, SR	1987–90
* NH 84 NE	Glengeoullie	CAA, AG, SR	1987–89
NH 84 SW	Riereach Burn	CAA, SR	1987–89
* NH 84 SE	Carnoch Burn	CAA, SR	1986–88
NH 85 NW	Wester Delnies	CAA, AG	1987–90
* NH 85 NE	Easter Delnies	CAA	1987–90
NH 85 SW	Cawdor	CAA, TPF, AG	1987–90
* NH 85 SE	Meikle Kildrummie	CAA, AG	1987–90

PREFACE

Pressures on present-day environments are increasing. In areas of population and industrial growth the need for a thorough understanding of the local natural resources and overall ground conditions are paramount. One aspect of these needs is a comprehensive knowledge of the nature of geological and geomorphological change and to this end the British Geological Survey is directing research towards regions of major population, as well as surveying neighbouring wilderness areas where little or no geological information is available.

This memoir chronicles the geology of the eastern environs of Inverness, the principal centre of population in the Northern Highlands of Scotland, covered by the western half of Scotland 1:50 000 Sheet 84. Following a complete resurvey of the surficial Quaternary deposits and basement rocks and selective studies of the largely masked Devonian sequence, the memoir supersedes information on the district previously published in the Lower Findhorn and Lower Strath Nairn memoir of 1923.

This scenically attractive district borders the south-western end of the Moray Firth, an area most notable for the former ancient seat of learning at Fortrose on the Black Isle, the Battle of Culloden and the variety of fortifications dating back over 2000 years.

The most prominent geomorphological feature of the area is the hollow of the Great Glen and its extension into the Moray Firth. This feature reflects the line of a major crustal fracture, the Great Glen Fault Zone, which here parallels closely the coastline of the Black Isle. Displacements, both lateral and vertical, in this zone have a long history and account for many of the differences in the bedrock successions on either side, as well as being responsible for some modern earthquakes.

The oldest rocks mainly comprise Neoproterozoic-age sediments deformed and metamorphosed during the Caledonian Orogeny. They are represented here by the Rosemarkie Metamorphic Complex and the Central Highland Migmatite Complex lying respectively to the north-west and south-east of the Great Glen Fault Zone; both successions are cut by igneous intrusions of Precambrian to late-Silurian age and together form a crystalline basement to the Devonian Old Red Sandstone sedimentary rocks.

The Devonian sequences on both sides of the fault zone were deposited along the south-western margin of the freshwater Orcadian Basin. They are assemblages of red-bed fluvial conglomerates and sandstones interbedded with lacustrine silty sandstones and thin grey calcareous mudstones, some of which are bituminous and contain fish-bearing concretionary limestone lenses and nodules.

In the past, it was the local Devonian rocks and their fossil fish that drew most attention and many very distinguished geologists have devoted special attention to them; these include Hugh Miller, the doyen of Scottish geology. Now, however, climatic and sea-level changes are the focus of modern environmental studies and in these respects the preserved Quaternary sequence in this district provides some of the most important evidence available in Britain. These deposits document a fluctuating glacial and interglacial history of sedimentation and sea-level change dating back over 430 000 years, and aid our understanding of the great variety of local topographical and subsurface features.

The memoir also describes the structural history of the district, some geophysical features and the economic geology.

Peter J Cook, DSc, CGeol, FGS
Director

British Geological Survey
Kingsley Dunham Centre
Keyworth
Nottingham
NG12 5GG

10 August 1995

ACKNOWLEDGEMENTS

NOTES

In this memoir, the chapters on metamorphic rocks and igneous rocks were written by Drs A J Highton and S Robertson, the chapter on geophysics by Mr K Rollin and the Quaternary chapter by Messrs J W Merritt and C A Auton. The remainder was largely written by Dr T P Fletcher, who was also responsible for the overall compilation. Notes on the offshore geology by Mr C C Graham and on the hydrogeology by Mr N Robins have been incorporated appropriately.

This memoir has been prepared by the authors mainly on the basis of the resurvey, but pertinent internal reports and manuscript maps prepared by other BGS colleagues, past and present, are here acknowledged. Notable among these workers are Professors A L Harris and J D Peacock, Drs M Armstrong, J S Coats, W Mykura and P A Rathbone and Messrs D Ball, A Grout and C W Thomas. Devonian fossil records have been verified by Mr P J Brand, who also collected from this district; Quaternary fossils have largely been identified by Mr D K Graham and Drs R Harland, B J Taylor and I P Wilkinson. The memoir was edited by Drs D I J Mallick and Mr J I Chisholm.

Collaborative contacts with academic researchers on all aspects of the local geology are also gratefully acknowledged, especially with Professor D Q Bowen and Drs G R Coope, M H Field, C R Firth, A Haggart, H Heijnis, J E A Marshall, H McKerrell, J E Robinson, D A Rogers, J R Underhill and M J C Walker. Additional information has been provided by Mr A Morey and Miss A Andrew of the Highland Regional Council. The Nature Conservancy Council is particularly thanked for organising and funding excavations at Ardersier and Clava.

Without the cooperation of farmers, landowners and quarry operators, fieldwork would not be possible. We acknowledge, therefore, the access and help given to us during the resurvey, and especially on the larger estates by the Earl Cawdor, The Mackintosh of Moy Estate, Mrs M E Macrae of Clava Estate and the Forestry Commission (Scotland).

Throughout the memoir, the word 'district' refers to the area covered by 1:50 000 Geological Sheet 84W (Fortrose). National Grid references are given in square brackets; all lie within 100 km square NH.

Numbers preceded by the letters C, D, MNS, PMS or TS refer to the Scottish BGS collection of photographs; those preceded by the letter S refer to the Scottish BGS sliced rock/thin section collection and by the letters T or GSE to the Scottish BGS fossil collection.

Non-BGS numbers refer to specimens dealt with by the following laboratories:
AA — University of Arizona, National Science Foundation, Accelerator Mass Spectrometry Unit,
GU — Glasgow University Radiocarbon Laboratory, East Kilbride,
OxA — University of Oxford Radiocarbon Accelerator Unit,
SRR and AMS — NERC Radiocarbon Laboratory, East Kilbride.

Lists of Quaternary fossils recorded from the district are given in Appendix 4.

Enquiries concerning geological data for the district should be addressed to the Manager, National Geosciences Record Centre, Murchison House, West Mains Road, Edinburgh, EH9 3LA.

ONE

Introduction

This memoir is an explanatory account of the geology of the district covered by the Fortrose 1:50 000 Geological Sheet 84W (Figure 1), an area east of Inverness and a small part of the Black Isle around Fortrose.

The original survey was carried out by Hinxman, Horne and Peach on the six-inch scale between 1875 and 1904, and was published as parts of separate Solid and Drift editions of One-inch Geological Sheet 84 (Nairn) in 1923. An explanatory memoir for Sheet 84 written largely by Horne (1923) accompanied these maps, which were later reprinted in 1954 and 1958. In 1978, a composite Solid and Drift edition of the Fortrose Sheet was published at the 1:50 000 scale by photographic enlargement of the 1923 Drift edition.

The resurvey of the Quaternary geology on the southern side of the Moray Firth was made on the 1:10 000 scale largely by Messrs Merritt, Auton and Ball between 1986 and 1990. The pre-Devonian rocks south of the firths were resurveyed by Dr Robertson between 1986 and 1990, and those in the Rosemarkie area were re-examined by Dr Highton in 1989 in the light of a 1969 survey by Dr A L Harris and subsequent published work. Because of the extensive drift cover on the southern side of the firths, revision of the Devonian successions there was carried out by selected traverses over known exposures by Dr Fletcher between 1988 and 1990. However, he completely remapped the Devonian rocks and Quaternary deposits on the Black Isle in 1989.

The district covers an area of about 568 km^2, of which the Inverness and Moray firths occupy 91 km^2 and the Black Isle 45 km^2.

This region has a documented history of settlement dating back to 60BC. Inverness grew to its present importance in the 19th century, as the gateway to the Northern Highlands and the principal centre of trade and services. Geologically related activites have supplemented the local economy and limestone, sandstone, metamorphic and igneous rocks have been variously quarried in addition to subordinate mining and extraction of manganese, baryte, sand and gravel. Notable among the newer quarries was that at Leanach, which supplied the attractive red sandstone in the Clava Railway Viaduct, and also those at Daviot and Littlemill.

It should be noted that this can be regarded as a seismically active region. Major earthquakes due to movements on local faults were experienced around Inverness in 1816, 1890 and 1901. Each caused considerable damage, especially the 1901 event which reached magnitude 5 on the Richter Scale and was followed by aftershocks between 18 September and 21 November felt well beyond this district (Browitt et al., 1976; Musson et al., 1987).

Topography

A four-fold physiographical subdivision is evident (Figure 2), and indicates the controlling influence of the Caledonoid (north-easterly) trend of the bedrock geology. This trend is no more marked than by the Inverness Firth, the cliff-lined northern side of which parallels the Great Glen Fault Zone lying just offshore (Figure 1).

In the south and east, the main upland areas largely correspond with the outcrops of the Central Highland

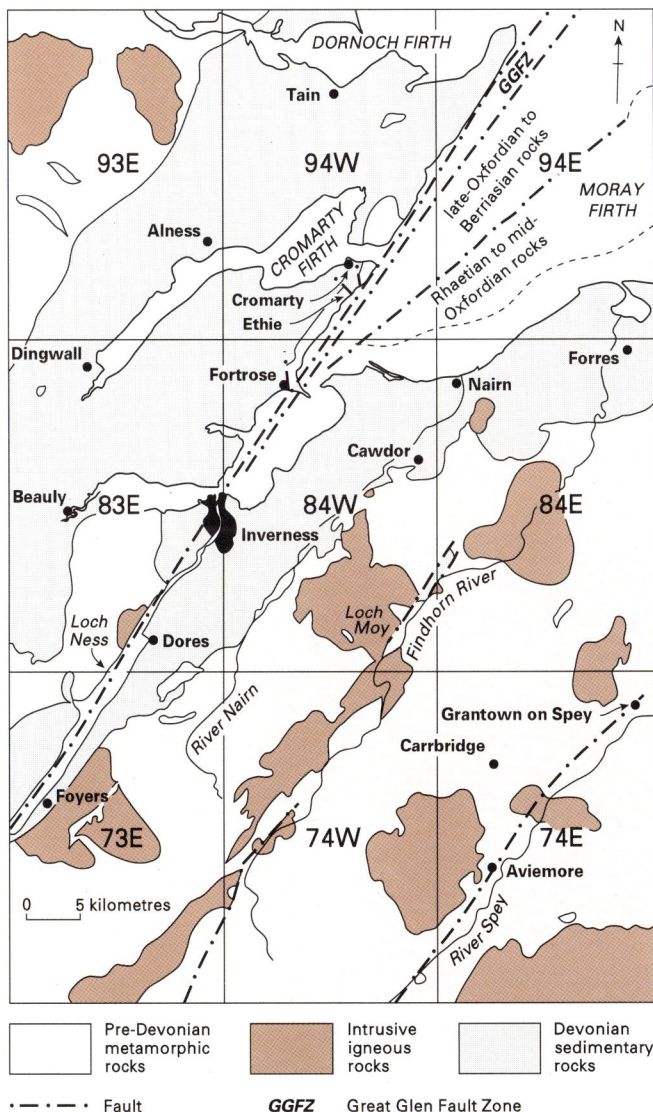

Figure 1 Map showing the location and regional geological setting of the district.

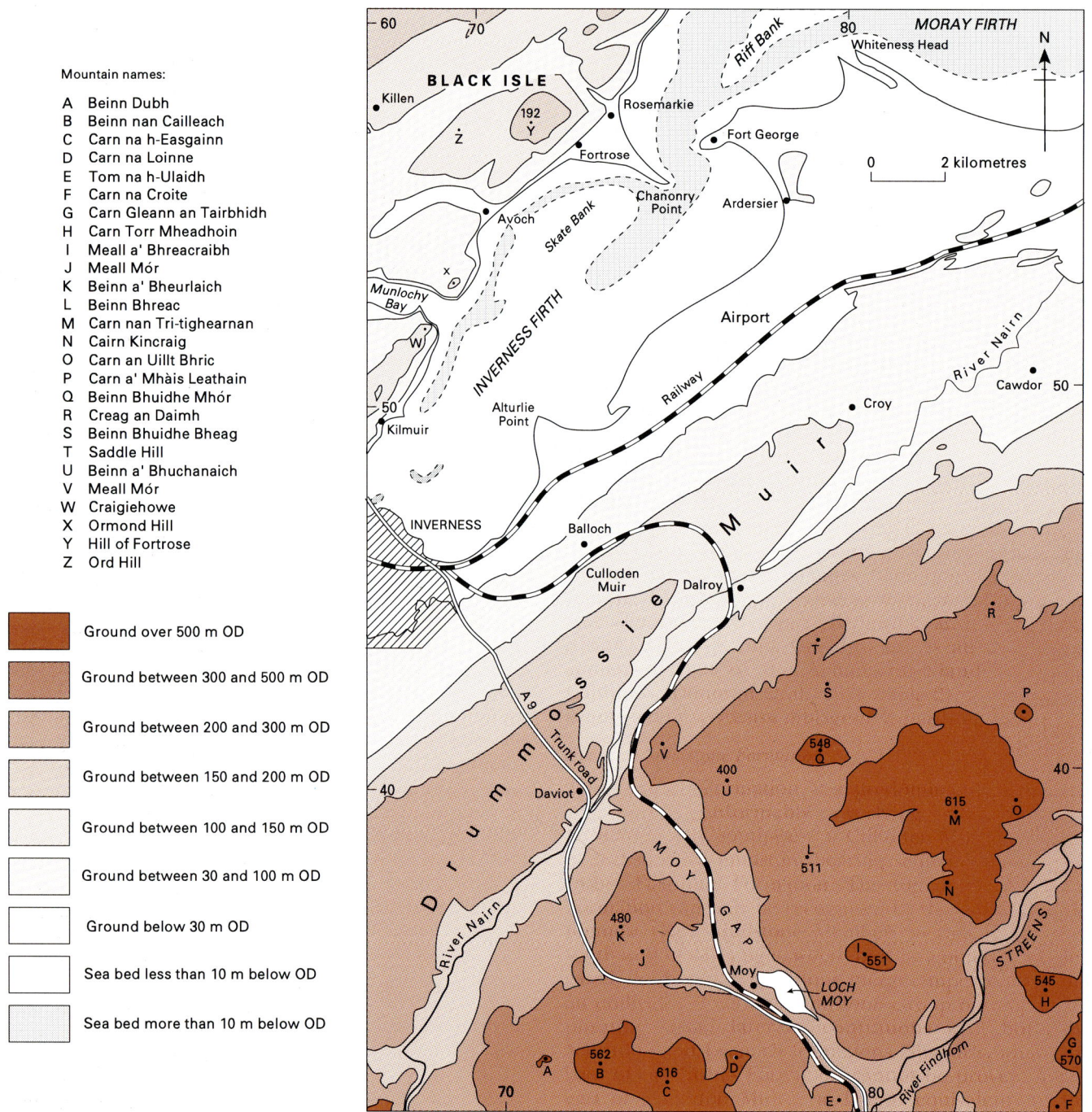

Mountain names:

A Beinn Dubh
B Beinn nan Cailleach
C Carn na h-Easgainn
D Carn na Loinne
E Tom na h-Ulaidh
F Carn na Croite
G Carn Gleann an Tairbhidh
H Carn Torr Mheadhoin
I Meall a' Bhreacraibh
J Meall Mór
K Beinn a' Bheurlaich
L Beinn Bhreac
M Carn nan Tri-tighearnan
N Cairn Kincraig
O Carn an Uillt Bhric
P Carn a' Mhàis Leathain
Q Beinn Bhuidhe Mhór
R Creag an Daimh
S Beinn Bhuidhe Bheag
T Saddle Hill
U Beinn a' Bhuchanaich
V Meall Mór
W Craigiehowe
X Ormond Hill
Y Hill of Fortrose
Z Ord Hill

Ground over 500 m OD

Ground between 300 and 500 m OD

Ground between 200 and 300 m OD

Ground between 150 and 200 m OD

Ground between 100 and 150 m OD

Ground between 30 and 100 m OD

Ground below 30 m OD

Sea bed less than 10 m below OD

Sea bed more than 10 m below OD

Figure 2 Map showing the principal topographical features of the district.

Migmatite Complex and intruded igneous rocks. These lie at the northern end of a high north-easterly sloping plateau covered by peat and till, dissected in this region by numerous immature valleys. In places, these cut into solid rock to form spectacular gorges (Plate 6) containing streams draining into the two principal rivers, the Nairn and the Findhorn. The upland area generally lies between 300 and 600 m above OD, with the highest points of Carn nan Tri-tighearnan (615 m above OD)

[8230 3905] and Carn na h-Easgainn (616 m above OD) [7435 3215] lying astride the glacially enhanced misfit Moy Gap through which the Findhorn originally flowed.

To the north-west, a second subdivision is virtually bounded by Strath Nairn, which is aligned roughly parallel with the contact between the migmatite complex and the Devonian Old Red Sandstone. This area is underlain by north-westerly dipping Old Red Sandstone and similarly contains high ground with an element of north-easterly slope, readily detected in terrain over 100 m above OD. Reference to Figure 2 and the 1:50 000 geological map clearly demonstrates the influence of the Quaternary cover here. Tills and moraines form a prominent scrubby upland belt from Drummossie through Culloden Muir to Croy, inland of a more fertile coastal spread of postglacial deposits, where the main settlements are located. Coastal departures from the Caledonoid trend, developed at Inverness, Alturlie Point and the headlands of Fort George and Whiteness, are ice-marginal in origin.

The Black Isle, in the north-west of the district, forms a third subdivision. Ridges of sandstone, conglomerate and gneiss parallelling the offshore Great Glen Fault Zone largely reflect north-westerly dipping Old Red Sandstone strata on the south-eastern limb of the Black Isle Syncline. As in the other subdivisions, this area has been heavily glaciated and the rocky ridges alternate with fertile valleys containing Quaternary clays, sands and gravels. A conspicuous feature of this region is the extensive terrace-sculpturing of the terrain, caused by ice movement.

The fourth topographical subdivision is the offshore area. Here, in the Moray and Inverness firths, the most prominent features are linear deeps, which form a system of channels extending north-eastwards (Figure 2). Similar deeps also occur at Kessock, immediately west of the district, and in the Cromarty Firth to the north (Hydrographic Office, 1981; 1989). They are either remnants of more-extensive, glacially deepened basins and subglacial drainage channels infilled by glacial postglacial deposits, or they were cut by subaerial drainage during the period of low relative sea level early in the Flandrian (Firth and Haggart, 1989). It has been inferred from seismic and bathymetrical data that such a drainage system once extended from the Beauly and Inverness firths across the Moray Firth as far east as Lossiemouth (Chesher and Lawson, 1983). The modern deeps are maintained by the scouring action of bottom tidal currents that are of sufficient strength to prevent any long-term deposition within them.

The largest deep within the district forms the narrow strait between Chanonry Point and Fort George (cover photograph). It is approximately 7 km long, 1 km wide and, in places, descends to more than 45 m below mean sea level, the maximum water depth within the district. Peak surface tidal currents in the strait attain speeds of up to one metre per second. At its northern end, in the Moray Firth, the deep joins two channels that run each side of the Riff Bank, a prominent north-east-trending bathymetrical high which shoals locally during low tides. Although the nature and composition of the bank are unknown, it is most likely to be either a glacial moraine feature or a late- to postglacial sandbank. The channel on the northern side is about 10 to 15 m deep and extends northwards beyond the district boundary, where it gradually diminishes as a distinct sea-bed feature. The southern channel turns eastwards to run parallel to the coast off Ardersier and is deepest off Whiteness Head, where water depths exceed 25 m.

Within the Inverness Firth, a series of deeps forms a narrow channel off the northern coast in water depths greater than 7 m. This extends from Kilmuir in the south-west to Chanonry Point in the north-east, where the channel opens into the main deep at the seaward entrance to the firth. The channel is approximately 300 m wide and contains several enclosed deeps that descend to depths of between 15 and 20 m, and to below 25 m off Fortrose. Elsewhere in the firth, the sea bed is relatively shallow and flat, except for several low, broad, banks with reliefs of about 2 to 3 m. The most prominent is Skate Bank off Fortrose, which is marked by shoals during extreme low tides.

Outline of geological history

Two distinct rock suites are recognisable in this district, a crystalline basement comprising mostly metasedimentary rocks, and an overlying sequence of non-metamorphosed sedimentary rocks.

Due to fracturing and movements along the Great Glen Fault Zone, which passes through the northern part of the district, two separate basement–cover rock successions are juxtaposed. The basement on the north-western side is differentiated as the Rosemarkie Metamorphic Complex and that on the south-eastern side as the Central Highland Migmatite Complex; the overlying sedimentary rocks on each side form different successions of more or less contemporaneous Devonian facies. Despite similarities of lithologies and tectonothermal history, correlation of the metamorphic complexes across the fault zone remains tentative.

Previous interpretations of the geological history of the Rosemarkie Metamorphic Complex suggested the presence of two tectonostratigraphical successions. An older assemblage of migmatitic gneisses, with distinctive bands of hornblendic lithologies, was equated with Lewisian rocks occurring on the Caledonian foreland farther west. It was interpreted as being tectonically intercalated with a younger assemblage of Moine metasedimentary rocks of Neoproterozoic age. In this account, however, all the rocks of the complex are tentatively attributed to the Moine Supergroup, and no Lewisian rocks are recognised.

The distinctive amphibolites and hornblende schists, rather than being Lewisian in origin, bear strong similarity to the suite of tholeiitic intrusions recognised in the Moine outcrop farther west, which was emplaced towards the peak of the early (late Proterozoic) metamorphic event of the Caledonian Orogeny. At Rosemarkie, this event induced localised migmatisation in the host metasedimentary rocks, and its closing phases were marked by the emplacement of a locally developed intrusive vein complex, comprising mostly leucogranites with minor pegmatite. These intrusions were deformed and recrystallised, and their host rocks subjected to further tectonothermal reworking, during an Ordovician phase of the orogeny. The peak of metamorphism, coinciding with the third, upright, phase of folding, achieved similar pressure and temperature conditions to the Proterozoic event. The end of orogenic activity in this part of the Scottish Highlands was marked by a phase of uplift and extensive faulting, focused largely on the Great Glen fault system. Within the Rosemarkie

Metamorphic Complex, fluid movement in these crustal fractures induced localised fenitisation, carbonate veining and brecciation.

South-east of the Great Glen Fault Zone, metamorphic basement rocks comprising migmatitic metasedimentary gneisses with minor intercalated non-migmatitic schists are here referred to as the Central Highland Migmatite Complex. Arenaceous and argillaceous protoliths were deposited, deformed and subjected to amphibolite-facies metamorphism with associated migmatisation during the late Proterozoic. Basic igneous rocks, recognised in parts of the district, were intruded at this time. Renewed deformation and migmatisation is thought to have occurred during the Ordovician. This was succeeded by the repeated emplacement of granitic rocks into the area to the north of Moy during the Ordovician and the late Silurian to early Devonian. The Moy Granite, which is the oldest major intrusion in the district, is composed of granodiorite and granite with associated marginal vein complexes. The Quilichan Vein Complex, south-east of the Findhorn Fault, has features in common with the Moy Granite. The Auchnahillin Monzodiorite has a very small outcrop, but an extensive area of hornfelsing and associated aeromagnetic anomalies suggests that it forms a large body at shallow depth, below and to the north-east of the Moy Granite. This intrusion is probably similar to the Findhorn Granite Complex, south-east of the Findhorn Fault, which encompasses monzodioritic and granodioritic lithologies as well as containing appinitic bodies; appinites are also apparently related to the Auchnahillin Monzodiorite. The youngest major intrusion in the district was emplaced into the central parts of the concealed monzodiorite intrusion; the Saddle Hill and Carn Odhar granites may represent the surface expression of cupolas related to this body. Minor intrusions include lamprophyres, microdiorites and quartz-feldspar porphyries.

The intensity of Ordovician deformation gave a strong north-easterly structural grain to the country, and fractures with this trend were utilised intermittently during subsequent periods of stress. During the later stages of the Caledonian Orogeny, sinistral stress along the Great Glen Fault Zone appears to have resulted in shear movements that created considerable fracturing. A permanent line of weakness was established and though it had little or no effect on Devonian sedimentation, dextral and compressive movements in late Devonian to Carboniferous times, and conspicuous extensional subsidence along its line in the Jurassic, Cretaceous and Neogene periods, illustrate its long-lasting effect.

The waning stages of the Caledonian Orogeny saw a welding of the North American foreland with northern Europe to form part of a large continental mass called Laurussia. By the close of Silurian time, differential uplift and erosion had exposed the roots of the Caledonian mountain chain, which in this district are represented by the basement suite. As erosion continued, concurrent crustal extension produced fault-controlled graben and half-graben basins that became the repositories for accumulating clastic detritus. Widespread red-bed continental sedimentation under tropical to subtropical conditions developed in Laurussia and gave rise to the Old Red Sandstone sediments that characterised Devonian deposition there.

Initially, tensional stress created numerous minor basins and a highly irregular topography. In northern Britain, a landscape of small hillocks remained as the vestige of the original mountains, and erosion of such uplands produced debris that was carried by vast fluvial systems to local fault-bounded depressions. The earliest Old Red Sandstone sediments in northern Scotland appear to have accumulated in the area south-west of this district, where tensional fracturing was also accompanied by volcanic activity. A variety of discrete fluvial and lacustrine regimes was established, which, in early Devonian times, were parochial in aspect. Conglomeratic spreadings began the process of hillock burial that continued into mid Devonian times. Thus, the early Devonian succession is replete with irregular unconformities, many of which were due to differential movements in scattered half-graben. In this district, only the Den Siltstone Formation on the Black Isle, largely formed in a marginal lacustrine setting, records events from this period.

By late Emsian (end of Early Devonian) time, north-eastern Scotland was dominated by a single large depression, the Orcadian Basin. This was largely occupied by a freshwater lake whose margins fluctuated in response to climatic changes and differing rates of subsidence. The Fortrose district lay at the south-western margin of the basin, where fluvial systems intermittently merged with lacustrine phases. In the southern part of the basin, the Early to Middle Devonian interval saw widespread coarse clastic sedimentation, with conglomerates overlapping earlier deposits, both discordantly and concordantly. In this district, the conglomerates form most of the basal Old Red Sandstone succession and irregularly covered the hillocky basement. The Kilmuir and Daviot formations represent parts of this basal spread, which accumulated as a piedmont alluvial fan system along the southern lake margin, thinning and fining northwards into the Orcadian Basin.

Periodic expansions of the warm-water lake allowed more-quiescent intervals of finer and limy sedimentation, especially in embayments controlled by the topography of the local protruding basement. Gradually, finer-grained fluvial regimes were established at the lake margins and intermittent exposure of bars and shoreline deposits allowed aeolian processes to rearrange the sands.

Fluctuating conditions throughout the Eifelian Stage of early Middle Devonian time were characterised by numerous cobbly, pebbly and sandy fluvial spreadings northwards into the lake, where changes in its fauna and flora were slowly evolving. Hence, the finer-grained succession above the basal conglomerates comprises a mixture of conglomerates and coarse to fine-grained sandstones exhibiting the full spectrum of fluvial and aeolian sedimentary structures, interfingering with lacustrine deposits—flaggy, calcareous siltstones, mudstones, and sporadic fossiliferous limestones and concretions.

North of the Great Glen Fault Zone, only one formation, the Raddery Sandstone, is recognised above the basal conglomerate. The contemporaneous sequence south of the fault zone contains more contrast, and four formations are recognised: Nairnside Sandstone, Leanach Sandstone, Inshes Flagstone and Hillhead Sandstone (Table 1, inside front cover).

Biological changes during the accumulation of the Orcadian Lake sediments are reflected by the seven fish assemblages proposed by Donovan, Foster and Westoll (1974). These serve as the principal means of correlating beds in the Orcadian Basin, but correlations generally are based upon relationships to the assemblage found in the Achanarras Limestone of Caithness. This is especially the case when dealing with palynological distributions, where only pre-Achanarras, Achanarras and post-Achanarras subdivisions are made (Richardson, 1965).

In this district, the Achanarras faunal elements are distributed through a major part of the Old Red Sandstone succession, with pre-Achanarras elements possibly limited to the pre-Clava Mudstone and pre-Killen Mudstone successions. South of the Black Isle, post-Achanarras assemblages occur in the post-Inshes sequence.

Towards the close of Eifelian sandstone deposition on the southern side of the firths, localised differential earth movements were accompanied by some erosion. Thus, in the Ardersier–Cawdor area, the Hillhead Sandstone with its post-Achanarras Eifelian fish was unconformably overlain by a mixed fluvial/lacustrine sandy sequence (the Nairn Sandstone Formation) containing a Givetian fish assemblage. Although there is no longer a record of younger Devonian strata in this district, sequences farther east indicate that Old Red Sandstone sedimentation continued, possibly with minor breaks, until early Carboniferous time (Peacock et al., 1968).

In other parts of the Moray Firth, Old Red Sandstone sedimentation was succeeded by marine and terrestrial Carboniferous deposition and by later Mesozoic sediments (Figure 1). However, no trace of rocks of Carboniferous to Neogene age has been detected in this district and events during this interval are represented by fault movements affecting the Devonian sequence and the evidence of vast erosion. Compression associated with tectonic stresses in late Carboniferous to early Permian time resulted in some reverse faulting and minor folding, but was largely accommodated by dextral movement along the Great Glen Fault Zone, sufficient to offset the Old Red Sandstone outcrop by as much as 29 km (Rogers, 1987). This movement accounts for the contrasting nature of the successions on each side of the Moray and Inverness firths.

Evidence amassed from hydrocarbon researches in regions to the north-east indicates that the Great Glen Fault Zone remained dormant throughout the late Triassic to Jurassic interval, and was briefly active as an extensional collapse lineament during a post-Early Cretaceous period of regional uplift. Major normal faulting followed in late Oligocene and early Miocene times. Splay faults disrupting strata on the Black Isle undoubtedly relate to movements along the Great Glen Fault Zone, but timings of such movements are uncertain. Reverse movements and minor lateral displacements are evident on some faults (Underhill and Brodie, 1993) and

probably account for areas of Old Red Sandstone (like the immediate hinterland of Fortrose and Avoch) that are folded and contain steeply dipping strata. Although not proven, faulted Jurassic sediments may occur within the firths in this district, continuous with those known to the north at Ethie (Figure 1). Thus, from the late Jurassic onwards until a few thousand years ago, a time span of over 100 million years encompassing the Cretaceous, Palaeogene and Neogene periods and more than half of the Quaternary Period, there is no local evidence of sedimentary deposition.

The growth of the Antarctic Ice Sheet in the mid Miocene and of the Greenland Ice Sheet by 3 million years ago suggests that global climates had begun to deteriorate long before the beginning of the Quaternary Period, 1.8 million years ago. The early Pleistocene probably witnessed periodic growth of glaciers in the Scottish mountains, but there is no evidence that the Fortrose district was glaciated until the 'Ice Age' intensified some 750 000 years ago. The marine record indicates that since then a rhythmic growth and decay of large ice sheets occurred in the middle latitudes of Europe, and ice sheets must have covered most of the Scottish mainland on several occasions. Cold episodes of ice-sheet growth (glacials) were halted by relatively short, warmer periods (interglacials) when climatic conditions rapidly became broadly similar to those of the present day. The last five glacial/interglacial cycles have each spanned about 100 000 years. They were mostly cold, but did include short warmer intervals with warm summers and cold winters. The colder periods within glacials are known as stadials, and warmer interludes interstadials; both probably became colder as each glacial cycle progressed.

The repeated glaciations considerably modified the pre-Quaternary topography of the district by widening, straightening and deepening pre-existing river valleys, breaching watersheds, and cutting meltwater drainage channels. Glacial erosion has largely removed evidence of Quaternary events prior to the last ice-sheet glaciation, that of the Late Devensian. However, the district contains a fragmentary earlier Pleistocene record. The oldest known deposit is a till that underlies biogenic material containing pollen of interglacial affinity at Dalcharn [815 452], near Cawdor. If the interglacial stage represented here is the Hoxnian, as suspected, the underlying till would be Anglian in age. Deposits in Rosemarkie Glen [7360 5785], on the Black Isle, at the Allt Odhar site [7978 3678], and in river cliffs of Allt Carn a'Ghranndaich [773 428] south of Clava, may also be of pre-Devensian age.

The last glacial episode, the Devensian, began about 120 000 years ago. The oxygen-isotope (^{18}O) record of the deep oceans (Ruddiman and Raymo, 1988; Imbrie et al., 1984) indicates that two relatively mild interstadials occurred in the early Devensian, between 120 000 and 80 000 years ago. One of these is represented by a bed of compressed peat between tills at the Allt Odhar site. The first major cold phase occurred about 70 000 years ago, when the district was probably glaciated, but the existence, extent and product of this 'early' Devensian glaciation are controversial. Deposits of shelly clay at Clava [7658 4411] were probably laid down in the Loch Ness Basin during a subsequent cool interstadial in the mid Devensian, perhaps at about 44 000 years ago. These deposits were rafted glacially to the Clava site during the build-up of the Main Late Devensian Ice Sheet, some 25 000 years or more ago. This ice sheet possibly reached its greatest extent in the Dimlington Stadial about 18 to 20 000 years ago and was responsible for the destructive erosion of older Quaternary deposits.

The Main Late Devensian Ice Sheet overwhelmed the entire district and the landforms and sediments it produced are ubiquitous. The high ground in the south and south-east of the district was first to become free of ice, some 13 500 years or more ago. High-level spreads of glaciofluvial deposits were laid down as meltwaters flowed around, or were ponded against, ice that remained on the lower-lying ground. As the ice sheet shrank, similar deposits, together with ice-marginal moraines and meltwater channels, formed at progressively lower levels. Meltwater became increasingly directed towards the major preglacial valleys of the Nairn and Findhorn, but the middle reaches of both remained choked with ice for some time, causing ponding upstream, particularly in the interconnecting Moy Gap. The final stages of deglaciation on land witnessed the formation of two splendid systems of eskers in the vicinity of Loch Flemington [810 520] and Littlemill [706 372]

Glacio-isostatic depression of the lithosphere caused by the Main Late Devensian Ice Sheet was sufficiently great to ensure that, although global sea levels remained low during deglaciation, local sea levels around the Scottish Ice Sheet were high. A series of late-glacial raised beaches and associated marine deposits consequently occurs up to 35 m or more above OD around the shores of the Inverness and Moray firths. Calving of icebergs into the sea caused glacial retreat to be more rapid in the firths than on the adjacent land, so that glaciomarine and beach deposits that accumulated adjacent to land-based ice now form relatively high ground on the coast, such as the Ardersier and Alturlie peninsulas. For a while, ice emanating from the Beauly Firth and the Great Glen probably coalesced in the Inverness Firth to form a tidewater glacier. Glacitectonically deformed glaciomarine sediments at Ardersier testify that a minor readvance of this glacier occurred, whereas subaqueous fan-delta gravels forming Alturlie Peninsula show that a still-stand occurred during its subsequent retreat.

All traces of the Main Late Devensian Ice Sheet in the district had probably disappeared by about 12 500 years before present (BP), during the relatively mild Windermere Interstadial. There is little known evidence of this warm interlude, nor of the subsequent Loch Lomond Stadial, during which glaciers returned to the mountains of north-western Scotland and tundra conditions generally prevailed. Relative sea level was probably a little lower than OD at this time, but after falling further, it eventually rose to about 8 m above OD at the climax of the Postglacial Transgression in the mid Holocene, when a series of postglacial raised beaches and associated marine deposits formed. These Flandrian raised beaches are commonly backed by the abandoned Main Flandrian Cliffline, which is particularly prominent along the southern shores of the Inverness Firth (Frontispiece). There was considerable erosion and alluviation in the valleys during the early Holocene, as the present drainage systems became established, whereas the late Holocene witnessed the extensive growth of blanket peat on high ground and lowland peat in waterlogged depressions.

TWO

Metamorphic rocks

The crystalline basement rocks in the district are composed predominantly of metasedimentary schists and gneisses with minor occurrences of metamorphosed basic and granitic rocks. Basement crops out in two areas, separated by the Great Glen Fault Zone. That to the north-west of the fault zone is restricted to a small inlier in the Rosemarkie area, referred to here as the Rosemarkie Metamorphic Complex, whereas that to the south-east, in the upland area south-east of Strath Nairn, is termed the Central Highland Migmatite Complex. The following accounts highlight many similarities, in terms of lithology and geological history, between the two areas of basement. Both are overlain unconformably by Devonian sedimentary rocks.

Terminology

In the following pages, the term gneissose describes rocks with a coarse metamorphic foliation and a discrete compositional layering, whereas the term migmatitic is used for rocks containing granitic segregations. Migmatites are gneissose, therefore, but a gneissose rock is not necessarily migmatitic. Psammites, micaceous psammites and semipelites are differentiated on the basis of their mica content. Semipelites broadly contain more than 20 to 30 per cent mica, whereas psammites contain less than 10 to 20 per cent mica.

ROSEMARKIE METAMORPHIC COMPLEX

Introduction

The Rosemarkie Metamorphic Complex is restricted to a small inlier on the north-western side of the Great Glen Fault Zone, adjacent to Devonian sedimentary rocks north and west of Rosemarkie (Figure 6). The metamorphic rocks are reasonably well exposed along the 3 km-shore section, both above and below the high-water mark, and in cliffs near Cairds Cave. Inland exposure is poor and largely restricted to road cuts along the A832 and crops in Rosemarkie Glen.

The inlier lies adjacent to the zone of brittle deformation associated with the Great Glen Fault Zone and is broadly coincident with an elongate positive aeromagnetic anomaly of 300 nT and a minor gravity anomaly centred close to Fortrose. The outcrop shows much evidence of brecciation and subsidiary fault movement, especially near Rosemarkie.

The status of the metamorphic rocks in this complex has not, as yet, been satisfactorily established. The memoir accompanying the original publication of Sheet 84 (Horne, 1923) makes no comment as to their affinities. However, Flett, commenting on petrographical aspects in the memoir, described the gneisses as 'the most remarkable series of rocks' in the district. In comparison with what was known then of the Moine Supergroup, they were regarded as being slightly anomalous.

Although brief reference was made by Garson and Plant (1973) to the gneisses of the inlier, the status of the component lithologies received little attention until Harris (1978) reported on findings noted during the revision of Sheet 94 (Cromarty) immediately to the north. This study was subsequently followed by the work of Rathbone (1980) and Rathbone and Harris (1980), who described the intercalation of Moine-like metasedimentary rocks with hornblende-bearing and acidic gneisses. These studies regarded any hornblende-bearing lithologies as *prima facie* evidence for the presence of Lewisian rocks.

The occurrence of Lewisianoid lithologies within the general Moine outcrop is mostly restricted to anticlinal cores of major regional-scale folds (Lambert and Poole, 1964) and to tectonic slices associated with ductile thrusts (Rathbone and Harris, 1979). The Rosemarkie outcrop, however, lies well to the east of the nearest occurrence of undoubted Lewisian rocks.

Unambiguous evidence for an unconformable relationship or tectonic discontinuity between the disparate units of the inlier is lacking (Rathbone and Harris, 1980) and the presence of Lewisian lithologies within the Rosemarkie Inlier is regarded here as unproven.

Metamorphic rocks

The metamorphic rocks of this complex are well exposed along the coastal section north-eastwards from Rosemarkie. The overall disposition of the component lithologies is problematical, due to the paucity of inland exposures and disturbance by late faulting. Figure 3 shows the distribution of lithologies and structures within the coastal strip, which mainly comprise weakly migmatitic feldspathic (?psammitic) gneisses with two distinctive layers of more typically Moine-like metasedimentary rocks.

Re-assessment of the geology of this inlier now indicates that many of the amphibole-bearing rocks, designated by previous workers as Lewisian lithologies, are similar to the metabasic suite recognised in the Moine to the west (Rock et al., 1985) and in the Cromarty Inlier farther north, where they were intruded during or after the earliest (D1) phase of deformation. Several of the larger hornblende-bearing masses within the complex, similarly appearing as intercalations within the Moine, retain the relict igneous textures typical of metagabbroic rocks from the Northern Highlands (Peacock, 1977) and the Central Highlands (see p.18 and Highton, 1992). The hornblendic gneisses of Rathbone and Harris (1980) are confined entirely to the zones of metabasic masses, and appear to pass outwards into the more feldspathic acid gneiss defined by these

Figure 3 Geological sketch map of the Flowerburn foreshore, north-east of Rosemarkie.

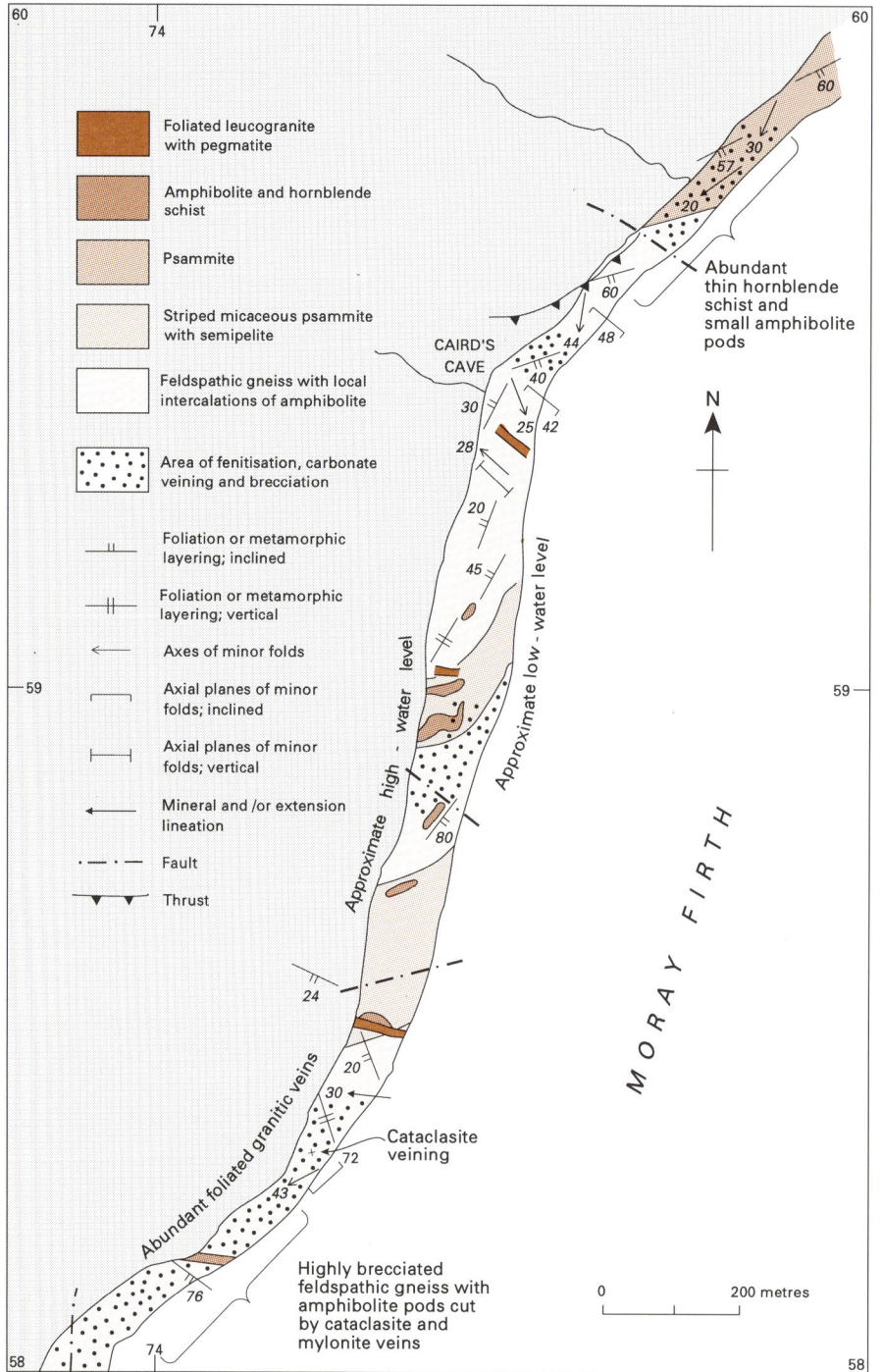

authors. Such lithologies may be of Lewisian origin, or may simply represent an interaction between the metabasic intrusive rocks and their host during regional metamorphism (Rock et al., 1985; Highton, in preparation).

In the absence of any geochemical information, the affinities of the amphibole-bearing lithologies, whether Lewisian or not, cannot be satisfactorily established.

METASEDIMENTARY LITHOLOGIES

Within the complex, two discrete lithological units of recognisable metasedimentary origin crop out: a layered psammite with semipelite, and a slightly gneissose psammite. The former, confined to the coastal part of the outcrop, is fine to medium grained, with little evidence of a macroscopic gneissose segregation. Slightly flaggy, gneissose psammites are exposed in road cuttings on the A832 west of Rosemarkie and also form the northernmost exposures of the coastal outcrop. The relationship between these occurrences could not be established, due to the paucity of exposure inland. The metasedimentary rocks of the complex show many similarities to lithostratigraphical units in the adjacent Moine and in the Central

Highland Migmatite Complex south of the Great Glen Fault Zone.

The psammites along the A832 are mostly medium to coarse grained, deeply weathered, pink to grey, with fine-scale jointing of less than 50 mm spacing. This lithology, in both the road and coastal exposures, is locally massive, but more commonly slightly flaggy. Discontinuous partings, picked out by thin micaceous laminae less than 2 mm thick, are spaced between 50 and 300 mm apart. Thin quartzofeldspathic segregations (lits) impart a weakly gneissose appearance. The rocks are not considered to be migmatitic, because they lack evidence of leucosome/melanosome development. The psammites typically comprise quartz, plagioclase, K-feldspar, biotite and muscovite. Accessory minerals include opaques, zircon, garnet, sphene and apatite, with secondary chlorite, epidote and carbonate; myrmekite is also common. Textures are mostly granoblastic, with the foliation defined by a preferred alignment of the micaceous phases.

Mixed units of psammite, micaceous psammite with thin intercalations of semipelite, rare quartzite and calc-silicate, form two distinct units within the gneiss outcrop on the coastal section (Figure 3). The component lithologies are fine to medium grained with little evidence of segregation. A grey, micaceous, slightly gritty psammite is predominant, forming discrete units up to 80 mm thick. In thin section (as in S 18381), this lithology is distinctly feldspathic, with plagioclase the slightly predominant feldspar, and quartz subordinate. Plagioclase grains are mostly granoblastic in form, of weakly zoned and twinned oligoclase. K-feldspar grains are more irregular, with evidence of subgraining. Myrmekite intergrowths are common. Small clast-like fragments of a poorly twinned K-feldspar and quartz crystals up to 0.3 mm in diameter are present. Biotite is the predominant mica, with only minor muscovite, occurring as irregular flakes interstitial to the quartzofeldspathic grains, and mica-rich laminae defining the compositional layering. Evidence for graded bedding in this lithology is equivocal. A green to cream-grey ellipsoidal calc-silicate pod up to 100 mm across was noted [at 7437 5869] in the micaceous psammite. The interlayered psammites appear compositionally similar, but are relatively mica-poor. Intercalations of schistose semipelite range from 10 to 300 mm thick. They are characterised by an abundance of macroscopic porphyroblasts of a purplish mauve garnet. The garnets, up to 5 mm long, are mainly elongate within, and wrapped by, the principal crenulation cleavage (S2) in these rocks. The schistosity is defined by muscovite/biotite-rich laminae separating microlithons of predominantly strained quartz with minor feldspar. Metamorphic index minerals were not noted in these rocks.

Contacts with the feldspathic acid gneisses range from sharp, with thin, highly strained, flaggy zones [7438 5873], to apparently transitional [7475 5969]. At the former locality, the gneissose lithology shows some evidence of localised high strain, with the quartzofeldspathic segregations streaked out adjacent to the contact. The non-gneissose lithologies, however, appear little disturbed. The latter locality is marked by non-gneissose metasedimentary rocks intercalated over a distance of one metre with thin layers of a K-feldspar-rich (?psammitic) gneiss in which thin amphibolite layers are present.

FELDSPATHIC GNEISSES

Much of the coastal outcrop, and the rare, small inland exposures at Hillock [7383 5862], Wester Balmungie [7381 5972] and Hillockhead [7440 6015] essentially comprise a unit of feldspathic acid gneisses with locally abundant intercalations of hornblende-rich (amphibolite) bands. The gneisses are pink, medium-grained feldspathic rocks, which bear close similarity to gneissose arkosic psammites both within the Northern Highland Moine outcrop and the Central Highland Migmatite Complex. The gneissose lits are dominantly feldspathic, with a perthitic microcline having a slight predominance over a poorly twinned low-andesine plagioclase. Quartz is a minor constituent of the lits, which appear to be mostly monzonitic in composition. In areas of low strain, the lits are texturally granoblastic. Where strains are higher, quartz, and to a lesser extent feldspar, recrystallise to a distinctive ellipsoidal (LS) subgrain shape fabric. The mafic laminae are fine-grained zones in which there is a localised abundance of a dark red-brown biotite, with quartz, minor feldspar and accessory minerals, such as apatite, zircon, sphene and allanite. These are much finer grained than the quartzofeldspathic lits (less than 0.2 mm). The biotite laths are typically elongate, parallel to the compositional banding in these rocks.

Locally, the gneisses become migmatitic with the development of small, medium- to coarse-grained granitic, or more rarely pegmatitic, segregations (leucosomes). The segregations, up to 10 mm wide, are mostly concordant with the compositional layering in the gneisses (as in S 72568). Mafic-rich melanosomes are poorly developed and mesosomes are only identifiable as fine-grained zones bounding the leucosomes and holding irregular, embayed flakes of biotite. These are essentially granodiorite in composition, weakly granoblastic with oriented biotite flakes defining the foliation.

Both the gneissose lits and migmatitic leucosomes show evidence of disruption, becoming boudinaged and/or cut by thin shear bands (as in S 18383). Here, subgrains of plagioclase develop a weak zonation from andesine (An_{20}) cores to rims of albite (An_{12}).

AMPHIBOLE-BEARING LITHOLOGIES

The amphibole-bearing lithologies within the outcrop vary in size and form from thin, diffuse domains within the feldspathic gneisses, to concordant sheets and large masses some tens of metres thick. Most are fine to medium grained, with a green to greenish grey colouration in hand specimen. All forms are cut by foliated microgranite sheets and carbonate veins. As previously indicated, the status of these lithologies both at outcrop and in the wider context of the tectonostratigraphy of the Moine is open to question, particularly the larger component masses within these gneisses. In the light of this uncertainty, and for the purposes of this description, the larger metabasic masses are discussed separately.

Interlayered amphibolites

The feldspathic acid gneisses are locally intercalated with numerous layers of amphibolite within zones in which sheets and masses of undoubted metabasic igneous lithologies crop out (see below). These intercalations occur on all scales up to several tens of metres thick. The thin sheets are mainly fine to medium-grained amphibolite/hornblende schist (Plate 1) mostly comprising a mid to pale green (?)actinolitic amphibole whose preferred orientation defines the foliation in these rocks. Quartz, feldspar and biotite form only minor components. Subidioblastic ilmenite is the most common accessory mineral, with minor apatite and sphene (as in S 18379). The internal fabric is generally parallel to both the margins of sheets and the regional fabric in the host lithologies. In the cores of minor folds formed during the D2 phase of deformation this fabric is crenulated, with a well-developed extension fabric parallel to fold axes.

Within the feldspathic gneisses, relatively amphibole-rich domains, up to several millimetres in thickness, may be developed adjacent to the metabasic layers (as in S 72572). These rarely become monomineralic, but may pass locally into the more substantial bands of amphibolite or hornblende schist. These domains within the gneiss are generally fine to medium grained, comprising plagioclase, amphibole, biotite, quartz and minor K-feldspar. Sphene is the principal accessory mineral, with apatite, zircon, allanite and ilmenite. Secondary minerals include epidote, chlorite and carbonate. The domains are dominated by irregular grains and aggregates of a dark green to pale greenish brown amphibole. Most are slightly elongate parallel to the principal foliation. Amphibole crystals are mostly inclusion-free, but may hold minor rounded sphene or apatite. Secondary alteration to a blue-green amphibole, or replacement by carbonate, is common. Accessory minerals in the gneiss are largely confined to these mafic-rich layers, and in particular the amphibole-rich aggregates. Within the amphibole-rich domains the quartzofeldspathic component is predominantly plagioclase-rich, with quartz and minor microcline. Here, plagioclase is mostly poorly twinned and sericitised, but, where fresh, compositions up to andesine (An_{44}) occur. K-feldspar is poorly twinned, its presence being largely defined by myrmekite. Grains are mostly granoblastic, up to 0.2 mm long. Quartz, as in the acid gneisses, forms pools of strained aggregates, elongate to the foliation.

The layered nature of the rocks, and in particular the presence of the fine-scale amphibole-rich domains, led Harris (1978) and Rathbone and Harris (1980) to suggest a lithological similarity with the hornblende-bearing gneisses of Lewisian inliers elsewhere in the Moine of the Northern Highlands. However, the occurrence of these diffuse zones bears many similarities to the unusual hornblende-bearing psammites reported by Rock et al. (1985) adjacent to metabasic masses within Loch Eil Group psammites of the Moine to the southwest of this area. These were interpreted as evidence for the interaction of the metabasites with their host during regional metamorphism.

Plate 1 Rosemarkie Metamorphic Complex: feldspathic ('acid') gneiss with intercalations of amphibolite/hornblende schist cut by subparallel foliated leucogranite veins in zone of high D2 strain. Outcrop fractured and cut by carbonate veins [7505 5996] (D4890).

Other metabasic lithologies

The larger metabasic masses form trains of boudins or pods up to 50 m across. Contacts are sharp and concordant with the foliation in the host rocks, although, overall, the boudins show evidence of a discordant relationship to the local tectonostratigraphy. The margins of these larger masses are everywhere fine grained and schistose, but pass rapidly into amphibolite, over distances of less than 0.5 m, reflecting the more competent nature of the bodies. Most are coarse-grained, melanocratic to mesocratic amphibolites with little variation. At several places [7433 5851, 7437 5870, 7437 5810 and 7443 5895], the 5 to 20 m-thick boudins preserve a relict subophitic texture. Although most of the larger masses are structureless, the coarse-grained amphibolite at one point [7443 5900] is cut internally by thin sheets of hornblende schist up to 50 mm wide. The origin of these internal structures is uncertain; they may represent either relict internal intrusions, tectonically modified, or minor shear zones.

Mineral assemblages and textures in the metabasic rocks are almost entirely metamorphic, comprising amphibole, plagioclase feldspar, quartz and biotite, with sphene, ilmenite, apatite, zircon and monazite as the more important accessory minerals, and rare garnet. Amphibole is the dominant mineral, which when fresh is pleochroic from green to mid-brown. It occurs mostly as subidioblastic crystals up to 1 mm long or as poikiloblastic aggregates enclosing mainly quartzofeldspathic phases (as in S 18378). Rounded inclusions of sphene, quartz, zircon and apatite are common. Biotite is uncommon and forms small irregular grains, mostly overprinting the amphibole. Plagioclase occurs mainly as granoblastic aggregates. Compositions are andesine, with weak normal zonation from cores of An_{40} to rims of An_{30}. Ilmenite is abundant in all samples, forming irregular grains intimately associated with the mafic aggregates. Although clearly of metamorphic origin, this mineral shows evidence of recrystallisation with thin growth rims of sphene.

In specimen S 72569, a relict subophitic texture is preserved. This comprises large, poikilitic plates up to 4 mm long and aggregates of mid-green to brown pleochroic amphibole. All carry abundant inclusions of ilmenite and sphene. Ilmenite is mostly anhedral, mainly forming small aggregates, partially replaced by sphene. Small patches of a green amphibole are preserved, in which ilmenite crystallites show a preferred orientation. This texture is similar to those described from metagabbroic rocks in the Moy area in the south-eastern part of the district (see p.18 and Highton, 1992), which preserve relict exsolution lamellae of ilmenite in hornblende after clinopyroxene. In the Rosemarkie Complex, euhedral apatite and rounded zircon are common inclusions, with rare monazite. Biotite is not significant, occurring only as small, randomly oriented crystals overgrowing the amphibole. The eu- to subhedral forms of original cumulus plagioclase crystals are recognisable, though now mainly replaced by granoblastic aggregates. Where fresh, however, compositions are mainly andesine (An_{38}), although a single, euhedral crystal enclosed within an amphibole plate was found to be labradorite (An_{54}). All lack significant zonation. Quartz is a minor component, forming small irregular grains or granoblastic aggregates with feldspar. Patches of micropegmatite are also present in the interstices.

Structure

Minor and intermediate-scale folds are common. However, due to the restricted nature of the outcrop, the disposition of the major structures affecting the rocks of the complex are not known.

Both at outcrop and in thin section, the Moine-like and amphibolite lithologies appear to have undergone a similar deformational history; evidence for early (pre-Caledonian) deformational structures and metamorphic assemblages in the interlayered amphibolites of Rathbone and Harris (1979) is absent. Three main phases of ductile deformation were recognised in the rocks of the Rose-

markie Complex. The similarities of these structures to ones in the adjacent Northern Highland Moine succession are such as to suggest that deformation in the complex is mainly Caledonian.

The principal fabric in the metasedimentary lithologies is a compositional layering. This is thought to represent a composite fabric (S0/S1), in which bedding has been modified during the first phase of deformation. This is well demonstrated in the sequence of weakly gneissose siliceous psammites exposed in the quarry south of Whinhill [7386 5841]. In the main face, prominently displayed rootless, isoclinal minor folds are intrafolial to the layering. These folds are picked out principally by thin feldspar-rich layers in the psammites and a mica fabric within thin semipelitic intercalations. The axial planar fabric to these earliest (D1) structures is essentially penetrative, and locally accentuated by thin quartzofeldspathic segregations. Similar relationships have not been recorded in the distinctly gneissose lithologies. Although the gneissosity is everywhere parallel with the compositional layering in the weakly gneissose lithologies, it may be representative of a multicomponent fabric. In their review of the structural history of the inlier, Rathbone and Harris (1980) described isoclinal folding of the purported Moine/Lewisian boundary at one locality just north of the district boundary. To this minor structure, these authors ascribe a deformational age of 'D2'. However, the structure shows strong resemblance to the D1 folds described here, and the 'D2' age cannot be proved.

Minor second-phase (D2) folds are numerous throughout the outcrop, with few of intermediate scale. All are tight to close, upright, asymmetrical structures, with a mostly north-easterly trend, plunging at moderate angles to the north-east. They are characterised by thickened hinge zones and attenuated limbs on which minor shear zones are developed. Components of the D2 fold pairs are distinguishable, with rounded synforms and angular antiforms. In the fold cores, an axial-planar crenulation cleavage (S2) is recognised in semipelitic lithologies, developed as a spaced segregation fabric in the psammites. An intense, mainly coaxial, intersection/extension lineation associated with these structures is present in most lithologies. Local divergence of the extension lineation from coaxiality reflects a curvilinearity of the D2 folds.

The D2 folds have been largely reworked and reoriented by later folds having a variable degree, but consistent direction, of plunge. On the south-west-dipping limbs of the third-phase (D3) structures, the D2 folds are almost indistinguishable from minor D3 structures. The later structures are recognisable only by the lack of a well-developed axial planar fabric and a generally contrasting sense of vergence.

The overall outcrop pattern of lithologies in the complex is controlled mainly by D3 folds. Both minor and intermediate-scale structures show a consistency, of plunge at moderate angles to the north-east, while axial surfaces are upright to steeply inclined, mostly to the south-east. Minor folds are more open in form than earlier structures, with little thinning on the limbs. There is generally no prominent axial-planar fabric, with only an open to tight crenulation in semipelitic lithologies. Interference structures with earlier D2 folds are common in the cores of intermediate-scale folds cropping out to the south of Cairds Cave [as at 7436 5862]. The sense of vergence on these folds indicates the presence of a large-scale antiformal structure to the east.

Metamorphism

The metamorphic history of the inlier is poorly known. Metamorphic index minerals are largely absent or poorly preserved, and the effects of retrogression are widespread. The preservation of regional metamorphic hornblende in the metabasic rocks and gneisses indicates that the Rosemarkie Complex has undergone middle amphibolite-facies metamorphism.

Garnet is present in all the weakly segregated semipelitic lithologies; porphyroblasts are mostly irregular, slightly elongate within the foliation and deformed in the D2 crenulation. Small porphyroblasts of muscovite and shimmer aggregate are common, overprinting the S2 foliation. By analogy with similar observations in the Moine of the Northern Highlands (May and Highton, in preparation) and the Central Highlands (Highton, 1986), these may indicate the former sites of an alumino-silicate phase.

Retrogression is common, with garnet, biotite and to a lesser extent amphibole, being overgrown and replaced by irregular masses of chlorite and/or carbonate. Feldspars are mostly replaced by sericitic mica and more rarely carbonate, and by irregular grains and/or aggregates of epidote.

Granitic intrusions of the Rosemarkie Complex

The metamorphic rocks of the Rosemarkie Inlier are cut by numerous veins and sheets of acid composition. They mostly comprise foliated, leucocratic microgranites, which locally may be slightly K-feldspar porphyritic, and which everywhere are characterised by a reddish colouration both on fresh and weathered surfaces. The intrusions range in thickness from 5 mm to 3 m. Most are parallel sided with sharp contacts, oriented at low angles, but clearly discordant to the compositional layering and/or gneissose foliation in the host lithologies (Plate 2). Where abundant, the granitic veins form ramifying networks, which locally are anastomosing, while the larger intrusions are discrete sheet-like masses.

These granitic rocks essentially comprise quartz, plagioclase and K-feldspar with minor muscovite, which appears to be primary (as in S 18384). Compositionally, they are monzogranites with plagioclase and K-feldspar in roughly equal proportions. Accessory minerals include zircon and magnetite, with rare apatite and monazite. Common secondary minerals include chlorite, zoisite, carbonate and, more rarely, a pale blue sodic amphibole (Garson et al., 1984). Evidence for recrystallisation is widespread in these rocks. Even where strains are relatively low, igneous textures and mineralogies are rarely preserved, with the notable exception of

partially recrystallised K-feldspar phenocrysts and rare small patches of micropegmatite. Elsewhere the rocks are mainly granoblastic, but rarely show equilibrium polygonal textures. Quartz occurs largely as small aggregates ranging from weakly strained, equant grains to elongate ribbons which, in areas of higher strain, are tabular in form. Plagioclase is mostly sericitised, with microcrystalline haematite exsolved parallel to the principal crystallographic axes. Where fresh, the feldspar shows a weak normal zonation from oligoclase (An_{20}) cores to albite (An_5) rims. K-feldspar is a poor to weakly twinned microcline, with slight alteration to mica products. The phenocrysts in the slightly porphyritic variants comprise subhedral crystals, up to 5 mm long, of coarsely perthitic microcline. These show only minor evidence of recrystallisation, with subgrain developments at the margins of the crystals. Mica phases are rare, with small flakes of muscovite less than 0.2 mm long. The presence of minor chlorite may in part indicate the former presence of biotite in these rocks (Rathbone, 1980). Both minerals are oriented, defining the ubiquitous foliation.

Rare examples of cross-cutting microgranite veins oriented at high angles to the regional fabrics show evidence of boudinage and are more commonly folded. Most veins and sheets are strongly foliated, the fabric being defined essentially by a quartz-feldspar shape fabric and oriented muscovites. The foliation in these intrusions is everywhere concordant with the principal cleavage in the host lithologies, identified as being a composite S1/S2 cleavage, axial planar to the minor D2 folds (Rathbone and Harris, 1980). The veins are clearly folded in the cores of these D2 structures on both small and intermediate scales and a pronounced extension lineation is developed in granitic rocks parallel to the D2 fold axes. This lineation is a shape fabric essentially defined by single elongate grains and ribbon aggregates of quartz. The strains associated with the development of this intense fabric are reflected in the axial ratios of the strain ellipsoid, 18:2.5:1, recorded by Rathbone and Harris (1980). These must represent minimum strain values for these rocks, with some uncertainty arising from inherent assumptions about initial grain shapes and the effects of subsequent recrystallisation. Where the intrusions are porphyritic, K-feldspar phenocrysts are elongate, and wrapped by the quartz fabric. The linear form of these augen is accentuated by the development of coarse-grained quartz pressure shadows in the low-strain areas (as in S 18384).

The relationships observed at outcrop indicate an age essentially later than the emplacement of the suite of metabasic rocks and the D1 deformational event, but pre- to possibly syn-D2. The veins and their internal foliation are clearly reworked by folds of D2 and D3 age. The only radiometric age determinations to date have been by U-Pb analysis of zircons reported by Rathbone (1980) from two granitic veins in the inlier just north-east of this district, which yielded ages of 2380 Ma and 384 Ma. Both were regarded as unrealistic, however, with the samples possibly having suffered lead loss.

Plate 2   Rosemarkie Metamorphic Complex: inclined foliated sheets of leucocratic microgranite cutting the tectonically modified layering of the psammitic ('acid') gneisses [7453 5937] (D4892).

CENTRAL HIGHLAND MIGMATITE COMPLEX

Previous research

The basement metasedimentary schists and gneisses south-east of the Great Glen, and within the Fortrose district, were originally grouped with the Moine Series of Sutherland and Ross-shire (Horne, 1923). Later, Johnstone (1975) defined the Moine on the basis of lithology as the succession of monotonous psammites occurring stratigraphically below the more diverse rocks of the Dalradian (sensu Harris and Pitcher, 1975). Johnstone differentiated the Moine into 'Older Moine' in the Northern Highlands and 'Younger Moine' in the Central Highlands on isotopic evidence that indicated a Neoproterozoic deformation in the former, prior to deposition of the latter. It was suggested that an unconformity between the two successions was either obscured by deformation or hidden by displacement along the Great Glen Fault Zone. Subsequent work in the Central Highlands apparently located such an unconformity, though modified by high-strain zones and locally referred to as the Grampian Slide; this separated rocks interpreted as having contrasting metamorphic and structural histories (Piasecki and van Breemen, 1979; Piasecki, 1980; Piasecki and Temperley, 1988). Migmatitic gneisses referred to as the Central Highland Division were considered to represent the basement to the largely non-migmatitic succession of the Grampian Division (Piasecki, 1980). This view has not gained universal acceptance; Lindsay et al. (1989) found no difference in structural history between the two divisions. They suggested that a migmatitic/metamorphic front separates the migmatitic and non-migmatitic rocks. Since the Grampian Group (equivalent to the Grampian Division) is included within the Dalradian Supergroup (Harris et al., 1978) on the basis of a sedimentary transi-

tion into the overlying Appin Group (Glover and Winchester, 1989), Lindsay et al. (1989) considered that the migmatites in the Central Highlands should also be included within the Dalradian Supergroup.

As there is still a lack of consensus on stratigraphical relationships in the Central Highlands, the migmatites in the Fortrose district are here included within the Central Highland Migmatite Complex. The use of this descriptive lithodemic term for a group of rocks dominated by migmatitic gneisses, but with associated intercalated non-migmatitic rocks, whether these be tectonic (Highton, 1992) or stratigraphical intercalations, avoids premature lithostratigraphical interpretation.

Metabasic rocks, which crop out mainly in the Findhorn Valley, are also included within the migmatite complex.

Lithologies

METASEDIMENTARY LITHOLOGIES

Gneissose psammite and micaceous psammite are generally the most abundant lithologies, comprising 70 to 80 per cent of the migmatite complex. However, coarse-grained migmatitic semipelites are widespread in the southern part of the district. Thin quartzites are a minor lithology in the south-west, comprising less than one per cent of the complex. Most of the rocks are migmatitic but, whereas semipelites are extensively migmatitic, the psammites have migmatitic and non-migmatitic variants. Exceptions are the non-migmatitic rocks in the Meur Bheòil and Gleann an Tairbhidh areas, which comprise about one per cent of the complex. Their relationship to the migmatitic rocks is uncertain and on present evidence the possibility that they are tectonic enclaves cannot be discounted. They have lithological similarities with Grampian Group metasedimentary rocks farther east, in the Aviemore–Grantown area. In this account, they are treated as components of the Central Highland Migmatite Complex.

The lack of exposure and near absence of sedimentary way-up structures, together with the occurrence of polyphase deformation and numerous intrusive rocks, hinder the establishment of a stratigraphical succession. Thus, for descriptive purposes, the district is subdivided into five regions as indicated in Figure 4; they are the Moy–Farr, Meall Mór–Dalroy, Dalcharn–Glengeoullie, Cairn Kincraig and Glenkirk–Ballachrochin regions. Since there is some correspondence between particular lithological associations in many of these regions, a provisional stratigraphical succession is proposed (Figure 5).

DETAILS

Moy–Farr region

Psammite and micaceous psammite with semipelite succession in the Moy area The area south of the Moy Granite is underlain by a variety of gneissose psammites and micaceous psammites, with migmatitic semipelites intercalated on scales ranging from a few millimetres to hundreds of metres. The lithological variation is apparent around Creag an Eòin [730 339], where psammite and micaceous psammite are interlayered with micaeous psammite–semipelite units up to a few tens of metres thick. Towards the south-west, psammite with some quartzite units up to 2 m thick becomes dominant.

The Moy area is dominated by flaggy, locally migmatitic psammite and micaceous psammite, with concordant, and in places pegmatitic, quartz-plagioclase leucosome segregations, interlayered on a scale of a few millimetres to few centimetres. Biotite laminae, with or without muscovite and with 3 to 20 mm spacings, are common. However, grey gneissose psammite and micaceous psammite without micaceous laminae are widespread, as seen also around Meall Mór. Garnet is locally developed within the psammites, especially in the micaceous psammites. In Daviot Quarry [719 392], the predominant lithology is massive grey gneissose micaceous psammite. A layering defined by more and less micaceous units several centimetres to several tens of centimetres thick is superimposed in places on a finer-scale layering defined by similar variably micaceous units. The grey micaceous psammite is in sharp contact with leucocratic migmatitic psammite containing some biotite laminae and a few migmatitic semipelite units up to 50 cm thick. The psammite is composed of quartz, microcline, plagioclase, biotite and a little muscovite.

Within the Moy area, Horne (1923, p.53) recorded the occurrence of a pebbly rock '1/2 mile east of Beinn a' Bheurlaich'. Re-examination of his hand specimen (S 18371) reveals that the rock is a psammite containing mafic laminae up to 1 mm thick separated by 2 to 5 mm bands of more leucocratic quartzose rock. Ovoid and rounded clasts are up to 7 mm in diameter and comprise 15 to 20 per cent of the rock. In thin section, the clasts are seen to be composed of coarse-grained quartz and quartz–K-feldspar aggregates. In contrast, the groundmass is composed of fine to medium-grained quartz, plagioclase, K-feldspar and biotite. Mafic laminae contain a higher proportion of biotite together with abundant opaque minerals (assumed to be largely ilmenite since the rock has low magnetic susceptibility), sphene and some zircon. This occurrence of pebbly rock was not located during the recent resurvey.

The migmatitic semipelites are typically medium to coarse grained with prominent quartz-plagioclase leucosome segregations. With increasing proportion of leucosome, the migmatitic semipelite grades into a rock of granitic aspect, both to the

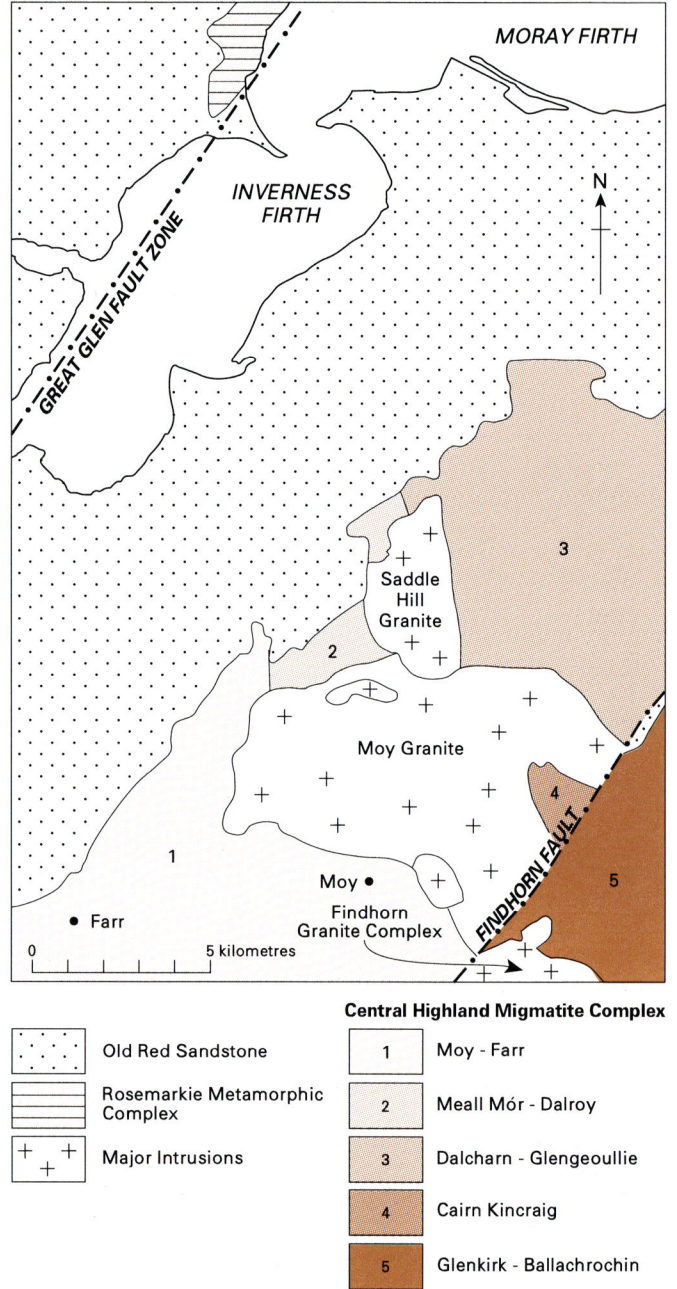

Figure 4 Sketch map showing the main regions used for the description of the Central Highland Migmatite Complex.

north-west of Beinn a' Bheurlaich [at 7240 3662] and on Carn na Loinne [7616 3284]. A 'lenticle of amphibolite' and 'small knots of hornblende-biotite rock' within semipelitic gneiss were recorded during the primary survey [at 7291 3542 and 7240 3659 respectively], but they are now obscured by dense forest. Lithological boundaries are, for the most part, transitional, such that migmatitic semipelite grades into gneissose psammite and micaceous psammite with increasing abundance of intercalated psammitic units.

Psammite and quartzite succession in the Beinn Dubh–Beinn nan Cailleach area The gneissose psammite/micaceous psammite

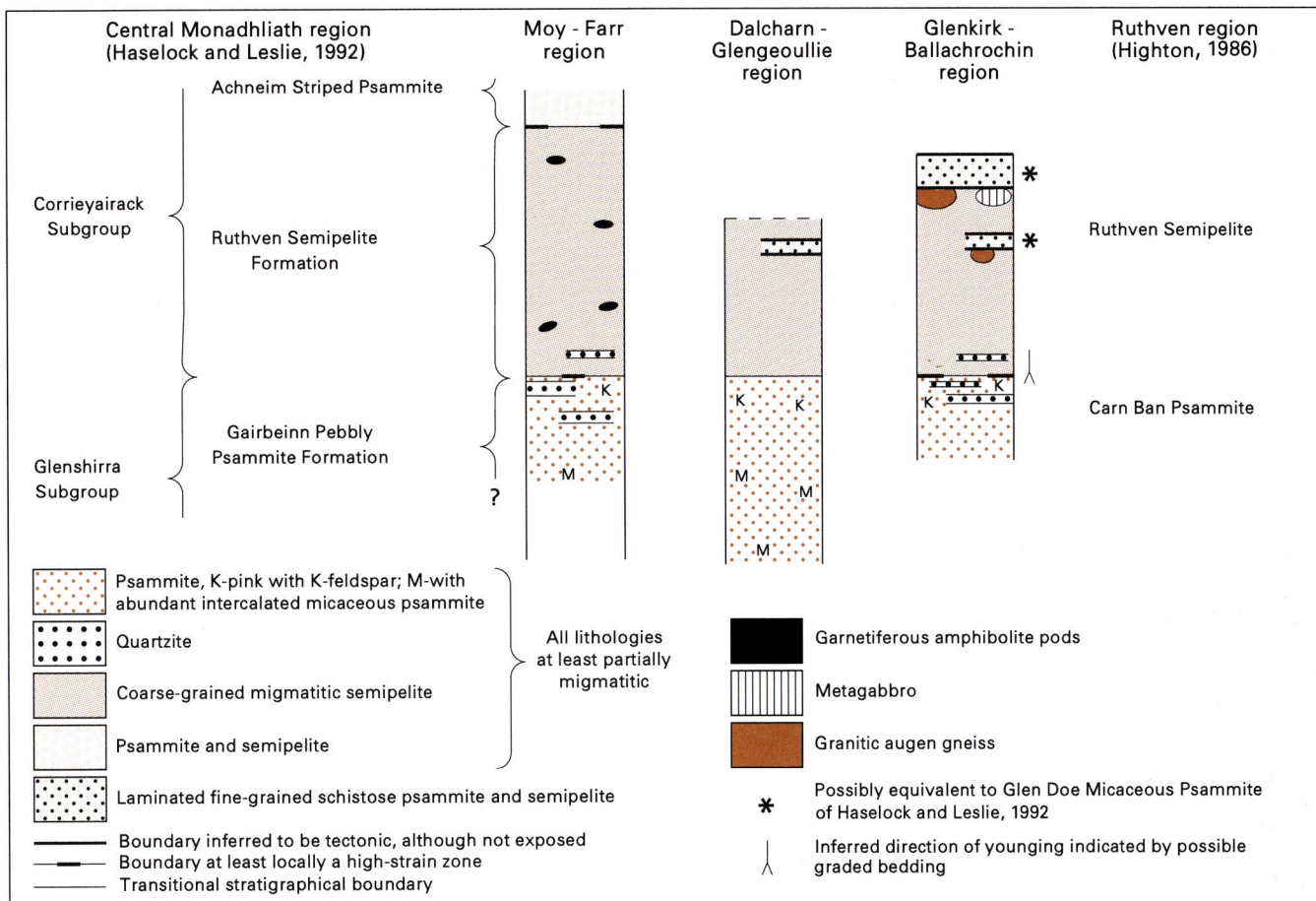

Figure 5   Inferred stratigraphical successions in the Central Highland Migmatite Complex of the district and possible correlations with adjacent regions.

with semipelite succession is represented to the south-west by psammite with intercalated quartzites. Psammites are typically striped, the striping being variously defined by biotite-rich lamellae mostly 3 to 10 mm apart, less abundant thin units of micaceous psammite and a colour banding in shades of grey. Thin-section examination indicates that the colour banding is compositionally controlled, at least locally, such that the more mafic layers comprise mainly plagioclase and biotite with subordinate quartz and microcline. In contrast the more leucocratic layers comprise quartz, microcline, plagioclase and biotite; some layers are composed almost exclusively of quartz and microcline. The psammite contains units of quartzite typically up to 1 m thick, although they are up to 8 m thick in places [as at 723 338]. The quartzites commonly contain thin biotite lamellae adjacent to contacts with psammite. To the east, quartzites are developed on Carn na Loinne [766 328], where they occur as units up to 2 m thick containing biotite-lamellae spaced 10 to 100 mm apart and having gradational contacts with the adjacent psammite.

Psammite succession in the Creag an Reithe area   Pink or pinkish grey strongly deformed gneissose psammites are well exposed on Creag an Reithe [663 337] and Creag Shoilleir [666 337]. They are composed of approximately equal proportions of quartz, microcline and plagioclase together with biotite, which is dark brown in thin section. The psammites are typically banded with biotite-rich laminae. In places, thin micaceous

psammite units separate quartzofeldspathic layers and rare quartzitic layers typically 5 to 10 mm thick (some up to 50 cm thick). The psammites also display a colour banding in shades of grey and pink mostly on a scale of 10 to 20 mm. They are correlated with the psammite–quartzite succession of the Beinn Dubh–Beinn nan Cailleach area.

Migmatitic semipelite succession in the Creag Bhuide and Meall Mór– Meall na Fuar-ghlaic areas   The contact between psammites with intercalated quartzites and migmatitic semipelites in the area north-west of Beinn Dubh [705 334] is inferred to be sharp, based on the absence of interlayering, although the actual contact is not exposed. Migmatitic semipelites are well exposed around Strath Nairn both within the uplands that contain extensive areas of bare rock and in Strath Nairn itself, where there are numerous *roches moutonnées*. The semipelites here are typically coarsely migmatitic gneisses composed of quartz, plagioclase, biotite and muscovite. Garnet is quite widely distributed, although not ubiquitous, and apatite is an abundant accessory mineral. Interlayered units of micaceous psammite are mostly less than 40 cm thick and are a relatively minor lithology within this unit. Quartz-plagioclase leucosome segregations are a prominent feature of the rocks. They developed at several stages in the structural history, as discussed in the following pages. They comprise both medium-grained concordant layers, typically 5 mm or so thick and coarse-grained layers and lenses up to several centimetres thick. The latter produce the charac-

Plate 3 Central Highland Migmatite Complex: migmatitic semipelite showing both fine and coarse-grained contorted leucosome segregation. Northern slope of Creag Bhuide [6640 3165] (D4349).

teristic coarsely migmatitic aspect of these semipelites (Plate 3). In places, the coarse-grained segregations coalesce to produce a rock of granitic appearance, albeit heterogeneous, with mafic biotite–muscovite-rich schlieren and patches of pegmatite. The complex structural and migmatitic history has produced rocks that are relatively homogeneous on a broad scale, but very heterogeneous at any particular exposure. The migmatitic semipelites contain inclusions or pods of coarse-grained garnetiferous amphibolite up to 1 m in diameter (p.18). Thin (50 to 200 mm thick) units of quartzite with sharp contacts against the migmatitic semipelites occur around Carn Bheithin [667 319]. Psammite units generally less than 50 m thick with thin quartzites are intercalated with the migmatitic semipelites in the Achvaneran [678 342]–Beachan [681 348] area.

West of Meall Mór [around 696 338], a westerly increase in the amount of intercalated micaceous psammite indicates a possible transition into the mixed successions of the Farr area.

Mixed psammite and semipelite successions in the Farr area Bedrock in the Farr–Inverarnie area of Strath Nairn [685 330] is very poorly exposed. The few localities where rock crops out suggest that the area is underlain by a mixed succession of gneissose psammites, micaceous psammites and semipelites lying within the fold core between the migmatitic semipelites of the Creag Bhuide and Meall Mór–Meall na Fuar-ghlaic areas. A stream section at Inverarnie reveals gneissose psammite and semipelite/pelite units interlayered on scales of about one to 50 cm. Psammite is somewhat more abundant than the semipelitic lithologies. Lithological layering is apparent on several scales such that individual colour-banded psammite units alternate with more and less semipelitic units.

A mixed group of interlayered gneissose psammites and micaceous psammites with thin semipelites exposed in the area south-east of Meall Mór [around 702 333] is correlated with the Inverarnie succession. The interlayered rocks are adjacent to highly strained migmatitic semipelites east of Meall Mór [704 335] and in Strath Nairn [672 317]. In the former area, the contact is not exposed and a gap 3 m wide separates the two exposed lithological units; thin psammite units up to 30 cm across are rare, even in the marginal parts of the semipelite. Similarly, in the latter area, no contact is visible alongside a broad zone (about 450 m wide) of highly deformed semipelitic gneisses.

Psammite succession in the Carn Dar-riach area Exposures beside a forestry track [6931 3164], 900 m east-north-east of Farr House, reveal grey migmatitic psammite with a lithological layering on about a 50 mm scale, defined by more and less feldspathic layers. Approximately 250 m to the south-east [6948 3145], psammite and decomposed migmatitic semipelite are interlayered on a scale of between 5 and 10 m.

Around Carn Dar-riach, [712 314] psammite and micaceous psammite, in part migmatitic, are interlayered with thin units of migmatitic semipelite up to 2 m thick, although, more typically, these are only a few centimetres thick.

Meall Mór–Dalroy region

Metasedimentary rocks exposed north-west of the Moy Granite consist predominantly of massive, medium-grained, granular psammites and micaceous psammites. They are in part gneissose with some quartz-plagioclase leucosomes. Psammites are quartz-rich with plagioclase and rare K-feldspar and although generally grey are pink in places. Many display a layering defined by biotite-rich laminae and thin layers of micaceous psammite. Units of migmatitic semipelite occur within the psammites, although their distribution is unclear because of poor exposure. A prominently striped rock (layers about 1 cm thick) composed of interlayered micaceous psammite and gneissose semipelite crops out 1 km east-south-east of Dalroy [at 7774 4419].

Dalcharn–Glengeoullie region

Migmatitic semipelite Migmatitic semipelite units in the Dalcharn–Glengeoullie region, occurring as intercalations within the gneissose predominantly psammitic successions, have many features in common. They are typically coarsely migmatitic with quartz-plagioclase segregations, lenses and layers, up to 2 cm thick, within essentially quartz-plagioclase-biotite-muscovite gneisses. Apatite is abundant. Garnet occurs in places, whilst thin sections of some specimens reveal biotite-muscovite intergrowths with rare andalusite and fibrolitic sillimanite. Muscovite-free semipelitic gneiss adjacent to Clunas Reservoir [at 8566 4572] contains granitic leucosomes composed of quartz, plagioclase and K-feldspar. Subordinate interbanded units of micaceous psammite mostly range from a few centimetres to 1 m or so in width.

Psammites   Psammites, which are migmatitic over extensive areas (Plate 4), range from 'clean', quartzofeldspathic rocks to massive, grey micaceous psammites. Pink microcline-bearing psammite, in part migmatitic, is exposed in a zone extending south-west from Glengeoullie to Creag an Daimh, but cannot be followed farther for lack of exposure. A layering of mica laminae is developed in places and migmatitic semipelite units (up to 1 m thick) occur close to Glengeoullie Bridge [8563 4761]. The contact between this psammite unit and migmatitic semi-pelite is exposed in Riereach Burn [around 8477 4446]. Here, gneissose psammite in units typically 10 to 20 cm thick, is inter-layered with gneissose micaceous psammite and semipelite with the proportion of psammite decreasing south-eastwards into the semipelite.

Elsewhere within the Dalcharn–Glengeoullie region, psam-mite and micaceous psammite are interlayered on all scales. The most extensive lithology is a grey micaceous psammite with or without a banding defined by micaceous laminae. Migmatitic quartz-plagioclase leucosomes are developed sporadically.

Schistose psammite and semipelite succession in the Meur Bheòil area
Grey, homogeneous to weakly laminated, fine to medium-grained schistose micaceous psammite and semipelite crop out in the Meur Bheòil [between 852 401 and 857 401]. The only indication of any migmatisation in these rocks is the occur-rence of rare, small quartzofeldspathic segregations. The rocks are composed of variable proportions of quartz, plagioclase and dark brown biotite. Some rocks are plagioclase-rich, whereas others are quartzose. Muscovite is less abundant than biotite and occurs as cross-cutting porphyroblasts. Garnet is common, but not ubiquitous.

Cairn Kincraig region

Metasedimentary rocks are exposed in a small area on the south-eastern side of the Moy Granite north-west of the Findhorn Fault. Relationships between lithologies are commonly obscured by extensive granite and pegmatite sheets and by the drift cover. Layered and locally migmatitic psam-mites in the north-east of the area, showing prominent mica-rich laminae 5 to 10 mm apart, are succeeded abruptly to the south-west by migmatitic, locally garnetiferous, semipelite. The migmatitic semipelite contains intercalated flaggy, grey, schistose micaceous psammite and semipelite [around 835 364]. To the south-west, the migmatitic semipelite apparently grades into massive, grey, locally garnetiferous micaceous psammite, although the exact nature of the contact is not clear. Micaceous laminae, about 5 mm apart, define the fabric in the micaceous psammite. Micaceous psammite, together with abundant granite and pegmatite sheets, forms the high craggy exposures of Creag a' Chròcain [828 358]. To the south-west, micaceous psammite is replaced by migmatitic psammite occurring as rafts and xenoliths within the granite vein complex west of Shenachie [around 823 349]. The nature of the contact between the micaceous psammite and the migmatitic psammite is not known.

Glenkirk–Ballachrochin region

The area south-east of the Findhorn Fault, in the south-eastern corner of the district, is underlain by interlayered gneissose psammite and semipelite with minor developments of

Plate 4   Central Highland Migmatite Complex: migmatitic psammite with prominent pegmatitic leucosomes and intrafolial isoclinal folds. Riereach Burn [8563 4765] (D4347).

quartzite, metagabbro and non-migmatitic, fine to medium-grained, laminated, schistose psammite and semipelite.

Gneissose psammites   Psammites are characterised by the devel-opment of a striping defined by biotite laminae, 1 to 2 mm thick, with or without muscovite, separated by quartzofelds-pathic layers 5 to 20 mm thick. A psammite unit containing thin quartzite layers on Cnocan Mór [836 348] is locally pink; other pink psammites within the district contain microcline. Quartz-plagioclase migmatitic leucosomes are widely developed.

Migmatitic semipelite   Coarse-grained quartz-plagioclase leuco-somes are ubiquitous within the migmatitic semipelites. Apatite is an abundant accessory mineral, whereas garnet is rare. Silli-manite (p.22) is developed in the Glenkirk area. A micaceous psammitic or semipelitic augen gneiss (p.23) occurs adjacent to both the psammite–migmatitic semipelite boundary and to metagabbro bodies along the eastern bank of the River Findhorn [8279 3359 and 8307 3437]. Similar K-feldspar-bearing migmatitic gneisses occur close to the migmatitic semipelite-psammite boundary on Tom a' Ghealagaidh and adjacent to laminated schistose psammite and semipelite in Allt Seileach [8346 3246].

The migmatitic semipelites are heterogeneous on an exposure scale as a result of at least two phases of migmatisa-tion (p.21), although they are more homogeneous on a broader scale, with only sporadic development of interlayered units of micaceous psammite 50 cm thick. Contacts between psammite and migmatitic semipelite are sharp, with no evidence for interlayering of the lithologies.

Quartzites   Quartzite units up to 2 m thick occur within mig-matitic psammite adjacent to migmatitic semipelite east and north-east of Shenachie [833 356 and 8355 3490]. Some indi-vidual quartzite units show transitions from feldspathic quartzite into mica-bearing quartzite. The implication, if this represents fining-upward grading, is that the migmatitic semipelite is younger than the psammite.

Migmatitic psammite and semipelite are separated by 5 to 10 m of quartzite in Carnoch Burn [8678 4026], immediately east of the district. A body of weathered amphibolite 50 cm thick, containing retrogressed garnets up to 2 cm in diameter, occurs close to the contact between the quartzite and the semipelitic gneiss.

Quartzite units up to 4 m thick, containing thin interlayered migmatitic semipelite units, are developed within migmatitic semipelite in a zone 50 m wide that can be traced along strike for about 1 km, north-east of Shenachie [840 360]. Quartzite units up to 5 m thick lie in a more restricted zone within an adjacent migmatitic semipelite [835 346]. The quartzites are feldspathic and in places migmatitic. Thin quartzite units also have been noted within a third migmatitic semipelite unit adjacent to a metagabbro body [8285 3376].

Laminated schists in the Gleann an Tairbhidh area   Laminated, fine- to medium-grained schistose psammite, micaceous psammite and semipelite crop out in a zone trending north from the southern margin of the district for about 1.5 km through Gleann an Tairbhidh [838 319] and Gleann Seileach [849 333]. The most striking feature of these rocks is the lack of migmatisation and the preservation of an apparent sedimentary lamination. Micaceous psammite is composed of quartz, plagioclase, muscovite, biotite and some K-feldspar. Poikiloblastic muscovite is abundant and occurs within thin (few millimetres thick) micaceous units. The contact between the non-migmatitic rocks and the adjacent migmatitic successions has not been seen.

METABASIC LITHOLOGIES

Two distinct suites of metabasic rocks are recognised within the migmatite complex. Garnetiferous amphibolites occur mostly around Strath Nairn, whereas metagabbros occur in the Findhorn Valley.

Garnet amphibolites   Garnet amphibolites, occurring as pods up to 1 m in diameter within migmatitic semipelites in the Moy–Farr region, are rare, but occur preferentially in a south-west-trending zone east of Creag Bhuide [around 669 316] and more extensively in another south-west-trending zone 3 km long between Coire Buidhe [662 320] and Beachan [around 682 343]. An exposure of a 1 m-thick strip of garnetiferous amphibolite is recognisable south-east of Strath Nairn on Beinn Dubh [7068 3254]. The amphibolite pods essentially comprise coarse garnets, locally up to 2 cm in diameter, hornblende, biotite and quartz. In thin section, the garnets contain biotite and sphene, and numerous inclusions of quartz that locally pass laterally into large quartz aggregates several millimetres across. The garnets are typically rimmed by a symplectic intergrowth of plagioclase, hornblende, biotite and opaques, up to 2 mm wide. The groundmass is composed of green hornblende, in part forming an equigranular polygonal mosaic, with less abundant reddish brown biotite. Rounded quartz together with sphene are common inclusions in the hornblende. There is evidence that interaction between the migmatitic semipelites and the amphibolite pods has resulted in hornblende being replaced by biotite and more abundant quartz. Ultimately, the pods are only represented by coarsely garnetiferous, more mafic, areas of the gneiss. The fabric in the migmatitic semipelite typically wraps around the amphibolite pods.

Metagabbros   Metagabbro bodies crop out north-east of Ruthven on the east bank of the River Findhorn [828 337 to 831 344] and on the west bank [821 335]. They are interpreted as tectonic intercalations within the metasedimentary rocks, thus forming part of the Migmatite Complex and as such they are described here rather than in Chapter Three. Metagabbro bodies occur in the form of lenses up to 80 m long composed of gabbro with ophitic texture passing into patches of gabbroic pegmatite. The gabbro is cut by gabbroic pegmatite veins 5 cm wide and in places by subhorizontal sheets of mesocratic metadolerite up to 20 cm thick [as at 8282 3367]. The gabbros are composed of subhedral plagioclase poikilitically enclosed by clinopyroxene. Pyroxene contains abundant ilmenite exsolution lamellae together with rare lamellae of orthopyroxene. These gabbros are amphibolitised to a greater or lesser degree and show a range from gabbro through metagabbro to amphibolite. A pale brown amphibole forms a rim to pyroxene in contact with plagioclase. Highton (in preparation) considers that this may be a late magmatic feature. With progressive amphibolitisation, pyroxene was first replaced by pale brown to pale green, commonly bladed amphibole, but the ophitic texture and the exsolution lamellae are still preserved. Such pale amphibole is in places replaced by a darker brown amphibole with ilmenite lamellae that coalesce to form discrete ilmenite grains with or without sphene. The brown amphibole is replaced by green hornblende. Subsequent recrystallisation involved replacement of the ophitic texture by a more equigranular, hornblende-plagioclase mosaic.

Locally, the metagabbros are foliated and deformed by small shear zones. Highton (1992) has shown that to the south-east of the district, gabbros occur within a zone of high strain in which non-migmatitic laminated schists are interleaved with migmatitic gneisses. The lensoid nature of the gabbro bodies within the Fortrose district is similarly interpreted to be the result of tectonic disruption. They occur close to the contact between migmatitic psammites and semipelites, where small exposures of grey, fine-grained non-migmatitic psammite and micaceous psammite, similar to the laminated psammite-semipelite of Gleann an Tairbhidh, are developed, although there is little evidence of high strain. The migmatitic augen gneiss (p.23) also occurs close to the same lithological boundary. Geochemical analyses of three specimens of metagabbro are presented in Appendix 2. They all fall within the field of subalkaline tholeiites.

Stratigraphy

It is apparent from the above lithological descriptions that certain associations of lithology recur in several parts of the district (Figure 5) and that the most extensive succession is exposed in the Moy–Farr region. Here, psammites and micaceous psammites grade into clean, layered, microcline-bearing and commonly pink psammites; pebbly psammite has been recorded previously at one locality and quartzite units lie within the psammite adjacent to coarsely migmatitic semipelite. The migmatitic semipelite, which contains thin quartzite units and pods of metabasic rocks in the form of garnet amphibolite, apparently grades into a mixed psammite and semipelite succession. The latter crops out at only two localities. There is no way-up evidence in the region.

A similar association of lithologies is recognised, apparently tectonically repeated, in the Glenkirk–Ballachrochin region. Clean, commonly pink, layered and migmatitic psammites pass northwards into thin quartzites apparently stratigraphically overlain farther north by migmatitic semipelite (an interpretation based on a single example of possible graded bedding). Possible graded and cross-bedding in similar psammites west of Drynachan Lodge [at 8647 3976], immediately east of the district, is consistent with this interpretation. Quartzites also occur within the migmatitic semipelite. Non-migmatitic flaggy schistose psammites and semipelites occur as intercalations within the semipelite, whereas augen gneiss, pods of metagabbro and flaggy schists occur close to the inferred stratigraphical top of this unit.

An association of psammite, quartzite, garnetiferous amphibolite and migmatitic semipelite lies immediately east of the district in Carnoch Burn. However, in the Cairn Kincraig region, no metabasic rocks have been recognised, although the migmatitic semipelite does contain intercalations of flaggy schistose psammites and semipelites.

North of the Moy Granite, neither quartzites nor metabasic rocks have been recognised and the lithological succession is less well constrained. However, psammite, coarsely migmatitic semipelite and flaggy non-migmatitic schistose psammite and semipelite form part of the association in the Meur Bheòil area. Farther north, scattered exposures of clean, partly pink, microcline-bearing psammite adjacent to migmatitic semipelite are all that is manifest of the association.

Since all or part of a similar association of lithologies is identifiable in a large part of the district (Figure 5), there may be considerable tectonic repetition of a relatively thin succession, perhaps a few kilometres thick, possibly with only one major unit of migmatitic semipelite present. In Strath Nairn, this unit is continuous along strike to the south-west with the Ruthven Semipelite of the Ruthven area (Highton, 1986), and lithologically comparable to it. Adjacent migmatitic psammites with thin quartzites, inferred to underlie the migmatitic semipelite of the Fortrose district (Figure 5), should therefore be equivalent to the Carn Ban Psammite of Highton (1986) and the Gairbeinn Pebbly Psammite Formation of Haselock and Leslie (1992). Whereas there are strong similarities with the former, correlation with the latter is less convincing. The Gairbeinn Pebbly Psammite is non-migmatitic and preserves well-developed sedimentary structures and pebble beds, features not recognised in the Fortrose district, apart from one unsubstantiated occurrence of a pebbly rock (Horne, 1923, p.53). Similarities, however, include the common occurrence of arkosic psammites and quartzites near to the top of the formation. The mixed psammite and semipelite succession of the Moy–Farr region, inferred to overlie the migmatitic semipelite (Figure 5), may equate with the Achneim Striped Psammite (Haselock and Leslie, 1992), although no

description of this is published. Laminated schistose rocks intercalated with the migmatitic semipelite, and having associated granitic augen gneiss and metagabbro, may be equivalent to part of a sequence originally described in Haselock et al. (1982), and later named the Glen Doe Psammite (Haselock, 1992). Large areas north of the Moy Granite are of unknown affinity. Here, the preponderance of grey, variably micaceous psammite is a lithological feature similar to that developed in the Moy–Farr region north-east of, and possibly stratigraphically below, the psammite-quartzite succession.

Structure

Previous work encompassing the structural history of the Central Highland Migmatite Complex has suggested that recumbent folding and penecontemporaneous migmatisation, together with the emplacement of metagabbros (Highton, 1992), predated deformation associated with the development of the Grampian Slide (Piasecki and Temperley, 1988). Deformation postdating the Grampian Slide produced north-north-east-trending upright structures throughout the Central Highlands (Piasecki and Temperley, 1988).

Three main phases of deformation are recognised within the metasedimentary rocks in the Fortrose district. A compositional layering and lamination (S1) thought to have resulted from the tectonometamorphic modification of original bedding, together with concordant stromatic leucosome segregations, developed in the metasedimentary rocks during the earliest folding event (D1). Isoclinal and commonly intrafolial folds related to this event are rare, but are recognised in laminated and layered psammitic rocks from widely spread localities. They are well displayed in Riereach Burn north-east of Glengeoullie Bridge and on Creag an Daimh [8331 4422]. The S1 layering is deformed by later fold structures (D2 and D3).

D2 folds have been re-oriented by the D3 structures in such a way that their original orientation is not clear, although it is thought that fold-axial surfaces trended north-west. Furthermore, where S1 strikes approximately north-east on D3 fold limbs, many D2 structures are indistinguishable from D3 folds, since they are similar in appearance and, in this structural setting, more or less coaxial. Large-scale D2 fold structures are recognised mainly in the southern part of the district by the disposition of migmatitic semipelite units that act as markers (Figure 6). North of the Moy Granite, similar structures have not been recognised, because of poor exposure. Both D2 and D3 strains were apparently quite low north of the Moy Granite, so that in places the later fabric (S2) is markedly oblique to the earlier (S1) in micaceous psammitic and semipelitic lithologies, as in Allt Dearg [around 816 431]. Close asymmetrical folds, with an axial-planar fabric, refold tight to isoclinally folded micaceous laminations in psammitic rocks in the Creag an Daimh–Braevall Wood area [835 450]. In places, this fabric is developed where no folds are present, whereas at one point [8331 4422] tight folds are recognised in a thin pegmatite vein and not in the composite S0-S1

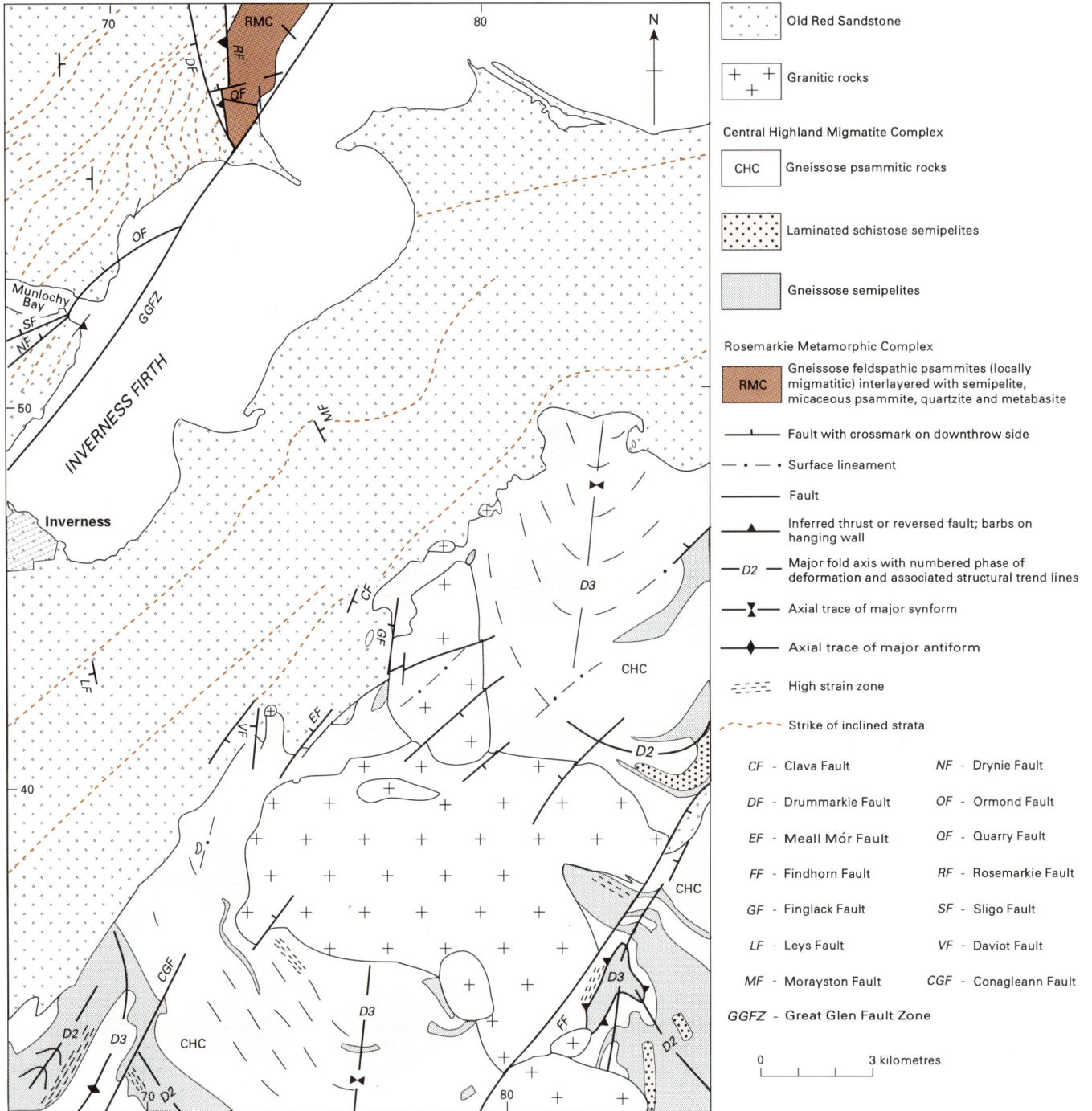

Figure 6 Simplified structural map of the district.

fabric. Whilst the folds and axial-planar fabric trend north-east here, they are thought to be the result of a re-orientation of D2 by D3 structures. Other north-east-trending D2 structures are well developed in Riereach Burn around Glengeoullie Bridge [856 476]. Coaxial intersection and rodding lineations are associated with D2 structures.

South of the Moy Granite, D2 strain is higher, resulting in generally tighter fold structures and local

zones of strong deformation and shearing on fold limbs. D2 fold limbs commonly show attenuation and in places have been excised, as in the psammitic lithologies in the Beinn nan Cailleach area [720 330]. The intense rodding lineation associated with D2 high-strain zones is particularly well developed in psammitic lithologies, especially in the Creag Shoilleir area [667 337]. Broad zones of high strain can be examined in several parts of the district (Figure 6), but the original trend of these zones is not known since they have been re-oriented by D3 fold structures. The extent of large-scale dislocation on slides associated with D2 fold limbs is not known, although the interleaving of migmatitic and non-migmatitic rocks is thought to have occurred during D2 folding (Highton, 1992). Metagabbro bodies considered to have been emplaced after the D1 folding are confined to the zone of D2 interleaving (Highton, 1992). Within the Glenkirk–Ballachrochin region, a repetition of the inferred stratigraphical succession of psammite (with some quartzite) overlain by migmatitic semipelite (with thin quartzites) may be the result of thrusting, since different parts of the succession are juxtaposed. The tectonic discontinuity is inferred to occur at a lithological boundary marked by the occurrence of augen gneiss and small exposures of non-migmatitic schistose rocks. Metagabbro bodies straddle this boundary, although there is little evidence of a significant high-strain zone. A larger intercalation of non-migmatitic laminated schists occurs in the Gleann an Tairbhidh area within the structurally higher migmatitic semipelite unit (allowing for the effects of the D3 antiform in the area). This may indicate that a thrust stack is present, at least in the south-eastern corner of the district. It may be inferred that intercalations of non-migmatitic rocks elsewhere within the district, also closely associated with migmatitic semipelite units, may be related to thrusting.

The inferred thrusts in the Glenkirk–Ballachrochin region are early D2 or even pre-D2 in age, since they are apparently folded around hook structures produced by the superposition of D2 and D3 folds.

D3 folds plunge gently to steeply towards the north or north-east and have steeply inclined axial surfaces trending between north and north-east. They range in amplitude from a few centimetres to several kilometres (Figure 6).

Small-scale D3 folds are not common in the northern part of the district. South and east of the Moy Granite, however, D3 strain is higher and small-scale open to tight fold structures are locally abundant and well displayed in north-west-striking zones east of Strath Nairn. It is not possible to distinguish D2 and D3 structures in the north-east-trending zone in and to the west of Strath Nairn. Structures with opposing senses of vergence testify to the presence of at least two generations of small-scale folds in this area. Axial-planar fabrics are not developed generally in D3 structures.

The best examples of superimposed fold structures produced by the three phases of deformation are to be found in psammitic lithologies in the Beinn nan Cailleach area [720 330], where a 3 mm-spaced S2 cleavage, axial planar to close D2 structures, is refolded by open D3 folds.

Migmatisation

The migmatite complex has experienced at least two phases of migmatisation, related to the D1 and D3 deformation events, which have produced leucosome segregations composed mostly of quartz and plagioclase. Migmatites are recognised throughout the outcrop, with the exception of the schistose intercalations around the Findhorn Valley. Structural relationships are best seen in the coarse-grained migmatitic semipelites in the south-western part of the district in the Farr area.

Stromatic leucosome segregations, typically 1 cm or less thick, are concordant with the regional D1 tectonic fabric. The leucosomes have been transposed both by D2 and D3 fold structures to an extent that in places segments of leucosome are re-aligned along axial surfaces of folds. At Glengeoullie Bridge [8562 4760], migmatitic leucosomes in psammites (Plate 4) are parallel to the composite S0–S1 tectonic fabric, whereas leucosomes in interlayered thin (1 m-thick) units of migmatitic semipelite have been transposed by D2 folds and parallel the D2 fabric. The thickness of these leucosomes is reduced to between 1 and 3 mm in high-strain zones associated with attenuated D2 fold limbs, as seen to the north-east of Creag Bhuide [between 668 315 and 677 323]. The intense fabric in these high-strain zones is cut by coarse-grained, thicker (up to 5 cm) leucosomes ranging in form from stromatic to lensoid. In places, neosome development has proceeded to such an extent that the rock has taken on a granitic aspect. In the Creag Bhuide area [665 317], several exposures demonstrate that the younger leucosomes are aligned along the axial surfaces of D3 folds. The cross-cutting nature, together with prominent discordant mafic selvedges with respect to the D1 or D2 fabrics, indicate that these are not transposed segregations. It is apparent that leucosome segregation is penecontemporaneous with D3 deformation. Elsewhere, coarse leucosomes are contorted and apparently folded (Plate 3), at least in part, as a result of selective segregation mimicking earlier folds.

Where D2 strain is low, there is a convergence in appearance between the two generations of leucosome, so that in many parts of the district it is not possible to differentiate the migmatites.

The schistose rocks around the Findhorn Valley show no evidence for migmatisation other than rare leucocratic lenses. They are also finer grained than the remainder of the migmatite complex. Tectonic interleaving during, or before, the D2 deformation, as suggested above, may account for the lack of D1 migmatisation. However, these rocks apparently have not been affected by either phase of migmatisation. This raises the question as to whether there is a compositional control on the distribution of migmatisation. It should be pointed out that Haselock (1984) recognised geochemical differences between the Glenshirra and Corrieyairack successions in the Corrieyairack area of the Central Highlands, that may equate with the migmatitic and non-migmatitic rocks recorded here.

Regional metamorphism

The metamorphic history of the Central Highland Migmatite Complex is not well known, since index minerals are not abundant. Furthermore, the coarse-grained nature of many of the metasedimentary rocks hinders interpretation of the structural chronology of mineral growth. Garnet occurs throughout the complex, whereas andalusite and fibrolitic sillimanite are only

a

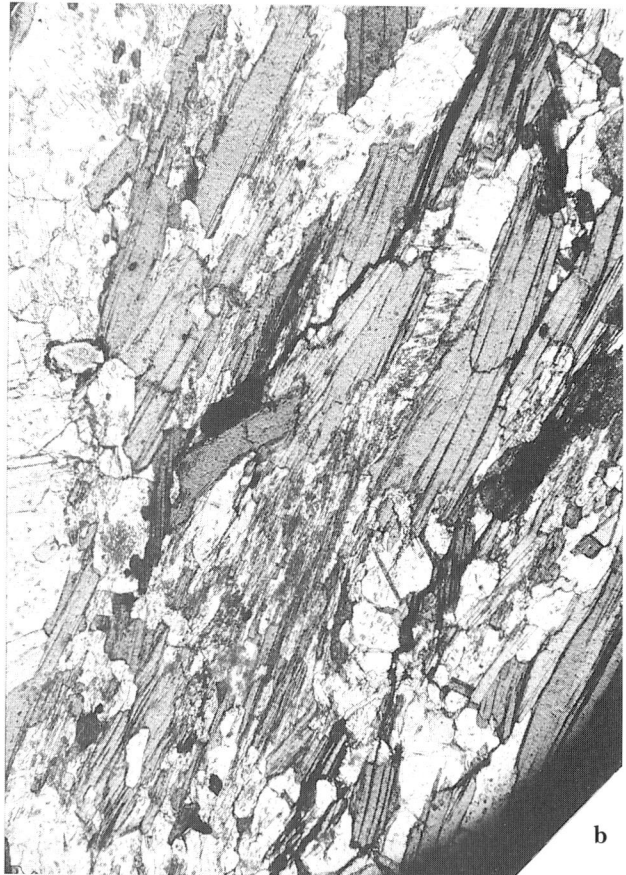

b

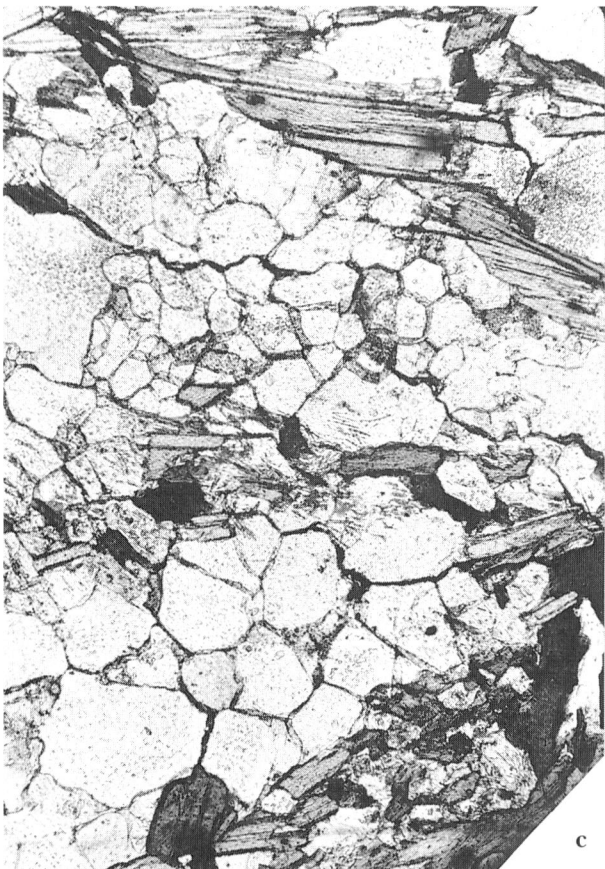

c

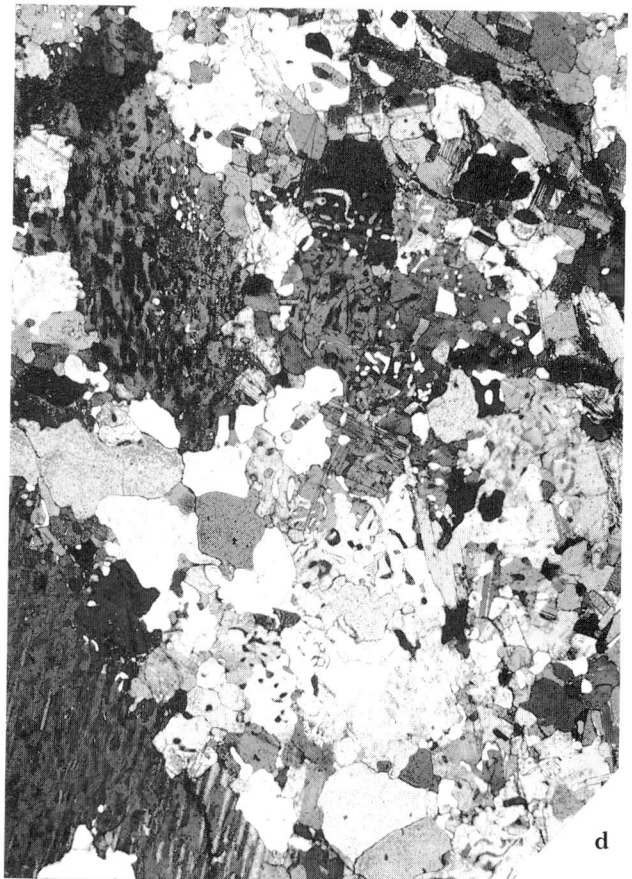

d

Plate 5 Photomicrographs
a) Granoblastic polygonal quartz with symplectite pseudomorphing biotite towards the bottom right. Hornfelsed Moy Granite, Carn Dubh Beag [7728 4042]. (S 9398 × 17.5, cross polarised light) (PMS 615).
b) Sillimanite-bearing hornfelsed semipelite. Sillimanite is intergrown with biotite (lower centre) and fibrolitic sillimanite occurs within quartz (top left). Tom a' Ghealagaidh [8292 3259]. (S 81919 × 35, plane polarised light) (PMS 616).
c) Sillimanite-bearing hornfelsed semipelite. Fibrolitic sillimanite occurs within granoblastic polygonal quartz, demonstrating the thermal metamorphic origin of the sillimanite. Tom a' Ghealagaidh [8292 3259] (S 81919 × 35, plane polarised light) (PMS 617).
d) Irregular textures composed of antiperthite and myrmekite in granitic gneiss. Allt Seileach [8347 3246] (S 81916 x 20, cross polarised light) (PMS 618).

developed locally in semipelitic lithologies, probably as a result of thermal metamorphism.

Metamorphic features recognised throughout the district in semipelitic lithologies include the occurrence of corroded and embayed garnets, in places only occurring as relicts, and the near-ubiquitous occurrence of muscovite porphyroblasts overgrowing biotite. Muscovite, typically discordant to the tectonic fabric (S1) defined by biotite, may overgrow D2 'high-strain' fabrics. It is mostly of a similar grain size to biotite, although in places coarse muscovite porphyroblasts up to 1 cm or more across are recorded. Muscovite porphyroblasts may contain inter-grown 'stubby' biotite, opaque ores and in places fibrolitic sillimanite.

Granitic augen gneiss, composed of quartz, plagioclase, K-feldspar and biotite, is developed locally within the migmatitic semipelites, particularly in the Glenkirk area, structurally immediately below the proposed thrust along which the metagabbros may have been emplaced. The rocks preserve heterogeneous textures further complicated by the effects of thermal metamorphism, as discussed below. Plagioclase is in part antiperthitic and contains abundant quartz 'blebs' together with myrmekite (Plate 5d). K-feldspar is commonly coarsely perthitic. Muscovite is absent, whereas allanite is a common accessory mineral occurring in partially metamict grains up to 1 mm in diameter. Relicts of garnet occur within plagioclase-biotite aggregates. Such rocks are interpreted as being the products of partial melting of micaceous psammitic or semipelitic metasedimentary rocks. The occurrence of two tectonic fabrics within some of these rocks, with at least the later fabric postdating migmatisation, suggests that metamorphism with associated partial melting occurred early in the history of the complex. The absence of muscovite may relate either to the original composition of the sedimentary precursor, or to the metamorphism in which K-feldspar was generated at the expense of muscovite. However, age relations with muscovite porphyroblasts elsewhere in the district are not known.

Contact metamorphism is widespread within the pre-Devonian rocks south of the Great Glen Fault Zone and is attributed to post-tectonic igneous intrusions. This metamorphism is discussed in Chapter Three.

THREE

Igneous rocks

Intrusive igneous rocks occur on both sides of the Great Glen Fault Zone and underlie almost 50 per cent of the area south-east of Strath Nairn within the Fortrose district. They were emplaced into the Rosemarkie Metamorphic Complex and the Central Highland Migmatite Complex before deposition of Devonian sediments, as indicated by the absence of intrusive or extrusive rocks within the Devonian succession and by the occurrence of a variety of lithologies of igneous origin preserved as clasts within the Devonian conglomerates (Chapter Four).

Pretectonic basic intrusions, the metabasites described in Chapter Two, are represented by amphibolites in the Rosemarkie area, metagabbro bodies in the Findhorn Valley and garnetiferous amphibolite pods in Strath Nairn. Deformed granite veins in the Rosemarkie area are described in Chapter Two. Several periods of late to post-tectonic granite emplacement are recognised, with successive intrusions in the area north of Moy. Minor intrusions of intermediate to acidic composition are widespread, although not abundant. Fenites in the Rosemarkie area are associated with faulting and are described in Chapter Six.

GRANITIC INTRUSIONS

The principal intrusions (Figure 7) include the Moy Granite, Auchnahillin Monzodiorite, Saddle Hill Granite, Findhorn Granite Complex and Balfreish Microgranite; minor bodies include the Quilichan Vein Complex and Carn Odhar Granite. The Moy Granite is late to post-tectonic, whereas the remainder are post-tectonic intrusions.

Moy Granite

The Moy Granite is the largest intrusive body within the district, underlying more than 55 km^2 between the Findhorn and Nairn valleys to the north of Moy. The granite was first described as a 'Newer Igneous Rock' by Horne (1923). Zaleski (1982, 1985) has examined its geochemistry and in a study of Speyside and Lower Findhorn granitic rocks, subdivided what she termed the Moy Intrusive Complex into a syntectonic intrusion of granodiorite and granite referred to as the Main Phase (here renamed the Moy Granite) and a post-tectonic leucocratic granite named the Finglack Alaskite (here renamed the Saddle Hill Granite). The Main Phase was considered to have been emplaced at about 455 ± 45 Ma, with an initial $^{87}Sr/^{86}Sr$ ratio of 0.7185 ± 1 based on a poorly constrained 'best-fit' line through Rb-Sr whole-rock data points. The age of Alaskite emplacement was thought to have been constrained by an Rb-Sr whole-rock isochron age of 407 ± 5 Ma with initial $^{87}Sr/^{86}Sr$ ratio of 0.70816 ± 20 (Zaleski, 1982, 1985). The Ardclach Granite, east of the district, is lithologically similar to the Moy Granite (Horne, 1923) and has yielded a concordant U-Pb monazite age of 475 ± 7 Ma (Zaleski, 1982).

Large areas of the Moy Granite are peat covered and unexposed. However, in places exposure is good, particularly in the vicinity of Beinn a' Bheurlaich [around 733 360] at the south-western corner of the intrusion and in stream and crag exposures on the western side of the Findhorn Valley between Carn an Uillt Bhric [838 391] and Ruthven [816 330] at its eastern margin.

Lithologically, the Moy Granite is a medium to coarse-grained (2 to 5 mm grain size), grey to pink, sparsely porphyritic to porphyritic, locally foliated granodiorite to granite. Despite this apparent variability, which is inferred to occur within a single intrusive phase since no boundaries between lithologies are available to scrutiny, individual exposures appear homogeneous, differing only in the abundance and size of feldspar megacrysts. Modal analyses are presented in Appendix 1.

Granodiorites and granites show no obvious pattern of spatial distribution; granodiorite is rather more abundant than granite (Figure 8). Almost all exposures of these rocks are K-feldspar-megacrystic, with megacrysts perhaps more sparsely distributed in the north-east. K-feldspar megacrysts are tabular and sub- to euhedral, typically showing a range of sizes up to 3 cm long. Megacrysts poikilitically enclose plagioclase in many places. The abundance of megacrysts varies on all scales from very sparse (1 per cent or less) to prominently megacrystic with 10 to 20 per cent megacrysts as seen in the Beinn a' Bheurlaich area [around 738 370]. There is no obvious correlation between the abundance and size of megacrysts; sparsely distributed megacrysts can be 6 cm long [as at 774 382]. Similarly, no correlation between abundance and size of megacrysts and composition of the host rock has been identified.

The groundmass consists essentially of medium-grained (2 mm grain size) quartz, plagioclase and biotite with K-feldspar commonly occurring only as an interstitial mineral. Biotite is included in plagioclase and quartz, but rarely contains plagioclase inclusions. It is typically straw brown to dark brown in thin section, although reddish brown biotites together with biotite + feldspar ± muscovite ± andalusite symplectites occur widely and are attributed to the effects of thermal metamorphism (see p. 33). The development of subgrains and/or recrystallisation of quartz throughout much of the intrusion is also attributed to this metamorphism. Plagioclase typically occurs as laths and tabular grains, commonly with irregular and embayed margins. Myrmekite is widespread

Figure 7 Map showing relationships between outcrops of igneous rocks, thermal meta-morphism (ruled lines) and total-field aeromagnetic anomalies (contours at 10 nT interval) in the southern part of the district.

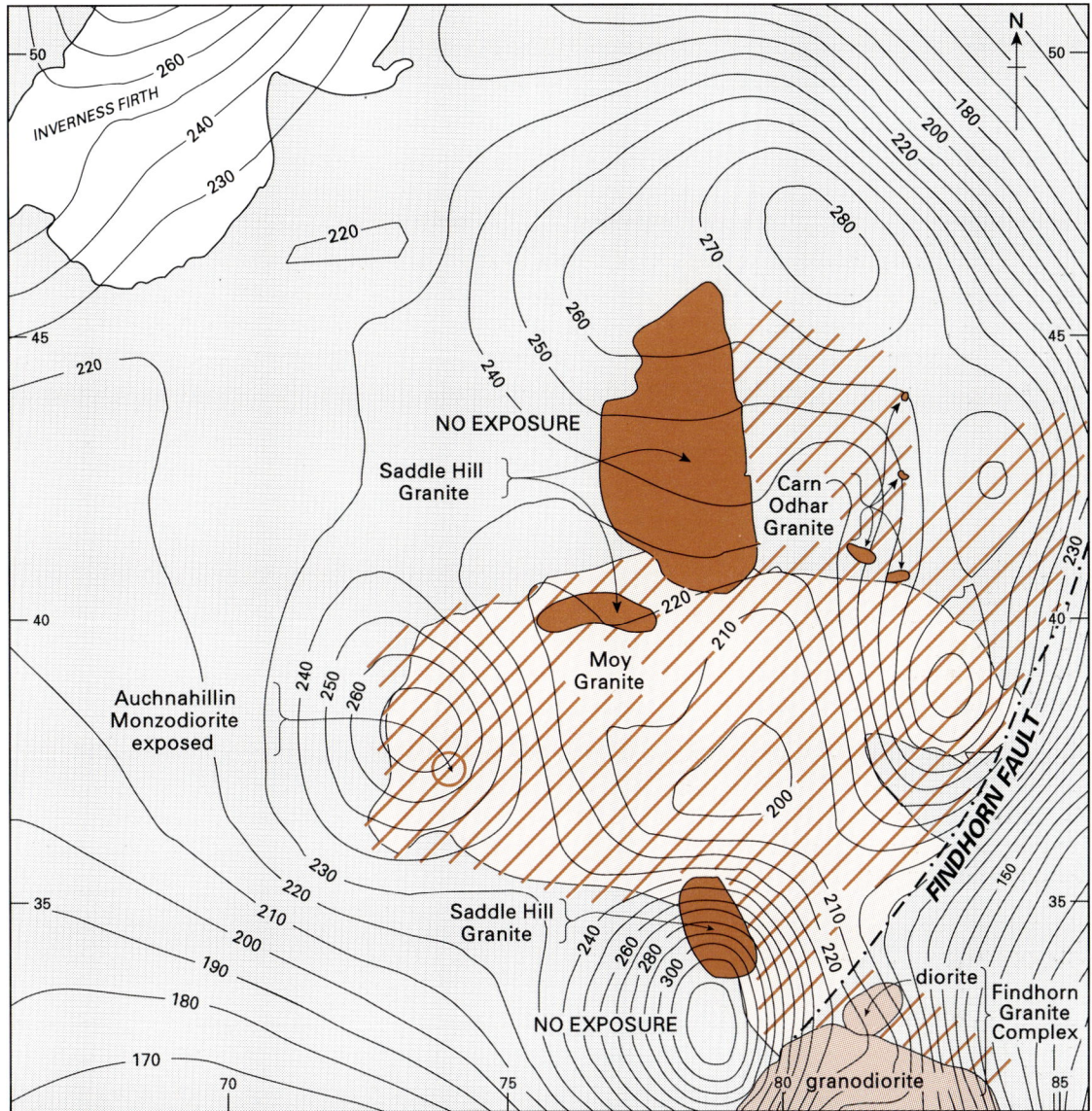

The annular aeromagnetic 'high' is attributed to the largely subsurface Auchnahillin Monzodiorite. The aeromagnetic 'low' in the centre also coincides with a prominent gravity 'low' (Figure 15) and is considered a reflection of the subsurface intrusions of the Saddle Hill and Carn Odhar granites.

at grain boundaries with K-feldspar. Apatite is a common accessory mineral.

Muscovite-bearing rocks are restricted to the more southern and western parts of the intrusion, particularly south-west of Moy. However, even in this area, muscovite is only sporadically developed and rarely exceeds 2 per cent of the modal composition (Appendix 1). It is a late-phase mineral, replacing biotite and occurring at feldspar grain boundaries where it forms ragged grains. No primary igneous muscovite is thought to be present. Muscovite is absent in the north-eastern part of the intrusion.

The granitic rocks are pink in various parts of the intrusion, but particularly so on the ridge between Carn

Dubh Beag [773 404] and Beinn a' Bhuchanaich [762 402]. The high craggy exposures [821 347] west of Shenachie, lying immediately north-west of the Findhorn Fault, are pinkish red. It is possible that the reddening was enhanced by the effects of the adjacent faulting. Elsewhere, K-feldspar is usually, although not every-where, pink, and accounts for the pinkish aspect of some exposures. Reddening is also a typical feature of joint surfaces and some areas of decomposed rock, as in Allt Breac [around 835 385].

The intrusion locally displays a foliation defined princi-pally by the preferred orientation of biotite. This is generally too poorly developed to measure with any accuracy, except in the area west of Beinn Bhreac [around

Figure 8
Plot of modal compositions of the major granitic intrusions in the southern part of the district.

Compositional fields (after Streckeisen, 1976):
a—granite,
b—granodiorite,
c—tonalite,
d—quartz monzo-diorite,
e—quartz diorite.

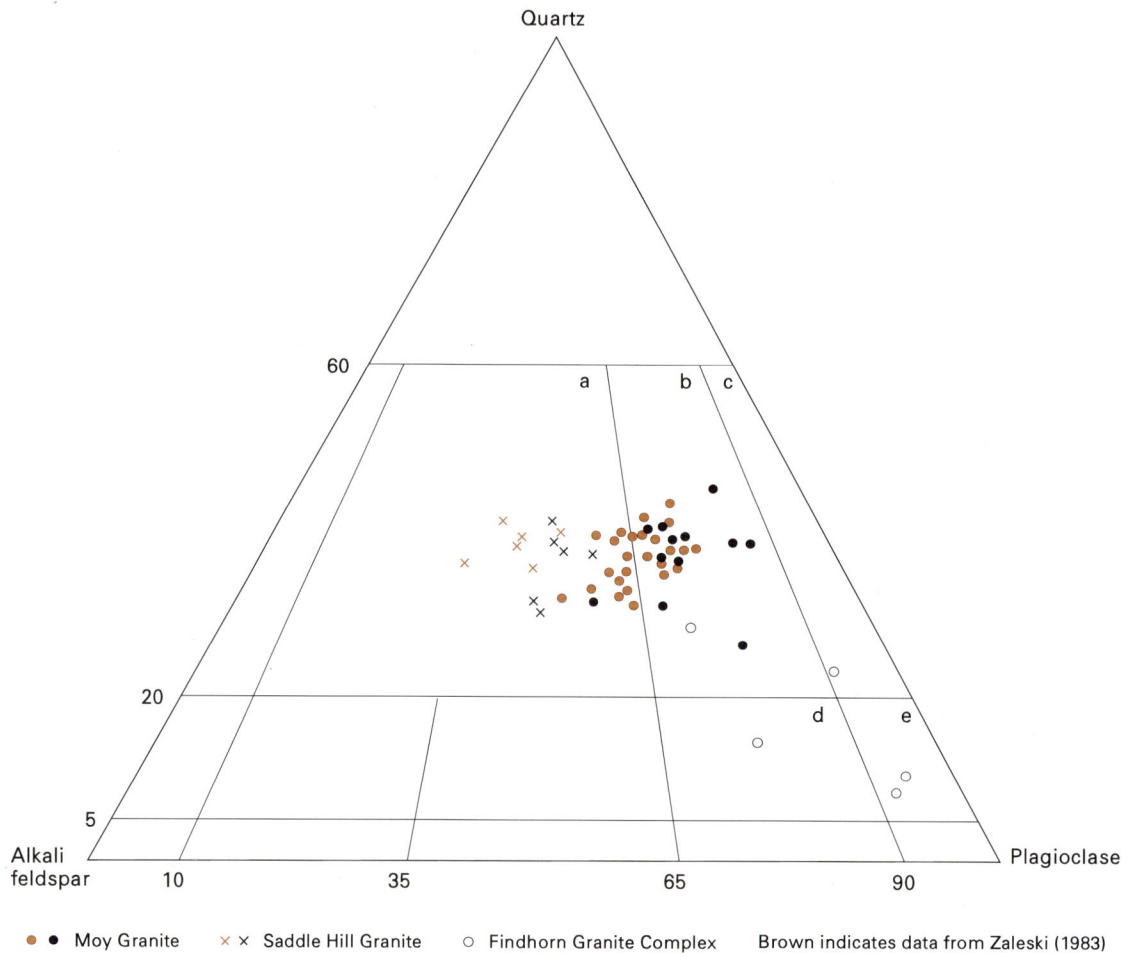

● ● Moy Granite   × × Saddle Hill Granite   ○ Findhorn Granite Complex   Brown indicates data from Zaleski (1983)

93TF14N

774 382]. Here, the feldspar megacrysts also show a weak alignment parallel to the biotite foliation that dips between north and north-north-west at 40 to 60°.

The eastern margin of the intrusion is well exposed, especially in the Ballaggan Burn area [837 372], where the contact is sheeted over a distance of 200 m. Granite and pegmatite veins and sheets up to 1 m thick, broadly concordant to the gneissose layering in the host metasedimentary rocks, become more abundant and are replaced by texturally heterogeneous granite with pegmatite pods containing screens and xenoliths of gneissose semipelite and psammite. These have been assimilated in part by the granite, as indicated by a range of contacts from coherent tabular bodies, through bodies with diffuse contacts, to irregular biotite-rich schlieren within the granitic rock; some are contorted and many contain coarse muscovite porphyroblasts. Assimilated xenoliths are reflected in places by a 'ghost gneissose layering' with concordant pegmatite veins.

A similar sheeted contact to the intrusion is exposed in the crags south-west of Creag a' Chròcain, as well as west and north-west of Shenachie, where micaceous psammite containing granite and pegmatite sheets passes to the south-west into an area of granitic rock containing screens and rafts of psammite. The host lithology in the Shenachie area is more psammitic than that in parts of the Ballaggan Burn area and assimilation of the metasedimentary rocks is less apparent. However, there remain areas where a 'ghost fabric' is preserved in the granite, in places close to coherent tabular rafts of psammite.

Elsewhere, the eastern margin of the Moy Granite is defined by the Findhorn Fault. The northern, western and southern margins of the granite are poorly exposed. Granite veining in these areas is apparently less intense than at the eastern margin, as seen in the Meall Mór area [737 356], although granite and pegmatite veins and sheets thought to be related to the Moy Granite are widespread, if not everywhere abundant.

Psammite and micaceous psammite xenoliths are very common and up to at least 20 m across on the ridge between Carn Dubh Beag and Beinn a' Bhuchanaich, possibly indicating a position close to the roof of the intrusion. A 2-m xenolith of grey, weakly foliated, medium-grained, sparsely feldspar-phyric tonalite is also present and contains plagioclase megacrysts up to 5 mm long. Although the contact between the tonalite and psammite is not exposed, both rocks are cut discordantly by pink coarse-grained granite to granodiorite, grading into pegmatite containing small psammite xenoliths and biotite-rich schlieren.

Elsewhere within the intrusion, xenoliths are rare. Biotite-rich schlieren and xenoliths occur in Caochan Odhar [at 8239 3757 and 8278 3823] adjacent to a large (about 0.5 km wide) raft of gneissose semipelite, and in an area [774 382] west-north-west of Beinn Bhreac. Lenses of pegmatite are almost invariably associated with granitic rocks containing schlieren and xenoliths.

The granite/granodiorite is cut by thin (mostly less than 1 m wide) microgranite sheets in the upper parts of the Moy Burn tributaries. Microgranite and fine to medium-grained granite sheets are quite abundant in the central parts of the intrusion in Allt na Beinne [between 800 382 and 811 390]. Most are less than 30 cm thick, although a steeply dipping sheet [at 8057 3873] is apparently about 5 m thick. A brick-red microgranite [at 8082 3884] contains biotite, quartz and feldspar phenocrysts up to 2 mm across and microgranite veins also occur in Caochan Odhar [824 371].

The Moy Granite together with included xenoliths preserves textures attributed to contact metamorphism. These textures are discussed on p.00.

New geochemical analyses of the Moy Granite are presented in Appendix 2 and plotted along with existing data for both the Moy and Saddle Hill granites (Zaleski, 1982; O'Brien, 1985) in a selection of variation diagrams (Figure 9a–c). CIPW norms of both granites are plotted in Figure 10a and b. A brief discussion of the geochemistry is given on p.00.

Auchnahillin Monzodiorite

A specimen (S 18373) of 'hornblende-biotite granite' was recorded by Horne (1923, p.63) from '5/8 of a mile west-south-west of Meallmore Lodge', in the south-western corner of the Moy Granite. Re-examination of the specimen reveals that the rock is a monzonite with a modal composition estimated to be quartz 2 per cent, K-feldspar 30 to 35 per cent, plagioclase 55 per cent, biotite 10 per cent and hornblende 2 per cent; the thin section is too small to permit a more precise modal analysis. The rock is coarse grained (2 to 5 mm grain size) with tabular plagioclase laths commonly up to 5 mm long, and with somewhat corroded margins. It also contains more-equant, although equally large and in part perthitic, K-feldspar and a small amount of interstitial quartz. Biotite, which is pleochroic from straw-brown to dark greenish brown is more abundant than green hornblende. Opaque ores are spatially associated with biotite and hornblende, as are sub- to euhedral lozenges of sphene up to 0.7 mm long. The hand specimen has a high magnetic susceptibility and has many features in common with lithologies within the Findhorn Granite Complex, described below. During the recent resurvey, a one-square metre exposure [7395 3776] of monzodiorite was examined, although it could not be ascertained whether this was in situ; blocks of monzodiorite lie nearby. Petrographically, the monzodiorite is similar to that described above, but with

poikilitic K-feldspars up to 6 mm across. A modal analysis of a specimen (S 95139) from this locality is presented in Appendix 1 (Table 5).

On the basis of the coincidence of thermal metamorphism in the Moy Granite and a significant aeromagnetic anomaly (Figure 7), the monzodioritic intrusion is interpreted as forming part of the roof of a large concealed body (see Chapter Seven). This body may be lithologically similar to the Findhorn Granite Complex, which contains monzodioritic rocks and granodiorites, has high magnetic susceptibility and is significantly younger than the Moy Granite. Since the aeromagnetic 'high' is annular in form, with a central 'low', it is thought that the central parts of the postulated intrusion were either much less magnetic than the marginal areas or that they have been intruded subsequently by a less magnetic rock. The Saddle Hill and Carn Odhar granites may conceivably be parts of such an intrusion.

An exposure [8428 3789] of hornblende-sphene-bearing granodiorite 2 m across crops out between Ballaggan Burn and Allt Breac. The rock is medium grained, grey to pink with scattered K-feldspar megacrysts, and shows lithological similarities with the Moy Granite, including the occurrence of recrystallised quartz and biotite. Major differences, however, include the presence of hornblende (modal analysis S 79364 in Appendix 1) in the form of dark bluish green grains and sub- to euhedral lozenges of sphene up to 0.5 mm across. The rock also has a magnetic susceptibility more than an order of magnitude higher than other specimens of the Moy Granite and occurs adjacent to the annular aeromagnetic anomaly. It may also be part of the Auchnahillin Monzodiorite.

Saddle Hill Granite

Zaleski (1982; 1985) differentiated a younger (407 ± 5Ma) intrusion, the Finglack Alaskite, from the older Moy Granite Main Phase on the basis of age and lithology. The alaskite is here renamed the Saddle Hill Granite since the intrusion is not an alaskite (defined by Le Maitre (1989) as a leucocratic granite consisting almost entirely of quartz and alkali feldspar) and it is not exposed at Finglack. Zaleski (1985) described contact-metamorphic effects in the Main Phase granite, which she attributed to the 'alaskite'.

The Saddle Hill Granite occurs in the form of three intrusive bodies; the largest is centred around Saddle Hill, with smaller bodies south of Carn Dubh Beag and east of Loch Moy (Figure 7). The intrusions, which are reasonably well exposed, particularly in crags on the southern side [789 433] of Saddle Hill, comprise essentially a medium to coarse-grained (3 to 5 mm grain size), non-porphyritic, pink, leucocratic granite. Biotite accounts for less than 3 per cent of modal compositions (Appendix 1). The granite is weakly quartz-phyric with subrounded grey quartz; feldspars are typically pink. Rare, small (10 cm across) pegmatite pods occur throughout the granite bodies.

Fine to medium-grained, pink, quartz-phyric granite exposed near Mid Craggie [735 388], close to the

Figure 9
Geochemical
variation diagrams
for the Moy and
Saddle Hill
granites.
a) Rb/Sr
b) K$_2$O/Rb
c) Y/Zr.

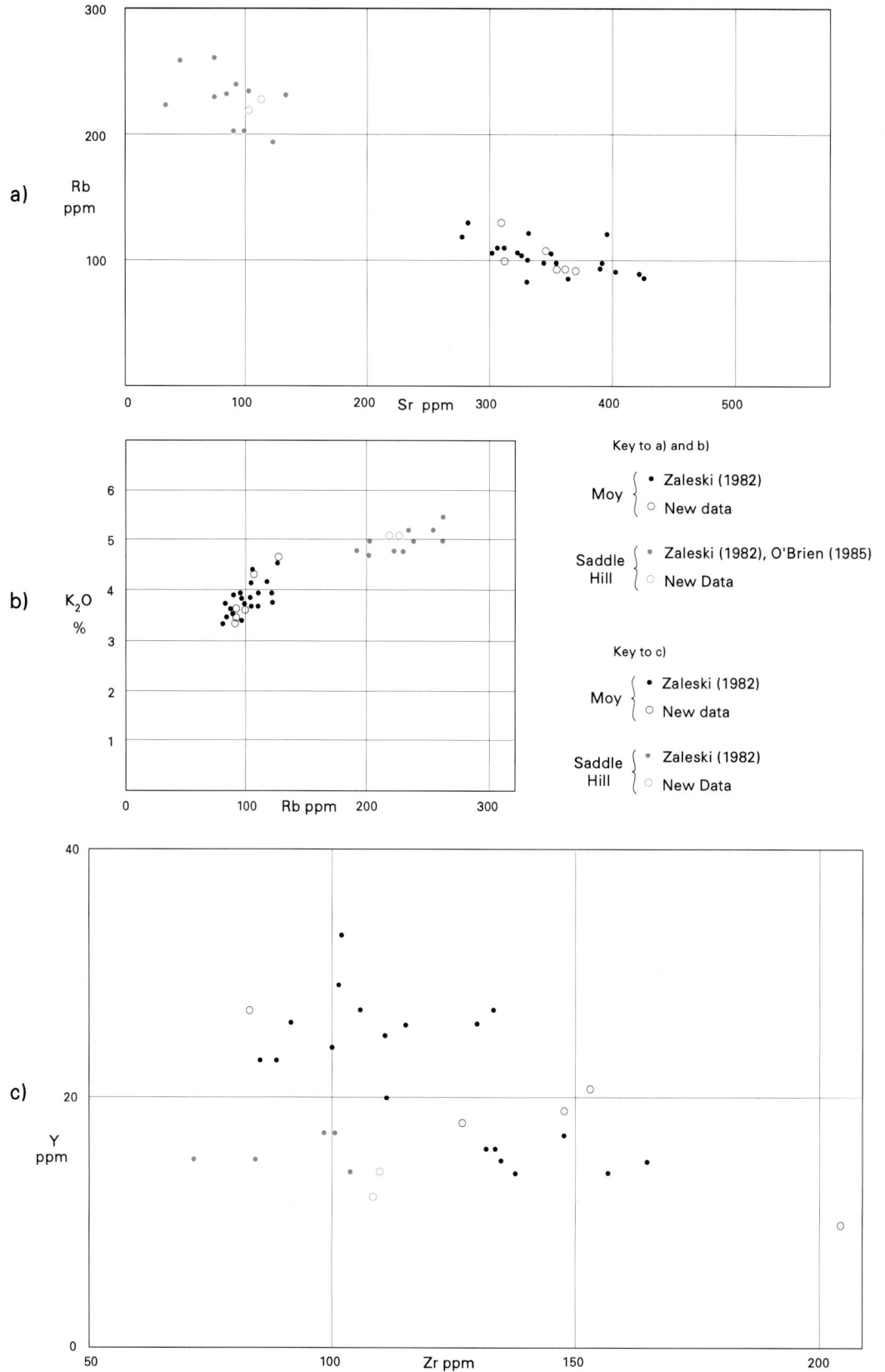

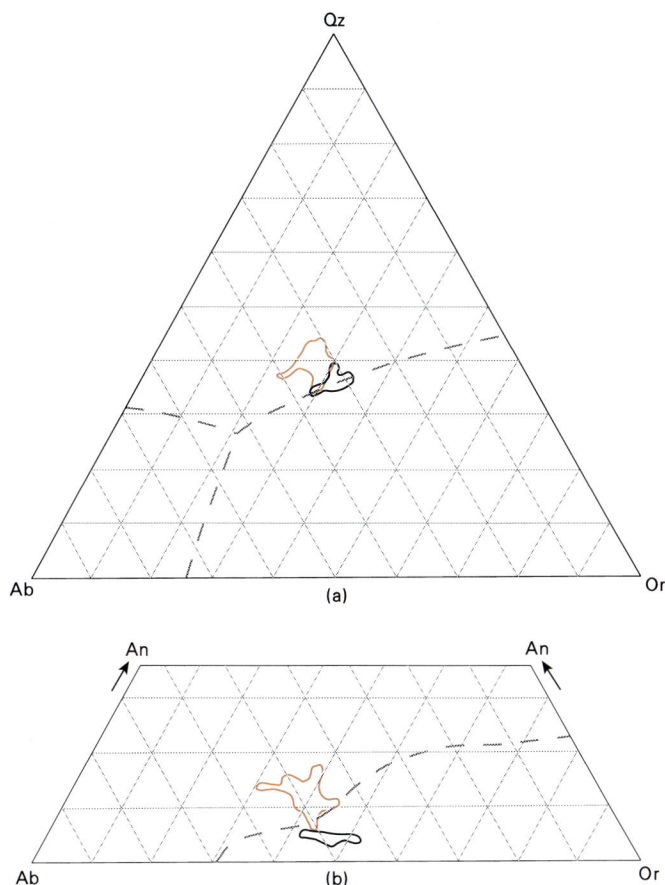

Figure 10 Diagrams showing the range of CIPW normative compositions of the Moy (brown) and Saddle Hill (black) granites, based upon data from Zaleski (1982) and new analyses: a) Qz-Ab-Or; b) Ab-Or-An. Cotectic lines (dashed) at 5 kb water vapour pressure after Winkler (1979).

inferred western contact of the Moy Granite, may represent the roof of a fourth intrusive body.

In thin section, the granite shows many features typical of coarse-grained leucocratic granites. These include prominently perthitic K-feldspars, irregular and in part embayed grain boundaries and rare accessory minerals (restricted to a few apatites). Biotite, usually altered to chlorite with spatially associated opaque ore, is included within plagioclase and both are included within K-feldspar.

The granite is cut by parallel-sided microgranite sheets and dykes (about 25 cm wide) around Beinn Bhuidhe Mhór [788 406] and in Allt Carn a' Ghranndaich [7753 4230]. In thin section, these are similar to the Saddle Hill Granite, although much finer grained, and show evidence for more-extensive replacement of feldspars by quartz.

Anastomosing, brick-red, very fine-grained haematitic veins a few millimetres wide occur in places, particularly around Beinn Bhuidhe Mhór and on Beinn a' Bhuchanaich [at 7624 4000]. They are broadly parallel to a set of joints trending 160°, but are somewhat

sinuous. They contain angular fragments of quartz and feldspar within an iron-stained, cryptocrystalline, apparently quartzofeldspathic groundmass. Such veins mostly have very sharp margins. However, in places, some veins pass into broader zones with a very fine-grained recrystallised, rather than brecciated, texture and have more diffuse contacts with the surrounding granite. Similar breccias have been recorded in association with zones of hydrothermal alteration from within the Monadhliath Granite (Highton, in preparation).

The outcrop pattern of the granite body south of Carn Dubh Beag [around 772 402] suggests that it is in the form of a subhorizontal sheet, whereas the outcrop patterns of other granite bodies are best interpreted as steep-sided cupolas. Contact relationships with the Moy Granite are not exposed, but contacts with the host metasedimentary rocks are exposed east of Finglack [at 777 439] and in Allt Cromachan [at 7707 4228]. At the former locality, the granite is fine to medium grained and grey or pink with pegmatitic segregations within 2 m of the contact. In places, this contact zone contains biotite distributed in such a way as to give the appearance of a 'ghost' banding, contorted in places. The contact is irregular where pegmatite has invaded the host metasedimentary rocks. At the latter locality, the granite contact is apparently sharp and steeply dipping. The contact zone is rather pegmatitic and has an irregular texture.

Two new geochemical analyses of the Saddle Hill Granite are presented in Appendix 2. The data are plotted in Figures 9 and 10 together with data from Zaleski (1982) and O'Brien (1985); a brief discussion is given on p.00.

Findhorn Granite Complex

The Tomatin (Findhorn) Granite was subdivided by van Breemen and Piasecki (1983) into late-tectonic granodiorites called the Glen Kyllachy Granite complex, and post-tectonic granodiorites with dioritic and appinitic variants, called the Findhorn Granite complex. The Glen Kyllachy Granite is considered to have been emplaced at about 443 Ma based on Rb-Sr muscovite ages for associated pegmatites and Rb-Sr whole-rock ages. The Findhorn Granite on the other hand is thought to have been emplaced at 413 ± 5 Ma based on Rb-Sr biotite and K-Ar hornblende ages (van Breemen and Piasecki, 1983). Initial $^{87}Sr/^{86}Sr$ ratios for the granites (van Breemen and Piasecki, 1983) of 0.7176 ± 4 and about 0.706, respectively, are similar to those for the Ordovician granites of Aberdeenshire and the late-Silurian granites throughout the Highlands, respectively.

The Glen Kyllachy Granite lies outside the present district, but the Findhorn Granite Complex occupies part of the south-eastern corner, south-east of the Findhorn Fault (Figure 7), in the Ruthven [814 330]–Balvraid [830 314] area. Exposure is restricted to a small area at Ruthven [8173 3300] and an area with small crops and loose boulders on the southern bank of the Findhorn [around 813 323]. On the basis of this limited evidence, the Findhorn Complex within the district appears to be a bimodal dioritic and granodioritic intrusion.

At Ruthven, medium-grained mesocratic quartz diorite with sporadic development of plagioclase phenocrysts contains veins up to 2 cm across and irregular patches up to 50 cm wide of leucocratic, coarsely porphyritic, quartz monzodiorite or granodiorite. Leucocratic patches contain abundant poikilitic K-feldspar megacrysts up to 1.5 cm long together with dioritic xenoliths. Contacts between the two lithologies are sinuous or lobate, and irregular, with bifurcating veins of quartz monzodiorite extending into the dioritic rocks. The south-western part of the exposure is dominated by quartz monzodiorite to granodiorite. Thin sections show that all lithologies contain hornblende replaced by biotite to a greater or lesser extent. Large dark brown biotite plates up to 2 mm long occur in places, but pyroxene is preserved in the cores of some amphiboles and, in one specimen, in biotite. Net-veined lithologies contain dioritic plagioclase-hornblende textures together with areas containing quartz, K-feldspar and plagioclase. Between these areas, dioritic textures are poikilitically enclosed by large K-feldspar megacrysts. The rocks, therefore, preserve macro- and microscopic features of magma mixing. Sphene is a common accessory mineral, both as discrete grains and as rims to opaque grains. Apatite is also prominent and abundant, commonly occurring as coarse needles which, in places, are concentrated in late interstitial K-feldspars (some of which are microcline). Some apatites contain 'cloudy' zoned cores thought to be produced by submicroscopic inclusions.

Pink porphyritic biotite-rich granodiorite crops out 700 m south-south-west of Ruthven on the southern bank [813 323] of the Findhorn. Tabular K-feldspar megacrysts up to 2 cm across are rather variably distributed, but may comprise up to 25 per cent of the rock. K-feldspar also occurs as interstitial grains. Quartz is less abundant than in the Moy Granite (see Appendix 1) and also occurs in part as interstitial grains. Hornblende has largely been replaced by biotite, which contains abundant apatite inclusions. Sphene is a common accessory mineral, both as discrete grains up to 1 mm across and as rims to ilmenite grains. The relationship between this pink porphyritic granodiorite and the granodioritic rocks at Ruthven is not known, due to lack of exposure.

Balfreish Microgranite

Small intrusions of 'binary granite' near Balfreish and Cantraydoune were considered by Horne (1923) to be apophyses of the larger granite intrusions. Several small intrusions of porphyritic microgranite and medium-grained porphyritic granite are here described together as the Balfreish Microgranite.

Microgranite exposed 300 m south-west of Cantraydoune [at 789 460] is composed of pale pink, porphyritic, fine to medium-grained granite. Quartz is the most prominent phenocryst mineral, occurring as grey grains and aggregates of grains up to 5 mm long. Both plagioclase and K-feldspar phenocrysts up to 1 cm long are recognised; plagioclase phenocrysts show partly sericitised cores overgrown by narrow fresh rims, whereas K-feldspar megacrysts are perthitic and poikilitically enclose plagioclase. Ragged, partly chloritised biotite also occurs as

phenocrysts up to 1 cm across. Loose blocks of pink, quartz-phyric microgranite are abundant 600 m west of Cantraydoune [at 785 461]. These are considered to represent a finer-grained variant of the porphyritic granite. A pink, quartz-biotite-phyric microgranite sheet striking north-north-east in the Burn of Cantraydoune [7924 4580] is at least 2-m thick and apparently dips steeply.

Similar quartz-phyric microgranite with quartz and some feldspar phenocrysts up to 4 mm across within a fine-grained pink groundmass is common as blocks and thin veins in psammite boulders in stone boundary walls around Balfreish [798 470]. In this area, a 2-m overgrown section [8002 4675] consists of microgranite passing into pegmatite. Biotite-rich patches there are assumed to be assimilated xenoliths.

Quartz-phyric microgranite south-east of Croygorston [around 773 450] is thought to represent a third body of the Balfreish Microgranite.

MINOR GRANITIC INTRUSIONS

Quilichan Vein Complex

Pink to red, biotite-muscovite granodiorite, grading into pegmatite, crops out in the form of veins, sheets and stocks south-east of the Findhorn Fault between Drynachan Lodge [865 397], immediately east of the district, and Shenachie [827 349]. Anastomosing granodiorite sheets are well exposed on the banks of the Findhorn at Poll Beag [849 372], where both low-angled and steeply inclined granodiorite sheets bifurcate and coalesce to form larger areas of granodiorite, several tens of metres wide. The granodiorite contains rafts of psammite as well as lenses of muscovite-bearing pegmatite. Larger bodies of granodiorite are well exposed in the river cliffs of Creag Ruadh, 500 m west-south-west of Daless [855 382] and in an excavated track section 500 m south of Quilichan [853 374]. On Creag Ruadh, where cataclasite related to the Findhorn Fault is extensive, the granodiorite is medium grained and non-porphyritic, whereas south of Quilichan it is rather inequigranular, carries a weakly developed foliation in places, and is medium to coarse grained with scattered megacrysts up to 3 cm long. It is cut by steeply dipping pegmatite sheets up to 20 cm thick and passes laterally into psammite cut by granodiorite and pegmatite sheets. Modal analyses of granodiorite specimens S 79363 and S 83509 from Creag Ruadh and south of Quilichan respectively are given in Appendix 1 (Table 5).

The granodiorites of the Quilichan Vein Complex are lithologically similar to the western parts of the Moy Granite and, in form of intrusion, with the eastern parts of that granite. Significant differences between the vein complex and the Findhorn Granite Complex are the presence of muscovite and absence of hornblende and sphene in the former.

Carn Odhar Granite

North of the Moy Granite in poorly exposed, peat-covered ground, pink to red, medium-grained leucocratic granite crops out in Allt Creag a' Chait [8234 4030] and near Allt Creag an Iomaill [8256 4213 and 827 435]. These bodies are from 40 m to at least 110 m across. The granite is cut by anastomosing quartz veins (a few millimetres thick) containing drusy cavities into which well-formed quartz crystals project. The quartz veins, which predominantly trend at about 020°, brecciate the granite so that in places granite and feldspar fragments lie within a quartz matrix. The Carn Odhar Granite has lithological similarities with the Saddle Hill Granite.

OTHER MINOR INTRUSIONS

Minor intrusions are sparsely distributed within the district. They include appinitic rocks, lamprophyres, microdiorites and quartz-feldspar porphyries. Geochemical data for a selection of minor intrusions in the district are presented in Appendix 2.

Appinites

Horne (1923) recognised several exposures of monzonitic rocks, together with basic rocks containing hornblende, biotite, a little pyroxene, quartz and alkali feldspar, immediately south-east of Strath Nairn. The rocks were considered to be part of the 'Newer Granites' and 'probably related to the foliated lamprophyres and syenites of Sheet 85'. Re-examination of these igneous rocks indicates that they show affinities with appinitic intrusions elsewhere in the Highlands. The appinites show compositional similarities, and an apparent spatial relationship, to the Auchnahillin Monzodiorite (Figure 7).

REHIRAN APPINITE

A medium to coarse-grained hornblende-biotite mafic igneous rock interpreted as an appinite crops out sporadically over an area of about 0.03 km² 500 m south-south-west of Rehiran, [around 8318 4590 and 8293 4590]. The intrusion ranges from an essentially hornblende-biotite rock to a more leucocratic, gabbroic lithology containing mafic 'clots' about 3 cm wide; both are cut by thin granitic veins and pods. Modally, the rock is dominated by green hornblende (about 50 per cent), in part replaced by biotite with sphene. Biotite is most abundant outwith the hornblendes, where it occurs together with rare colourless pyroxene. Pyroxene also forms cores to some amphiboles. Perthitic K-feldspar, plagioclase and quartz are generally interstitial to the mafic phases, although in places feldspar poikilitically encloses subhedral hornblende. The proportion of quartzofeldspathic minerals is variable and typically less than 30 per cent. Apatite is a significant accessory mineral in the form of euhedral grains up to 0.3 mm across. Although hornblende has recrystallised in places, there is little evidence for metamorphism or deformation of the rock. At one place [8302 4584], appinite is apparently intersheeted with a biotite psammite, whereas nearby [8318 4590] a 'raft' of micaceous psammite is enclosed within the appinite. The micaceous psammite

displays evidence of thermal metamorphism, including the development of 'mats' of white mica up to 4 mm across. These are associated with muscovite up to 1 mm across containing pseudomorphs after fibrolite and ragged opaque grains up to 1 mm; biotite is extensively chloritised.

DALTULICH APPINITE

Exposures of monzonite and syenite in dense coniferous woodland [7384 4183] immediately east of Mains of Daltulich are interpreted as representing an appinite. The intrusion is thought to occupy an area about 250 m in diameter, although external contacts are not exposed. The rock is typically medium grained and examination of thin sections indicates a composition dominated by blue-green hornblende, K-feldspar partly in the form of microcline, and plagioclase. Colourless to pale green pyroxene forms the cores to some hornblendes. In places, however, subhedral pyroxene phenocrysts, typically 1 to 2 mm long and partially replaced by blue-green hornblende, are abundant. Feldspars are interstitial to the pyroxene, with plagioclase in part poikilitically enclosed by K-feldspar. The pyroxene-rich rocks also contain abundant dark brown biotite in grains up to 2 mm across. Apatite is a common accessory in all lithologies and in places forms prisms up to 1 mm long. Quartz is not abundant. The appinitic rocks contain leucocratic, feldspathic patches up to 3 cm across together with pink segregations and veins up to 2 cm across composed almost exclusively of K-feldspar.

ALLT DEARG APPINITIC DYKE

A medium-grained appinitic diorite/monzonite dyke exposed in Allt Dearg [8131 4468] is composed of subhedral amphiboles, pale olive-green in thin section and poikilitically enclosed by feldspars; both K-feldspar and plagioclase are present. Amphiboles are partly replaced by dark brown biotite, whereas some rims have altered to chlorite. The dyke apparently trends north-west and is no more than 5 m wide, although contacts are not well exposed.

CREAG BHUIDE APPINITIC DYKE

A medium-grained appinitic diorite dyke is visible under loose blocks 300 m north-north-west of Creag Bhuide. The dyke apparently has a chilled margin 1 cm wide. In thin section, the dyke is seen to be composed of medium-grained (up to 1 mm grain size) amphibole and plagioclase. Amphibole is orange-brown and occurs in the form of laths commonly with euhedral basal sections; several of these have plagioclase within their cores. The amphibole is poikilitically enclosed by plagioclase, which occurs mainly in the form of laths as much as 2 mm long. Plagioclase is slightly more abundant than amphibole. The rock contains mafic phenocrysts up to 4 mm in diameter, now represented by chloritic aggregates partly replaced by carbonate and opaque ore minerals. The phenocrysts are partially rimmed by the brown amphibole and evidently represent the earliest phase in the rock. There is no direct evidence as to their original composition, but they appear, on the basis of their form, to have been pyroxenes. The dyke is at least 1 m thick, although the orientation is not known. A similar rock, at least in terms of field appearance, 190 m to the east-south-east, occurs in the form of a sheet dipping 25° towards the north.

Lamprophyres

Lamprophyres are rare within the district. A weathered and foliated dyke, 20 cm wide and composed largely of biotite, occurs in the River Nairn under the A9 road bridge at Daviot [7200 3382]. This dyke lies within a granite-pegmatite sheet, with which it has a lobate contact. Both the lamprophyre and the granite occur within a shear zone. The fabric in the dyke is consistent with that expected if it was syntectonic with respect to the shearing.

A lamprophyre 1.5 m thick exposed in Allt na Seanalaich, 240 m south-west of Shenachie [8252 3479], contains chloritic aggregates up to 5 mm across pseudo-morphing an unknown phenocryst phase. The chlorites are partly replaced by carbonate and occur within a fine-grained dioritic groundmass composed of plagioclase laths and chlorite. Plagioclase laths are strongly aligned and define a prominent flow foliation that wraps around the phenocrysts.

A lamprophyre dyke 5 m wide is exposed in Allt Dearg [8266 4655] 500 m south-west of Wester Barevan and strikes 020°; it dips at 55° towards the east. Two phases of dyke intrusion are recognised; a 1 m-wide central zone is fine to medium grained and has a sharp contact with the remainder of the dyke, which is very fine grained to cryptocrystalline. Patches of the fine- to medium-grained rock occur within the cryptocrystalline lithology. Pale sulphide is widely disseminated. In thin section, the central zone is composed of brown to olive-green, subhedral amphibole, poikilitically enclosed within pla-gioclase. In the absence of staining, K-feldspar has not been positively identified. Apatite needles are quite common within the plagioclase.

Daviot Breccia Pipe

A breccia zone was exposed in the north-eastern face of the main bowl of Daviot Quarry [719 391] in 1989 and was only visible in the two-dimensional quarry face. It is near vertical and thins upwards from its 2 m-wide base. It contains subangular clasts of psammite and micaceous psammite, similar to the adjacent undisturbed metasedi-mentary rocks, together with granite and pegmatite clasts, within a green micaceous matrix. The clasts are matrix-supported with no size sorting. The lowest 4 m of the eastern part of the breccia pipe contains fine-grained, purplish brown, rather porphyritic microdiorite that is itself brecciated in part. The microdiorite contains embayed plagioclase and some ragged chlorite pheno-crysts up to 1.5 mm across, together with rare rounded quartz phenocrysts. The groundmass is fine grained and dioritic with extensively sericitised plagioclase laths. The microdiorite is cut by thin calcite veins, but it is not clear whether the breccia pipe is related to the microdiorite dyke or whether the dyke occurs within a fault zone.

Microdioritic dykes

A suite of microdioritic dykes occurs throughout the district. The rocks range in composition from diorites through to granites and at outcrop are typically fine grained and purple to purplish brown. Most are less than 5 m wide, although a dyke in Allt Dearg [8231 4621] is up to 30 m wide. The dykes commonly bifurcate and, in places, include screens of the host metasedimentary rocks. They trend between north-east and north-north-west and are generally steeply inclined, except for two low-angled sheets cutting the Moy Granite in Allt Breac [8347 3823]. Another dyke, north-east of Shenachie [at 8340 3593], cuts biotite-muscovite granite sheets of the Quilichan Vein Complex. It is apparent from examina-tion of thin sections that dioritic compositions are the most abundant, although all members of the suite have certain common features. These include the occurrence of plagioclase phenocrysts and, to a lesser extent, chloritic aggregates as pseudomorphs after a mafic phe-nocryst phase. The latter are up to 5 mm across and referred to as xenocrysts, since they are commonly embayed and in part replaced by carbonate. The more granitic compositions also contain rounded and embayed quartz xenocrysts. The groundmass is typically brown (as a result of extensive alteration of igneous min-eralogy) and composed of plagioclase in the form of laths, chlorite, opaque ore minerals and locally sphene. Quartz occurs in the groundmass of the more granitic lithologies, which generally lack well-developed plagio-clase laths.

The range of lithologies encompassed by the microdi-orite suite includes characteristics overlapping with both the quartz-feldspar porphyries and the lamprophyres. The former case is shown by a microgranite, exposed in Allt Dearg [8231 4621], which contains xenocrysts of rounded and embayed plagioclase, quartz, K-feldspar and chlorite, encapsulated by a graphic intergrowth of quartz and K-feldspar. The remainder of the groundmass has a more granular texture. An overlap with the lampro-phyre suite is suggested by a dyke exposed north-east of Shenachie [at 8340 3593], which contains coarse chloritic or biotitic xenocrysts up to 5 mm in diameter within a fine- to medium-grained groundmass of plagio-clase (in the form of laths), biotite and quartz.

Quartz-feldspar porphyries

Felsites and quartz-feldspar porphyries occur sporadically throughout the district. They are typically pink with some brick-red variants. Felsites are fine grained and apparently granitic in composition, although the fine grain size precludes an accurate assessment of the mineralogy. Por-phyries contain rounded and/or embayed xenocrysts and xenoliths composed of quartz, plagioclase, K-feldspar, rare chlorite and granite within a fine-grained granitic groundmass. Dykes are typically up to 20 m wide and steeply dipping. They show a range of orientations parallel to prominent joint sets in the host metasedimen-tary rocks. They postdate the Moy Granite, as shown by the porphyry dykes within the granite north of Beinn a'

Bheurlaich [around 738 375], although relationships with the microgranites, which also cut the Moy Granite, are not exposed. A dyke 20 m wide in the A9 road section south of Meall Mór [at 7361 3472] is parallel to and separated from a microdiorite dyke 2 m wide by a screen of metasedimentary rock less than a metre wide. The existence of a close genetic link between the felsites and the porphyries is supported by the occurrence of a megacrystic central zone (with quartz and feldspar xenocrysts), 1 m wide, within a felsite dyke in Allt Dearg south-west of Wester Barevan [at 827 466]. There is no apparent intrusive contact between the two lithologies.

GEOCHEMISTRY OF IGNEOUS ROCKS

New geochemical analyses of granitic rocks, minor intrusions, metabasic rocks and a metasedimentary rock from the Fortrose district are presented in Appendix 2; they supplement those of Zaleski (1982), who discussed the geochemistry of the Moy and Saddle Hill granites; only a brief discussion is presented here.

The marked geochemical contrast between the Moy and Saddle Hill granites is shown by a selection of variation diagrams (Figures 9a–c). A granite ternary diagram (Figure 10) indicates that the CIPW normative composition of the Moy Granite extends from the three-phase cotectic line into the plagioclase field at 5 kb water pressure. The Saddle Hill Granite is more evolved than the Moy Granite, as indicated by higher concentrations of SiO_2, K_2O, Rb, Th and Ce, and lower concentrations of Sr, Ba, V and Y. Data for the Saddle Hill Granite (Figure 10) plot close to the three-phase cotectic line at between 2 and 5 kb water pressure (see Winkler, 1979).

The small amount of data from the minor intrusions (Appendix 2, Table 10) hints at geochemical differences between the appinite–lamprophyre suite of rocks and the microdiorite suite, although there are insufficient data for further speculation.

There is a marked geochemical difference between a single specimen of garnetiferous amphibolite and three specimens of metagabbro, but such a limited amount of data must be interpreted with caution. The former has lower Al_2O_3 and higher Fe_2O_3 than the latter for an equivalent concentration of SiO_2. Furthermore, V, Zn, Rb, Y, Zr, Ba, La and Ce are substantially higher in the garnetiferous amphibolite and Sr is markedly lower. Some of the differences, for example in Rb, may be the result of metasomatic modification of the garnetiferous amphibolite, since it occurs in the form of pods, generally less than 1 m in diameter. Other differences suggest that the two lithologies were derived from different source rocks.

THERMAL METAMORPHISM

Many rocks in parts of the south and east of the district (see Figure 7) show features characteristic of thermal metamorphism. In the semipelitic rocks, these changes include recrystallisation of quartz with resultant grano-blastic polygonal textures (Plate 5c), and the replacement of biotite by fine-grained muscovite-biotite symplectites; in places the latter contain andalusite and more rarely fibrolitic sillimanite (S 79381 and S 78354). In Riereach Burn [8137 4228], andalusite-muscovite intergrowths are more extensive and may have replaced muscovite porphyroblasts; they occur together with biotite-muscovite-andalusite symplectites. Discrete andalusite porphyroblasts overgrowing biotite are developed within flaggy schistose semipelites in Allt Carn a Mhàis Leathain [8559 4018]. About 1 km east of Saddle Hill [at 7986 4337], symplectites composed of biotite-muscovite-fibrolite and, in places, andalusite, are recognisable and fibrolite is intergrown with andalusite (S 79381).

Biotite in many specimens from this area is reddish brown, another feature characteristic of thermal metamorphism, and contrasts with the brown biotites in metasedimentary rocks elsewhere in the district. A migmatitic semipelite, 800 m east of Carn a Mhàis Leathain [at 8499 4093], contains coarse-grained (5 mm grain size) reddish brown biotite. Fibrolite and coarser-grained sillimanite occur within the biotite and muscovite. Garnet is embayed and replaced by biotite.

Granitic gneisses in the Glenkirk area, which are interpreted to have been partially melted, preserve evidence for later thermal metamorphism. This includes the occurrence of granular biotite-K-feldspar-fibrolitic sillimanite (Plate 5b), and possibly cordierite symplectites, as well as a development of granoblastic textures in some of the quartzofeldspathic domains. Fibrolite is intergrown with this granoblastic texture (Plate 5c), which is finer grained than the coarse-grained regional metamorphic/migmatitic textures. Cordierite is also developed outwith the symplectites. Only small relics of garnet remain, mostly enclosed within feldspars.

Thermal metamorphism in the semipelites is not spatially related to the surface exposures of the Moy or Findhorn granites. Indeed, some specimens collected from within 1 m of the Moy Granite contact show little evidence of recrystallisation.

Thermal metamorphic textures, described in detail by Zaleski (1985), also occur in many parts of the Moy Granite, as shown by the development of fine-grained biotite + andalusite + plagioclase/K-feldspar ± muscovite symplectites apparently pseudomorphing biotite (S 79400) and by the recrystallisation of quartz to a grano-blastic polygonal mosaic (Plate 5a). These textures are well displayed to the east of Beinn a' Bhuchanaich [around 765 403], where biotite-pseudomorph symplectites occur within both the Moy Granite and inclusions of tonalite and psammite. The development of muscovite within the Moy Granite may also be related to this recrystallisation.

Zaleski (1985) attributed the recrystallisation to thermal effects associated with the emplacement of the Saddle Hill Granite, on the basis of an apparent spatial distribution of recrystallised rocks within about 1000 m of the granite contact. However, semipelitic metasedimentary rocks within 20 cm of the contact of the Saddle Hill Granite in Allt Cromachan [7707 4228] show only patchy recrystallisation of quartz and do not contain any symplectites pseudomorphing biotite or muscovite; muscovite porphyroblasts overgrowing biotite are preserved. Similarly, micaceous psammite exposed closely adjacent to the Saddle Hill Granite contact and cut by granite veins on Creagan Breac [at 7775 4389] shows no evidence of mica recrystallisation. Within the Moy Granite, moreover, rocks showing evidence of thermal metamorphism are not restricted to areas adjacent to the Saddle Hill Granite. On the contrary, they are recognised throughout most of the exposed parts of the Moy Granite. Thermal metamorphic textures are comparable to those in metasedimentary rocks north and east of the granite, although muscovite is more abundant in the symplectites of the metasedimentary succession.

The areas affected by thermal metamorphism coincide with an annular aeromagnetic 'high' which roughly encircles the Moy Granite (Figure 7). The central parts of the granite coincide with an aeromagnetic 'low'. Diorite at Ruthven, which represents part of the Findhorn Granite Complex, crops out at the edge of the annular 'high' and the single exposure of the Auchnahillin Monzodiorite overlies a peak in the annular 'high' in the south-western part of the Moy Granite. These dioritic rocks have high magnetic susceptibilities. It is thought that the magnetic anomaly relates to a large subsurface dioritic intrusion. The thermal metamorphism of both the Moy Granite and the metasedimentary rocks resulted from the emplacement of this intrusion.

FOUR

Devonian rocks

INTRODUCTION

Rocks of Devonian age form about 38 per cent of the Solid outcrop in this district and cover the north-western half of the sheet. All are sedimentary in origin. They rest unconformably on an irregularly eroded basement of metamorphic and granitic rocks, Precambrian to Lower Palaeozoic in age, representing the worn-down remnant of the Caledonian mountains (Plate 6). Most of the sediments are arenaceous and were largely deposited in alluvial plains at the south-western limit of a major intra-continental basin. This tectonic depression, the Orcadian Basin, was located on the northern side of the Grampian Highlands and to the east of the Northern Highlands. Mainly centred on the Orkney region, this was one of many basins that developed in the Old Red Sandstone supercontinent of Laurussia (Ziegler, 1988).

All the Devonian rocks are regarded as components of the Orcadian Old Red Sandstone Supergroup, and subdivisions are based upon relative proportions of conglomerate, sandstone, siltstone or mudstone. Due to the juxtaposition of two separate parts of the Devonian succession by translocation along the north-east-trending Great Glen Fault Zone, two different sequences are recognised. On the north-western side of the Great Glen Fault Zone, only one group is developed, named the Black Isle Sandstone Group; this comprises three formations and is about 4800 m thick. To the south-east, two groups are recognised in this district, a lower Inverness Sandstone Group of five formations about 1750 m thick, and an upper Forres Sandstone Group with only about 200 m of the lowest Nairn Sandstone Formation preserved.

The unconformable cover of Mesozoic rocks, exposed outside of the district at Ethie and Brora to the north and at Burghead to the east, may extend into the district beneath the sea bed of the Moray and Inverness firths, but this has not been proved.

Study of the Devonian sequences on both sides of the firths is greatly hindered by the scarcity of exposure. On the south-eastern side, Quaternary sediments mantle most of the outcrops. Here scarce, spate-prone stream and river sections are not always accessible and the scattered quarries are largely overgrown or flooded. On the north-western side, the drift cover is more patchy, but ice sculpturing of the rock surface has largely obliterated the natural surface expression of mappable hard and soft stratal alternations.

PREVIOUS RESEARCH

As noted by Horne (1923, pp.17–23), Devonian rocks in this district have drawn attention since the earliest days

Plate 6 Basal Devonian unconformity in Glengeoullie Gorge [857 478] near Cawdor.

The crystalline basement (Central Highland Migmatite Complex) is overlain by bouldery polymict conglomerates of the Daviot Conglomerate Formation at the base of the Inverness Sandstone Group (D4345).

of natural history studies. Our main understanding dates back to 1828, when Sedgwick and Murchison (1835) outlined the overall character of the Old Red Sandstone succession. Great impetus to geological studies was provided by the findings of fossil fish by Hugh Miller (1841) at Cromarty on the north-western side of the Moray Firth and by Martin (in Gordon, 1859) on the south-eastern side. Probably the first comprehensive investigations were made known by Malcolmson (1859) in his 1838 address to the Geological Society of London and these led to other significant contributions by Agassiz (1844–45), Murchison (1859), Gordon (1859) and Miller (1858). Subsequent progress has been largely the result of fossil-fish studies initially by Traquair (1892–1914) and later by Watson (1908–1937), Westoll (1937, 1951, 1979) and Miles (1968), as well as of fossil-plant studies by Reid and Macnair (1898). However, the contributions of Geological Survey officers are significant; Horne (1923), in particular, was mainly responsible for the original six-inch geological maps that are the basis of the various published maps covering Sheet 84W.

The most recent observations on Old Red Sandstone rocks in the vicinity are those of Stephenson (1977) and Mykura (1983) on sequences to the south-west of this district, Armstrong (1964, 1977) on sequences to the north-west, and Rogers (1987) on Devonian correlations, environments and tectonics across the Great Glen Fault Zone in this and neighbouring areas. Also worthy of mention is the work of Richardson (1965, 1967), which concerns the correlation potential of palynological elements.

ORCADIAN OLD RED SANDSTONE SUPERGROUP

Area north-west of the Great Glen Fault Zone

Only one group is distinguished north-west of the Great Glen Fault Zone in this district; it is here named the Black Isle Sandstone Group.

BLACK ISLE SANDSTONE GROUP

The Black Isle Sandstone Group incorporates all the newly proposed Old Red Sandstone formations of the district on the Black Isle. It is predominantly a coarse-grained, sandy red-bed succession of lensing fluvial units deposited along the margin of a freshwater lake; two finer-grained lacustrine units are recognisable, the Den Formation and the Killen Member. The Den Formation includes laminated silty mudstones and thin limestones, and is unconformably overlain by a variable conglomeratic sandstone sequence differentiated into Kilmuir Conglomerate and Raddery Sandstone formations, the latter containing the Killen Member (Table 1).

All formations in this area lie on the south-eastern flank of the Black Isle Syncline and generally dip north-westwards at between 5 and 10°. In places, where affected by faulting, dips as high as 75° are developed, especially in a narrow coastal zone parallel to the Great Glen Fault Zone.

Den Siltstone Formation

The Den Siltstone Formation is the oldest Black Isle unit exposed in the district and is confined to one relatively small outcrop in Rosemarkie Glen alongside Drum-markie. Its base lies in the subsurface and the oldest beds at outcrop are faulted against rocks of the Rosemarkie Metamorphic Complex; it is overlain unconformably by arkosic conglomerates assigned to the Kilmuir Conglomerate Formation.

The Den sequence is tectonically disturbed and the exposed thickness is difficult to assess; as much as 150 m may be present. Most of the succession is drift covered or masked by vegetation, and some exposures noted in the past are no longer visible.

On the evidence of a former exposure on the hillside [7302 5344] and of debris thereabouts, the oldest beds in the formation are red breccio-conglomerates that probably represent the local basal Devonian beds; if so, they are equivalent to the Kilmorack Group farther west (Johnstone and Mykura, 1989, fig. 30). The bulk of the sequence comprises siltstones and silty sandstones generally younging westwards. The best exposures lie in the Ross-shire County Highways Depot at Rosemarkie Quarry [7292 5833]. Here the rocks are disrupted by several small-scale reverse faults and dips gradually increase westwards from 74° on the eastern quarry face to vertical; younger beds on the immediate glen slopes are conspicuously overturned [as at 7280 5333 and 7272 5350]. Most are greenish grey and planar laminated, but red and maroon beds are developed. Although not recognised in the quarry sequence, sporadic green micaceous shaly interbeds in the lower part of the formation are indicated by recordings of former occurrences on the hillside [7289 5365].

On the whole, the Den succession appears to represent the marginal sediments of a playa lake, where sheet floods produced sandy interbeds, some of which formed channel fills, as in a stream section [at 6266 5359].

Only one contact with the overlying Kilmuir Formation is presently exposed. Just below the lowest waterfall on the right bank [7263 5359], fine-grained sandstones dipping at 40° to the west lie concordantly below conglomerate. However, the surface trace of the conglomerate base transgresses the regional dip of the Den Formation on both sides of the glen and clearly much younger Den beds occur higher up the slope [7247 5341], where Horne's fieldslip indicates 'thick ribs, dark shale partings and grey limestone ribs'.

As yet, palynological studies have given no indication of age. However, the low stratigraphical position of the Den Siltstone, together with its unconformable relationship with the Kilmuir Conglomerate, suggest a possible correlation with strata in the Strathpeffer Group, farther west, which are assigned to the Lower Devonian (Richardson, 1967; Johnstone and Mykura, 1989, fig. 31C).

Kilmuir Conglomerate Formation

The Kilmuir Conglomerate Formation is a prominent conglomeratic unit lying between the Den Siltstone Formation below and the Raddery Sandstone Formation above. It is most thickly developed in the south-west, near Kilmuir, where it may be 400 m thick and contains relatively thick sandstone wedges. A thinning to the north-east is indicated by the thinner sequence around Rose-markie Glen and by its inferred continuation to the well-known thin conglomerate resting on metamorphic basement at Cromarty.

In Horne's memoir (1923, p.77), this formation was referred to as the 'Ord Hill conglomerate', but here it is renamed because two other hills in close proximity [6975 5735 and 6681 5378] bear that same name and lie on an outcrop of different conglomerate. The name 'Ord beds' was previously used by Horne and Hinxman (1914, p.59) for olive shales and fetid calcareous bands in the older Lower Devonian Strathpeffer Group on the western limb of the Black Isle Syncline, where the Knock Farril Conglomerate (Johnstone and Mykura, 1989, figs. 30 and 31C) appears to be the correlative of the Kilmuir Formation.

The base of the formation is exposed only in Rosemarkie Glen [6263 5359]. The basal contact with the Den Formation is considered to be an unconformity, since there are no signs of a fault there to account for the apparent cross-cutting nature of the conglomerate as it passes over the Den sequence on to the metamorphic basement on Callachy Hill. The top of the formation is markedly diachronous, with several intercalations of sandstone. Thus the surface trace of the contact with the Raddery Sandstone Formation is somewhat zig-zagged.

The best exposures are those on the shores of the Inverness Firth, astride the entrance to Munlochy Bay [6958 5307]. At Kilmuir, the oldest exposed beds form the rocky shore below St Mary's Church [6786 5012] and contain granite boulders. Most of these are well rounded and include pink biotite granites that may be equigranular or porphyritic with large K-feldspars. Slightly higher in the sequence, the clast-supported cobbly beds contain conspicuous thin cross-bedded sandstone lenses and quartzite boulders up to 0.6 m long, as exposed for instance below Craigbeck [6709 4938] and nearby [at 6850 5108].

Gritty sand in the matrix occurs throughout and many of the smaller clasts are barely rounded. Metamorphosed sandstone and quartzite are the main components, but at certain levels felsite, vein quartz, pegmatite, gneisses and onion-skin-weathering porphyritic andesite [as at 6876 5174] may accompany granite and rare mudstone fragments; granite boulders up to 1 m long are not uncommon. Due to the numerous beds involved, no adequate analyses have been made to establish a trend of change in clast composition through the sequence. Superficially, the proportion of quartzose metamorphic rocks and mudstone clasts appears to increase northwards, and in the Drummarkie area clasts are generally smaller.

Away from the coast, the formation forms a relatively drift-free high ridge alongside the Inverness Firth between Kilmuir and Ormond Hill [6967 5358]. In most places, the bedding is evident, especially where easily weathered thin sandstone interbeds occur. Around Loch Lundie [670 503], the relatively soft nature of such sandstones is manifest as a wide valley between the prominent conglomerate ridges of Girra [6505 5009] and Creag a'Chaisteil [6696 5065]. In this depression, maroon and red flaggy and pebbly sandstones are exposed on the northern shore of the loch [6717 5048], where steeply dipping strata have been shattered by faulting.

Due to the proximity of the Great Glen Fault Zone on its eastern side, the Kilmuir sequence in the south is affected by several subsidiary faults, most of which mark sandstone–conglomerate boundaries. Notable among these is Ormond Fault [676 517 to 699 538] marking the

Kilmuir–Raddery formational junction in the Craigiehowe to Wood Hill area.

Sedimentologically, the formation represents part of an alluvial fan fed by waters generally flowing from the south towards the main basin. In this district, the outcrop coincides with the north-western sector of the fan, which probably merged with a more localised debris-spread on the flanks of an upstanding inselberg of Rosemarkie Metamorphic Complex rocks.

A stratigraphical position similar to the Knock Farril Conglomerate and Cnoc Fynish Conglomerate on the north-western side of the Black Isle Syncline indicates the possibility of a merging of two parallel fan systems in the axial region of the Black Isle Syncline, which probably influenced the alignment of the structure.

Raddery Sandstone Formation

The bulk of the Old Red Sandstone sequence on the Black Isle comprises thin lensing units of conglomerate and silty mudstones in a thick sandstone succession. The main lensing units are regarded as members of one sandstone formation and lie in the western part of the isle; in the north-eastern part the formation is undivided (Table 1).

Avoch Sandstone Member   The Avoch Sandstone Member constitutes about 75 per cent of the Raddery Formation and conformably overlies the Kilmuir Conglomerate Formation, although the contact is diachronous and of an interdigitating nature. Similarly, its upper contact with the Drumderfit Conglomerate Member is diachronous, with sandstones interdigitating with conglomerates. The outcrop lies between Sligo in the south-west [670 515] and north Fortrose in the north-east [728 574].

The Avoch Member contains a variety of sandstone grades varying in colour from dark red, maroon and purple to dark brown and pale goethitic ochre. Generally, the finer-grained sediments have the darkest hues. Due to folding, faulting and highly variable dips, it is difficult to assess accurately its thickness; an estimate of 3000 m is made. Four subdivisions can generally be recognised.

In this district, the basal contact with the Kilmuir Conglomerate is not exposed. The oldest beds seen occur on the Platcock hillside [7267 5723] of northern Fortrose and around Sligo. These lowest sandstones are very coarse grained and arkosic. They are distinguished as very pink rocks with pink feldspar pebbles [7248 5733], in places sufficient to produce conglomeratic beds [7273 5738] that can be mapped separately [as at 676 517 and 661 516]. They are succeeded by a generally less pebbly subdivision having some red fine-grained micaceous siltstones [7201 5618] and marl seams [7184 5686 and 6713 5167]. A coastal section of this part of the succession between Fortrose and Avoch exhibits a variety of ochreous and reddish arkosic sandstones. Their somewhat indurated nature may be due to migratory-fluid cementing from the nearby Great Glen Fault Zone. Siliceous and quartzose sandstones also occur, together with silty mudstone interbeds [7179 5599]. Inland, some exposures show that parts of this sequence are particularly flaggy [7136 5679] and pebbly [7061 5589], features also to be observed farther south along the northern shore of Munlochy Bay [between 6875 5297 and 6775 5328]. One of the best sections of this interval of the Avoch Sandstone is in Bay Farm Quarry [6780 5325]. Unfortunately, the vertical walls do not allow close scrutiny, but rock quarried from here was used for building stone at Fort George [765 567], where wind-etched surfaces provide a plethora of fine sedimentological detail of fluvially deposited lithologies (Plate 7).

A third subdivision, of distinctive darker brown finer-grained sediments, is recognisable throughout the outcrop. Although largely comprising fine-grained sandstones, several intervals of rather shaly sandstone intercalated with ripple-bedded chocolate brown siltstones and silty mudstones are developed; some of the mudstones contain small concretionary carbonate nodules, as in

Shaltie Burn [6985 5081] (Horne, 1923, p.77). An exposure [7069 5170] in Deer Park indicates that the lowest beds of this subdivision include thin purplish shaly mudstones. The general fining-upward trend of this third subdivision is manifest by better-laminated to planar-bedded, commonly rippled, shaly silty sandstones in the Carse of Raddery Burn gorge section [7019 5113 to 6966 5047], Avoch Burn [6975 5028 to 6965 5034], on Tourie Hill ridge [6935 5453 to 6905 5435], at Corrachie Quarry House [6795 5429], along the Bay Wood scarp [6745 5326 to 6728 5335] and in Sligo Burn [6734 5202 to 6737 5211]. Some of the best examples of such lithologies can be examined in the stones of the estate wall alongside Dorroch Bog Wood at Corrachie Crossing [6780 5465], which probably derive from the infilled Corrachie Quarry. The highest part of this subdivision is

Plate 7 Avoch Sandstone: building stones in Fort George [7640 5680] hewn from Bay Farm Quarry [6780 5325] on the northern shore of Munlochy Bay on the Black Isle.

a) Small-scale trough cross-bedding within a planar-laminated arkosic sandstone (MNS 5696/9).
b) Soft-sediment deformation in red fine-grained arkosic sandstone; note also granite clast in channel above. Coin is 2 cm across (MNS 5696/8).

a

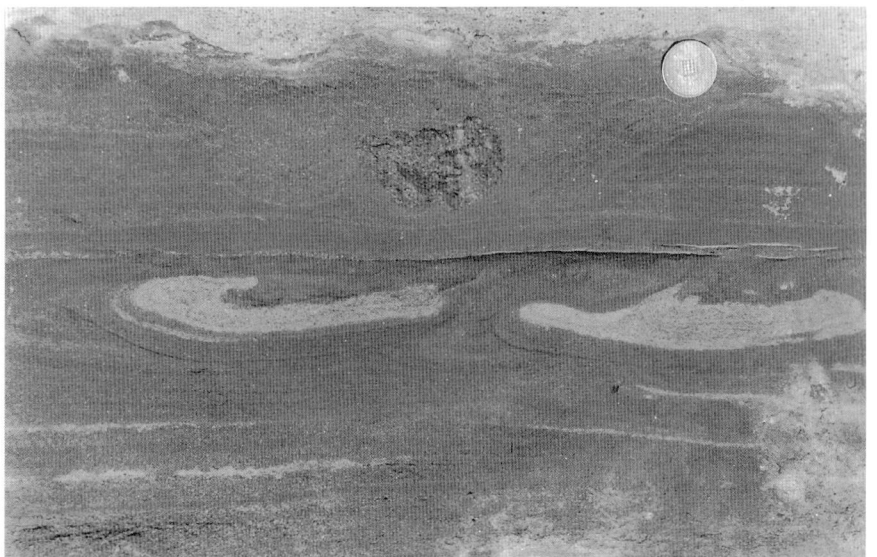

b

marked as a striped sequence of bright red thinly flaggy sandstones, the only exposures of which occur in Arkendeith Quarry [6930 5592] and its adjacent scarp [6926 5579 to 6902 5584]. In this sequence some striping is marked by green laminae [6920 5586] and there are also mudstone chips in the bases of some sandstones [as at 6928 5591].

The uppermost Avoch subdivision is barely exposed, but it is characterised by a gradual overall coarsening that probably heralds the prominent conglomeratic interfingerings of the overlying Drumderfit Member. Exposures of this transitional sequence are rare. A conspicuous knoll just north [6982 5644] of Arkendeith exhibits a channel-fill sandstone similar to those in Rosehaugh Quarry [6846 5099] which lie within a sandstone finger between the two lowest lenses of Drumderfit conglomerate. This quarry section and ditch exposures [7030 5729] west of Wester Craiglands both show that red and yellowish flaggy sandstones and rare micaceous siltstones also occur in the upper levels of the Avoch Member.

Drumderfit Conglomerate Member The Avoch Sandstone is succeeded by the prominent conglomeratic Drumderfit Member, extending into this district from its type area just south of Munlochy Bay (Horne, 1923, p.77). At its thickest, at Munlochy Bay, the member is about 240 m thick.

The best exposures of this cobbly sequence lie around Rosehaugh, near its north-eastern limit, where the outcrop splits into three main lenses, the highest of which is mappable as far as the northern end of Ord Hill ridge [7035 5818].

The north-easterly lensing of the reddish purple boulder-cobbly conglomeratic succession also accompanies a gradual fining, for boulders over 85 cm long are common in the south-west (seen just beyond the district boundary on Munlochy Bay [at 6640 5347]) whereas, on the Ord Hill ridge, the sequence is more pebbly, with clasts less than 25 cm long. No detailed clast analysis has been made, but quartzose metamorphic rocks and granite predominate, with andesite clasts significant in the north-east.

In addition to the exposures on the track along Ord Hill ridge, conglomerates of the lowest lens can be examined on the Arkendeith Wood track [6858 5603] and nearby scarp [6848 5594]. Medium to coarse conglomerates of the middle lens are exposed on the hillside [6824 5598], where large clasts also occur in associated red cross-bedded arkose. Only one other exposure of this member was encountered during the resurvey, that forming the falls [6720 5575] on Rosehaugh Burn.

Goose Sandstone Member The Goose Sandstone Member conformably intervenes between the Drumderfit Conglomerate Member and the overlying Killen Mudstone Member. It is about 440 m thick and extensively exposed in the tributaries of Rosehaugh Burn.

It is distinguished as a red, very coarse-grained sandstone unit with numerous pebble and conglomerate lenses. It is cross bedded virtually throughout and innumerable intersecting channel sets are developed; among these isolated pebbles occur. Some slump structures are evident, as are sporadic planar-bedded flagstones. A feature of many sets is the variable disposition of beds, indicative of a highly mobile switching fluvial regime. Such examples are well exposed in gorge-like walls in both Goose Burn and Killen Burn, especially at their confluence [6726 5605]; contorted cross-beds are notable at one place [6721 5703]. The top of the member is drawn at the base of the first mudstone in the Killen Burn section, which lies north-west of Coulnagour.

Away from these burns, the outcrop is masked by drift, but very small exposures of typical pebbly gritty sandstone occur, as on the slopes of Bog of Shannon Wood [6809 5663 and 6828 5676] and in a ditch [at 6948 5798].

Killen Mudstone Member The Killen Mudstone Member is a conspicuous interval of platy to flaggy, fine to medium-grained sandstones in which laminated, silty, carbonate-bearing mudstones occur. It lies conformably between the coarse-grained Goose and Millbuie sandstone members. The sequence is largely exposed in Killen Burn and is about 40 m thick. Its lower limit is taken at the base of the first mudstone [6702 5735] and its upper limit at the top of a marly mudstone with pedogenic carbonate concretions [6680 5747]. The latter locality lies just west of the district, in Sheet 83E.

Horne (in Horne and Hinxman, 1914, p.63) has given an outline of the basal part of the Killen Member, which is not now well exposed:

(1) yellow sandstone;
(2) light-coloured bluish shale or mudstone, 24 ft;
(3) dark bituminous beds with green partings and lime stone nodules, which contain fish remains, 24 ft;
(4) light bluish calcareous shale with plants.

This sequence rests on '(5) yellow sandstone', here referred to the underlying Goose Sandstone Member. Between the two bounding mudstone sequences flaggy sandstones and platy siltstones, some of which bear prominent sole structures (Plate 8), are still well exposed.

An important fossil-fish fauna has been recognised in the nodules near the base. The presence of *Cheiracanthus* sp., *Coccosteus cuspidatus* Miller ex. Agassiz (Plate 9), *Gyroptychius microlepidotus* (Agassiz) and *Osteolepis macrolepidotus* Agassiz, allows a correlation to be made with the Achanarras Fauna in the more lacustrine facies of the Orcadian Basin. In addition, a plant fragment akin to an early form of pine found in the Cromarty Fish Bed was collected by Tait (Horne, 1923, p.77). Fragments of *Cheiracanthus* and *Diplacanthus* have been recorded in the uppermost nodules of this member (Horne and Hinxman, 1914, p.63).

The Killen outcrop trends north-eastwards through the Barnside area and is poorly exposed in the headwater streams of the burn as laminated marly siltstone and shaly silty mudstone [6881 5933], pale grey-green shaly mudstone [6910 5956], and as thin-bedded, platy to flaggy sandstone with pale ochre or green marl

Plate 8 Killen Mudstone: interbedded greenish grey silty mudstones and sandstones with sole structures. Left-bank cliff of Killen Burn [6702 5737] (D4888).

seams 3 to 80 cm thick [between 6943 5964 and 6903 5928]. The last lithology is also exposed in the upstream area [6990 5995, 7003 6004 and 7017 6017].

As far as these meagre exposures allow, field evidence indicates that north-eastwards the mudstone content of this member diminishes considerably and its continuance as a mappable member beyond a point north of Muiryden [702 597] is not proven. Certainly, it is unlikely that there is any stratal continuity with the Cromarty and Ethie fish-beds, which bear related Achanarras fossils. The Killen fauna probably represents the youngest part of the Achanarras faunal range, in contrast to older representatives at Cromarty and Ethie.

Millbuie Sandstone Member Previously the name 'Millbuie' has been loosely applied to different intervals of the Old Red Sandstone succession in Ross-shire. Here it is formally applied to all strata in the Black Isle Sandstone Group above the Killen Mudstone Member, on which it rests conformably.

Exposures of the Millbuie Sandstone Member in the district are few, but this is a unit that has produced some of the finest building stones in the region and good quarry sections lie just beyond the district boundaries [at 6680 5759] in Sheet 83E and [at 6936 6084] in Sheet 94W. In this, district about 400 m of the lowest strata come to crop.

The lowest beds in the member are red, coarse-grained and pebbly arkosic sandstones. The Killen Burn exposures and adjacent quarry [6680 5759] just over the district boundary exhibit an intensity of intersecting channel fills. To the north-east, sporadic exposures along the Killen valley indicate that flaggy beds are developed at Barnside [6853 5909] and rare marl seams not far away [6719 5795]. A feature of these basal beds is the intermittent leaching of colour, with gingery hues

present at one place [6889 5950] and whitish, cross-bedded, coarse-grained sandstone nearby [6949 5992].

Higher up the Millbuie sequence, a mixed succession of leached cross-bedded pebbly and finer-grained flaggy beds is evident in predominantly bright to deep reddish sandstones, the best exposures occurring on the forestry tracks [6726 5897 to 6718 5911 and 6765 5938 to 6757 5945]; slightly younger planar-laminated sandstones on trough cross-bedded sandstones can be examined in an overgrown quarry [at 6728 5958]. The youngest part of the member is well exposed in a generally inaccessible flooded quarry [at 6730 6003]. Here a 5-m section shows planar, massive, trough cross-bedded and thinly bedded channel sandstones containing some volcanic clasts. The building stones of Fortrose Cathedral probably were hewn from this quarry; they are considerably wind-etched and clearly illustrate a full spectrum of fluvial sedimentary structures.

Raddery Sandstone Formation—undivided Within the north-eastern part of the Old Red Sandstone outcrop in this district, the continuity of the Drumderfit Conglomerate and Killen Mudstone members has not been established and, in the Fortrose–Raddery–Whitebog area, the sandstone outcrop above the Kilmuir Conglomerate cannot be subdivided. Hereabouts, the oldest sandstones are dark maroon and in part siliceous. On the northern slopes [7283 5978] of Flowerburn Wood, they rest on the Kilmuir Conglomerate, which here forms the base of the Old Red Sandstone succession. Similar micaceous flaggy sandstones lie along Rosemarkie Burn [7242 5979], and not far away [7298 6033] in Sheet 94W.

On the downthrown side of Drummarkie Fault, the youngest exposed beds appear to crop out in the Craiglands area, where red thin to medium-bedded sandstones and marl seams are notable in a 12-m shattered

Plate 9   Selected arthrodire fish fragments:

a) *Coccosteus cuspidatus* Miller *ex.* Agassiz, marginal plate (GSE 15084/T 3389a × 1). Killen Mudstone Member, Raddery Sandstone Formation, Killen Burn [6701 5737] (MNS 5750).
b) *Homosteus milleri* Traquair, dorsal view of skull roof (GSE 1209 × ¹/₃). Hillhead Sandstone Formation, Hillhead Quarry, Dalcross [7774 4983] (MNS 5751).
c) *Millerosteus minor* (Miller), nuchal plate (GSE 15085/T3426d × 6). Hillhead Sandstone Formation, Hillhead Quarry, Dalcross [7774 4983] (MNS 5752).

quarry section [7188 5785]. As recognised in the coextensive Avoch Sandstone Member, a change to finer-grained, darker red and maroon strata is evident and exposures in Rosemarkie Burn [at 7216 5934 and 7207 5924] may be laterally equivalent to part of the Carse of Raddery gorge section (see above).

A similar darkish, fine-grained interval is succeeded by bright red coarse-grained pebbly sandstones in the Den Burn of Raddery [7157 5934 to 7145 5986]; these probably represent the basal Goose Sandstone beds, since an older conglomerate in a quarry [7144 5912] is possibly related to the Drumderfit beds to the south-west. Certainly, the predominant cross-bedded pebbly nature of the lower beds in Den Burn is reminiscent of the Goose sequence in Killen Burn, especially in the presence of rotten granite clasts [as at 7145 5959]. On the evidence of green marl seams and thinly laminated sandstones in the upper part of the Den Burn section, and of bright red flaggy micaceous sandstones [7045 5978] near Muiryden, a fining-upward trend can be perceived as the level of the Killen Mudstone is approached.

Area south-east of the Great Glen Fault Zone

According to observations made during the original geological survey by Horne (1923, fig. 2) in the adjacent Nairn district, a prominent unconformity occurs within the Old Red Sandstone sequence on the southern side of the firths, separating two suites of fish-bearing sandstones. Due to masking by Quaternary deposits, there is no exposure of such a discordance in the Fortrose district. However, it is accepted here as marking the boundary between two distinct stratigraphical units, the Inverness Sandstone Group, below, and the Forres Sandstone Group, above; Rogers (1987) has interpreted this boundary as a faulted junction.

INVERNESS SANDSTONE GROUP

In this district, the Inverness Sandstone Group forms the basal Devonian sedimentary sequence on the southeastern side of the Great Glen Fault Zone. It comprises a mixed succession of fluvial and lake-margin clastic sediments subdivided into five formations (Table 1). The group is of Middle Devonian (Eifelian) age, being more or less equivalent to the Kilmuir and Raddery formations on the north-western side of the Great Glen Fault Zone. No representatives of older Old Red groups, such as the Lower Devonian Crovie Sandstone Group (Read, 1923), are preserved.

Daviot Conglomerate Formation

The Daviot Conglomerate is the basal formation of the Inverness Sandstone Group and rests unconformably on worn-down metamorphic and igneous rocks of the Caledonian mountain chain. Due to the paucity of exposure, the relationship with the overlying Nairnside Sandstone Formation is somewhat obscure. It appears to be unconformable, though the contact seems gradational because of reworking.

The main outcrop forms a relatively narrow northeasterly oriented strip stretching from Blàr Buidhe [675 345] in the south-west to the environs of Cawdor [860 500], but within the metamorphic and granitic hinterland, faulted outliers incised by the burns of Meur Bhèoil [850 402] and Allt Breac [850 385] testify to its former extensive development to the south-east.

Essentially a reddish imbricated polymict conglomerate of metamorphic and granitic clasts, red arkosic sandstone lenses are also common, with bedding indicating variable directions of fluvial flow. The cementing matrix is generally silica, but significant amounts of sparry calcite or baryte are patchily recognisable. In places [as at 757 422 and 804 476], fossiliferous limestones, some oolitic (Parnell, 1983, pp.199–200), occur and indicate original carbonate, in contrast to those where baryte is developed as a replacement matrix.

The formation has been interpreted as a bajada or piedmont spread of coalescing alluvial-fan sediments deposited along the line of active fault scarps paralleling the Great Glen Fault Zone (Johnstone and Mykura, 1989, fig. 32). However, the faulted outliers on Meur Bhèoil and Allt Breac clearly indicate that fan and intervening valley deposits probably covered a larger area, stretching back to the Cairngorms.

Generally, the Daviot sequence dips between 8° and 10° to the north-west, higher dips being associated with faulting. The present outcrops represent only the distal parts of the former piedmont and the variable outcrop widths are largely an indication of lateral thickness changes that reflect a buried hilly basement terrain. However, some wide outcrops indicate zones of low dip. Where the sequence is thinnest, upstanding mounds of basement can be inferred, some of them having remained proud until covered by younger Nairnside sands, as at Creag Shoilleir [670 346] and the Knoll at Daltulich [738 420]. Conversely, where the sequence is thickest, depressions, possibly coinciding with major outflows from the mountains, may be indicated. Thick sequences preserved between Meall Mór [745 410] and Finglack [770 440], and between Old Newton [825 495] and Riereach [855 478] may also include some downfaulted components. Accurate measurements of thickness have not been made, but estimates of 46 m for the thinner River Nairn section 2.5 km downstream of Daviot House and 76 m on nearby Meall Mór (Horne, 1923, pp.65–66) are sufficient to demonstrate the marked lateral changes. Thicker successions are preserved in Allt Breac, where over 250 m of strata have been noted (see also Horne, 1923, p.68).

No overall stratigraphical pattern is recognisable, but each exposure provides some measure of upward fining. Basal sections are invariably more brecciate than overlying sections, with high proportions of large angular clasts of local rocks, some over a metre long. Typically, however, the coarsest beds are composed of imbricated 30 cm clasts in a red gritty sandstone matrix. No major up-sequence changes in clast composition have been detected, but gritty sandstone lenses are more abundant in the higher levels, most of which are cross-bedded and about 70 cm thick. A feature of stream sections in the conglomeratic beds is their precipitous nature, none more so than the spectacular Glengeoullie Gorge [857 478] upstream of Cawdor (Plate 6).

Interbedded limestone is an unusual facies in this formation. Three lenticular bodies are known. A thin limestone near the top of the local Daviot sequence in Easter Town Burn [at 757 422] has yielded palynomorphs (Dr J E A Marshall, personal communication). Other exposures of fish-bearing limestones were first examined by Malcolmson (1859) near Balfreish [797 468] and at Wester Galcantray [8020 4735]; he recognised a 3 m bed at the former location, which is partially oolitic (Parnell, 1983) and a basal 1 m lens at the latter. Additionally, there is a calcareous shaly mudstone development with nodules near the top of the formation in the River Nairn section [7337 4172].

Good exposures of the basal unconformity occur in the River Nairn [7344 4125], in Dalroy Burn [767 448] and in Glengeoullie Gorge [8565 4780]. Contact with the overlying Nairnside Sandstone Formation is exposed at several localities. All appear to signify a sudden change to much finer-grained sandstone, although heralded by several fining-upward sandstone units in the uppermost conglomeratic levels. There are, however, problems in the interpretation of such a marked change when one considers the overall relationship of the various Nairnside members with the Daviot conglomerates. In this account, a break is recognised at the top of the Daviot Conglomerate, which has been masked by a reworking of the relatively unconsolidated topmost conglomeratic beds at the onset of transgressive Nairnside sedimentation (see Nairnside Sandstone Formation).

Fossil palynomorphs in the Daviot limestones indicate an age no older than Eifelian (Dr J E A Marshall, personal communication), thus the Daviot Conglomerate is probably coeval with the Kilmuir Conglomerate on the Black Isle.

Nairnside Sandstone Formation

The Daviot Conglomerate is overlain by a red and green muddy arkosic sandstone sequence distinguished by intervals of dark grey calcareous siltstones and mudstones that in places contain fossil fish remains and silty limestone nodules. This succession, here named Nairnside Sandstone Formation, is thickest in the area between Mains of Daltulich [740 200] and Clava Viaduct [763 448], where two named members allow four subdivisions to be established; a maximum thickness of 180 m is likely:

Clava Mudstone Member
Sandstone (unnamed)
Easter Town Siltstone Member
Sandstone (unnamed)

Elsewhere, the lower two subdivisions are not developed and the younger two subdivisions rest unconformably on the Daviot Conglomerate Formation.

Sandstone below the Easter Town Siltstone Member About 80 m of greenish grey sandstones and siltstones with thin red and purple gritty bands and one significant interval of fossiliferous laminated siltstone are developed in the area of greatest formational thickness. This sequence is drift covered and exposed only in Easter Town Burn, upstream of the railway bridge [at 7515 4305]. Three main sections are exposed.

DETAILS

Contact with the older Daviot Conglomerate is obscured by till and there is no exposure between the limestone interpreted as lying within the Daviot sequence [757 422] and the oldest recognisable Nairnside beds [7569 4275]. At the latter locality, green slightly wavy-laminated, platy sandstones with mud-cracked hexagons can be examined at low water and are overlain by grey, platy, fine-grained sandstones [7567 4288].

The next section [between 7555 4280 and 7553 4282] comprises pale, relatively thick-bedded, flaggy, medium-grained sandstone with pink feldspar grains, pale greenish grey platy sandstone, and sandstone with gritty patches. These are overlain by 1.5 m of dark grey, bituminous, laminated, muddy siltstones with small fossiliferous calcareous nodules and a prominent basal crystalline limestone, 10 cm thick, incorporat-

ing some quartz grit fragments. This distinctive fish-bearing unit is overlain by greenish grey fine-grained sandstone.

Farther downstream a thicker section is exposed between [7528 4285 and 7518 4295], where red and purplish fine-grained micaceous sandstones with pink ?dolomitic clasts at [7525 4825] are overlain by cross-bedded, lensing sandstone beds associated with thin gritty lenses and thin papery shale seams. These in turn are succeeded by greenish grey, thinly wavy-bedded, micaceous, silty sandstones with fish scales [at 7522 4291].

The junction with the overlying Easter Town Siltstone Member is best exposed under the railway bridge and is drawn at the base of a blackish, shaly silty mudstone 20 cm thick, where it rests on a fine-grained sandstone containing pyrite blebs. This exposure is a 'window section' revealed by bridge excavations; the mapped junction lies upstream nearer the road [at 752 430].

On the basis of identified fragments of *Coccosteus*, *Dipterus* and *Osteolepis* in nodules [7555 4280], the basal subdivision of this formation is considered to be of Middle Devonian age. The lithologies and the fish fauna both indicate a marginal lacustrine setting for this part of the succession. It seems not to have been a natural gradation from the coarse alluvial-fan environment of the underlying Daviot sediments.

Easter Town Siltstone Member The Easter Town Siltstone Member is a conspicuous sequence of dark grey, fine-grained, muddy sediments in the lower part of the Nairnside Formation. It conformably overlies the basal sandstone subdivision of the formation and is conformably overlain by grey flaggy sandstones. It is developed only in the 'embayment' between the Mains of Daltulich and Clava Viaduct, and is exposed in the gorge section [752 430 to 7493 4346] alongside Easter Daltullich (Easter Town on older maps). Detailed measurement shows it to be about 56 m thick, comprising nine units of mudstone, siltstone or sandstone (Figure 11).

DETAILS

The basal Unit 1 is 7.25 m thick and mainly siltstone. A black shale 20 cm thick at the base is successively overlain by 1.07 m of sandstone, a fining-upward shaly sequence 74 cm thick and a sandstone 42 cm thick. These are followed by laminated shaly siltstones with discoidal calcareous nodules containing *Osteolepis* remains.

The overlying Unit 2 is essentially of pale grey, fine-grained sandstones and is 6.50 m thick. There are rare shaly mudstone partings in the lower part, where a cross-bedded sandstone 30 cm thick forms the basal bed and a prominent shaly mudstone 11 cm thick lies 3.30 m above the base. The latter is overlain by a coarsening-upward, wavy-bedded sandstone with some cross-bed lenses beneath the top 1.3 m of interbedded thin mudstones and sandstone with a gritty top.

Unit 3 is of dark grey laminated silty mudstone a little over 16 m thick. Limestone nodules are well disposed in the lowest 2 m, but become fewer and incipient in the next 2 m, where numerous fish scales and spines are associated with pale silt wisps. Two very thin micaceous sandstones occur at 4 m and 8 m respectively above the base. Between the 8 and 14 m levels in this unit, the section is covered by wet till and here assumed to be of mudstone.

Thickness
in metres

Dark laminated silty mudstone

31

30

Section

29

covered

28

by

27

Till

26

25 — UNIT 3 — Thin sandstone

24

23   Dark grey silty mudstone

22

21   Micaceous thin sandstone

20   Dark silty mudstone with
silt wisps and incipient
19   small calcareous discoidal
nodules with fish fragments,
18   especially scales and spines

17   Limestone nodules
Very prominent gritty bed
16   Interbedded thin sandstone
and mudstone

15   Coarsening-upward unit of
wavy-bedded fine-grained
sandstone with some cross-
14   bedded lenses
Shaly mudstone

13   Pale fine-grained sandstone with
rare shaly mudstone partings

12

11

10   Cross laminations in fine-grained sandstone

Dark shaly siltstone with
9   calcareous discoidal nodules

8   Laminated siltstone muddy
with calcareous discoidal nodules

7

6   BP   Very pale laminated siltstone
BP   Laminated shaly siltstone
5   Shaly mudstone
Prominent bedding plane
Shaly mudstone
4   Prominent bedding plane

3   Blackish shaly silty mudstone
Pyrite blebs
BP   Fine-grained sandstone

2   Fine-grained sandstone

1   Pale fine-grained sandstone

0   Fine-grained sandstone

UNIT 2
UNIT 1

Top of waterfall beneath railway.
Burn flows on top of this sandstone
under railway bridge

Thickness
in metres

61   Flaggy fine-grained sandstone
with soft ochreous 'galls'

60

59   Shaly mudstone

58   UNIT 9   Obscured
section

57   Laminated siltstone
Laminated siltstone

56   'Paper shale'

55   UNIT 8

54   Laminated siltstone

53   Sandstone
Pyrite and clay balls
Shaly calcareous siltstone
Laminated siltstone

52   Shaly mudstone

UNIT 7

51

50   Planar-laminated
siltstone

49

48   Fining-upward sandstone-siltstone,
some cross bedding, pyrite lenses
47   and rafts

Dark grey laminated
muddy siltstone

46   Very pale grey, very finely-
laminated siltstone

UNIT 6

45

44   Pale laminated siltstone

Discontinuous limestone lenses and
43   incipient small discoidal nodules below

Dark shaly siltstone

42

41   Prominent dark grey-blackish shaly
silty mudstone, some slickensiding,
40   bituminous patches and brown
calcareous 'skins' along bedding

UNIT 5

39

38

37   Fine-grained calcareous sandstone

36   'Paper shale'
Fine-grained calcareous sandstone
Pyrite in shaly mudstone with thin
35   siltstones at base

Well-laminated siltstone with some
34   harder calcareous layers

UNIT 4   Prominent bedding plane
33   Siltstone

UNIT 3   Dark laminated silty
32   mudstone

31

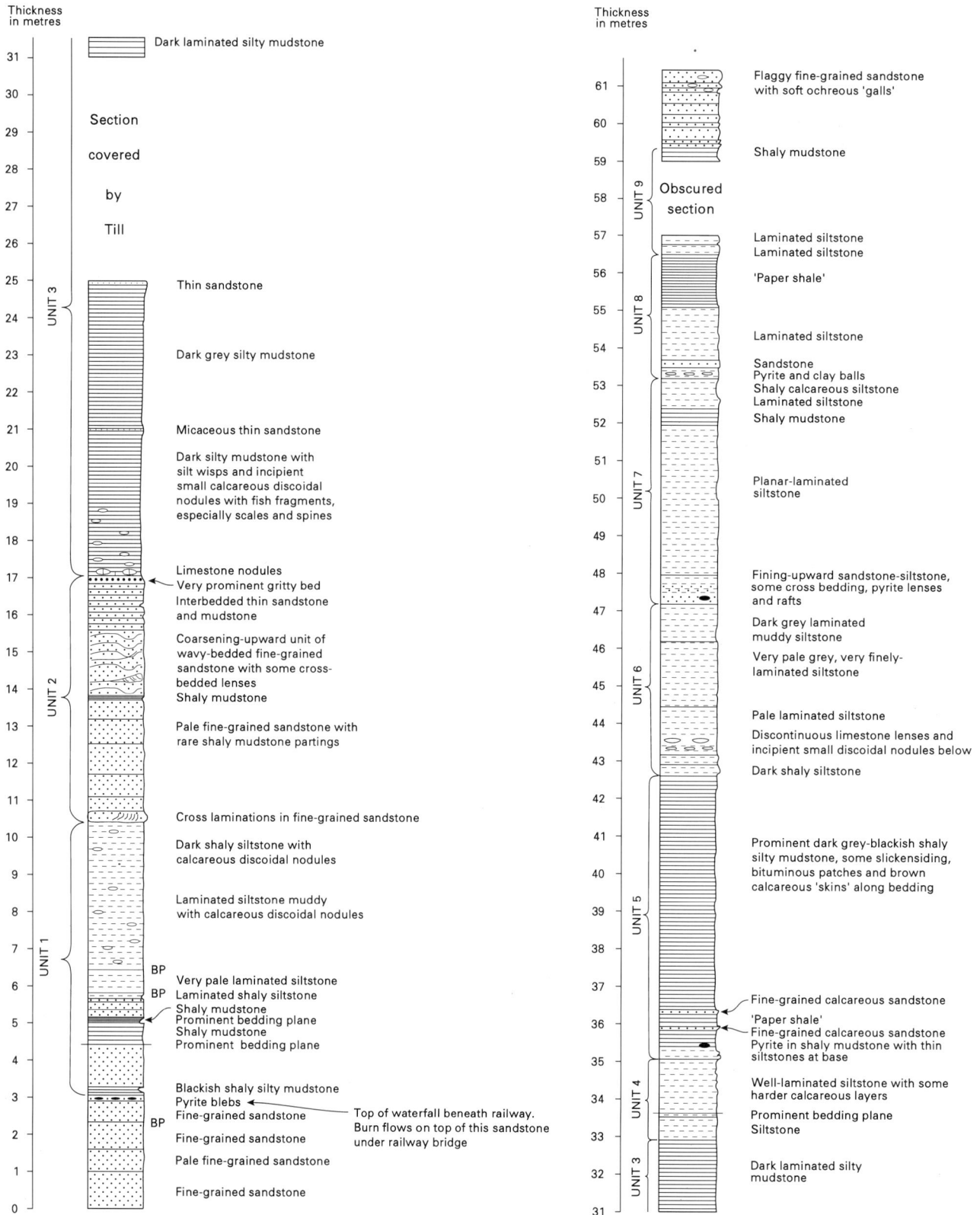

Figure 11   Type section of the Easter Town Siltstone Member (units 1–9) at Easter Daltullich
[between 752 430 and 749 434]. BP—bedding plane.

Unit 4 is 2.03 m thick and comprises well-laminated siltstone with thin hard calcareous layers and a distinctive separation or bedding plane 38 cm above the base. It is conformably succeeded by Unit 5 measuring 7.45 m thick, of blackish shaly silty mudstone with some bituminous patches, slickensided bedding planes and brown calcareous tufa 'skins' on bedding planes. The basal 1.12 m of this unit is particularly well laminated and pyritous, and two thin calcareous fine-grained sandstones 0.61 and 1.06 m respectively above the base are separated by paper shale.

Unit 6 is 4.59 m thick and forms a fining-upward siltstone sequence. The basal 1.85 m are of pale laminated siltstone, a little darker and more shaly near the base and with prominent limestone lenses and subjacent tiny incipient discoidal nodules about 75 cm above the base. The succeeding 1.77 m are very pale, very finely laminated and overlain by dark grey laminated muddy siltstone.

Unit 7 is 6.38 m thick and is marked by conspicuous pyrite lenses and small rafts at the base of a fining-upward cross-bedded sandstone 83 cm thick. This is successively overlain by 3.98 m of planar laminated siltstone, 50 cm of shaly mudstone and 1.07 m of slightly coarsening-upward, interbedded shaly mudstone, laminated siltstone and calcareous shaly siltstone with pyritous clay balls at the top.

Unit 8 similarly comprises a fining-upward sequence 3.08 m thick, having a sandstone 23 cm thick at the base. This is overlain by 1.4 m of laminated siltstone beneath a prominent sequence of paper shale at the top.

The topmost Unit 9 is partially obscured by vegetation and is about 2.5 m thick. It is a fining-upward sequence of laminated siltstone and shaly mudstone. The conformable junction with the overlying thick sandstone subdivision that intervenes between the Easter Town and Clava members is exposed at the foot of the gorge [7493 4346].

The rocks of this member represent lacustrine deposits. However, unlike the overall processes affecting sedimentation in the main Orcadian Lake, where coarsening-upward rhythms prevail, the Easter Town sequence generally comprises fining-upward units, and does not include clearly defined laminites or thick cross-bedded sandstones (Donovan, 1980, pp.37–40). The sequence probably developed in a more stagnant lagoon-type setting in an embayment fringed by exposed Daviot conglomerates. The end of Easter Town sedimentation more or less saw the filling in of the embayment, but the lateral extent of the silty mudstone facies is not known. In this account, it is assumed to pass sharply into sandstones towards the margins of the embayment, the junctions being shown on the geological map as zig-zag lines.

Palynomorphs are abundant in this member and indicate stratigraphical positions at or below the level of the Achanarras Limestone of northern Caithness (Dr J E A Marshall, personal communication). *Osteolepis* has been identified at one locality [7511 4311], and from mudstones of Unit 8 a plant (BGS specimen T3403d) comparable with *Psilophyton thomsoni* Dawson has been collected.

Sandstone between the Easter Town and Clava members   A sandstone sequence conformably intervenes between the two fine-grained members of the Nairnside Formation in the area between the Mains of Daltulich and Clava Primary School [758 440].

The basal beds are exposed at the bottom of Easter Daltullich Gorge and are grey, flaggy, fine-grained sandstones with soft ochreous galls. Higher similar beds occur in the River Nairn bank [at 7465 4365] and contain the small crustacean *Asmussia* (*Estheria* of Horne, 1923, p.67). North-east and south-west of the embayment, lateral equivalents form reddish basal beds to the Nairnside Formation, where they have been reworked with, and probably largely derive from, the immediately underlying Daviot conglomerates.

In the Mains of Daltulich river section, the base is taken at the top of a prominent fining-upward sandy conglomerate unit on the bend [7339 4173]. Between this point and the base of the Clava Member [7337 4191], there is a succession of reddish ochreous gritty coarse-grained sandstones about 40 m thick. Most of the sandstones are planar bedded, in units less than 25 cm thick (Figure 12). They largely display some degree of fining upward and may have cross-bedded conglomeratic bases.

To the north-east, this subdivision is greatly attenuated and just downstream of Clava Viaduct wedges out between the Clava Mudstone and the Daviot Conglomerate. Farther downstream it re-appears as a gradually expanding unit north-east of Dalgrambich [792 471], eventually merging (by reworking) with the prominent sandstone lens at the top of the Daviot Formation just east [8605 5050] of Cawdor.

At its thickest, this sandstone subdivision is about 80 m thick and, according to the literature, the anomalous presence of *Asmussia* suggests a stratigraphical position above that of the Achanarras Limestone in the northern part of the Orcadian Basin—a conclusion not supported here by faunas in the overlying sequences; clearly, the complete range of *Asmussia* is yet to be established. The lithofacies indicate fluvial activity at the margin of a body of standing water. South-west of Allt Lugie, this subdivision is not distinguished and equivalent strata form an undivided outcrop also involving the lateral equivalents of the Clava Member and the Leanach Formation.

Clava Mudstone Member   The Clava Mudstone is a sequence of mudstones, concretionary limestones and fine-grained sandstones, which forms the topmost subdivision of the Nairnside Formation. It represents a period of lacustrine sedimentation following the essentially fluvial episode that was responsible for the deposition of the underlying sandstones. It is conformably overlain by the Leanach Sandstone Formation.

Due to widespread mantling by Quaternary deposits, its full extent has not been determined and, at the time of survey, no trace of it was found south-west of the River Nairn–Allt Lugie section [7337 4191]. However, Rogers (1987, p.42) has reported the presence of lacustrine beds [at 7074 3850] that may indicate a south-westward continuation into the Lochan an Eoin Ruadha area (see section in Horne and Hinxman, 1914, fig. 3).

The outcrop shown on the geological map is mainly an extrapolation between exposures sparsely spread north-eastwards into the Cawdor area. The principal sections occur between the mouth of Allt Lugie and old quarries [7366 4235] to the south of Nairnside House, at a bluff

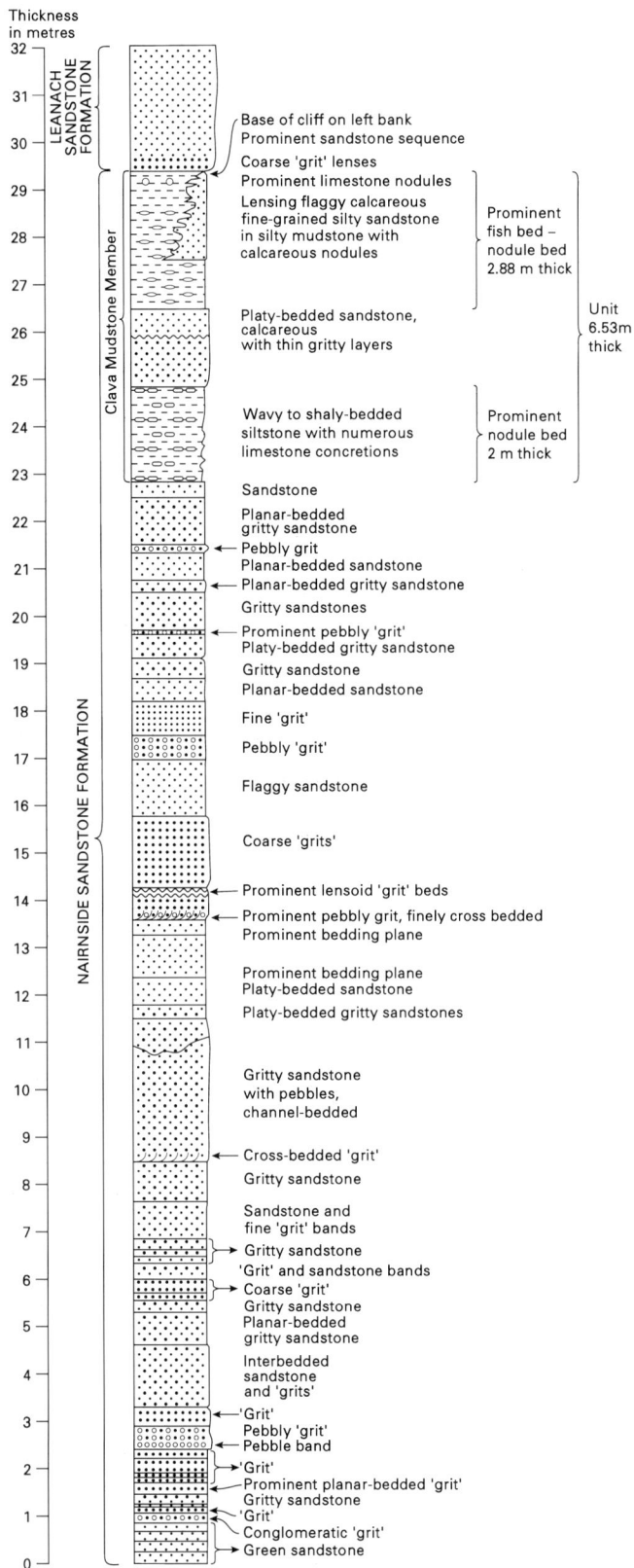

Figure 12 Section of the Nairnside–Leanach sandstone junction in the River Nairn at the mouth of Allt Lugie [733 419] near Mains of Daltulich.

[7548 4458], under Clava roadbridge [7592 4483], opposite the mouth of Dalroy Burn [at 7645 4505] and downstream of Cantray Bridge [at 8009 4817]. All but the bluff section are similar in sequence, comprising fish-bearing dark laminated silty mudstones with thin tabular limestones and/or calcareous nodules separated by one or two planar-bedded sandstones. The best exposures are just below the mouth of Allt Lugie (Figure 12) and opposite the mouth of Dalroy Burn (Figure 13). Both sections contain two significant sandstone intervals, although the upper one at the former locality is a discontinuous body representing the edge of a persistent channel system; the member is 6.53 m thick at the former locality and about 11 m thick at the latter.

At the above-mentioned bluff section on the left bank of the River Nairn, 8.07 m of strata are exposed and appear to represent a deeper-water facies, being a much more homogeneous succession of finely laminated muddy siltstones, in part pyritous, and with sporadic limestone occurring as small lenticles. The facies is more like the laminites at the bases of the Orcadian Lake cycles of northern Caithness and suggests that the influence of the earlier-noted lagoonal embayment in this area possibly diminished during Clava deposition.

The Clava Member represents a lacustrine phase and the fish fauna, which includes *Cheiracanthus murchisoni* Agassiz, *Cheirolepis* sp., *Coccosteus cuspidatus*, *Diplacanthus* sp., *Dipterus valenciennesi* Sedgwick and Murchison, *Mesacanthus* sp. and *Osteolepis* sp., is an assemblage similar to that of the Achanarras-Niandt Fauna of Caithness (Donovan et al., 1974, table 2), a comparison supported by spore assemblages (Richardson, 1965).

Leanach Sandstone Formation

The Leanach Formation is a predominantly sandstone succession conformably overlying the Nairnside Formation and, south-west of Culloden Muir, is overlain by a distinctive mudstone-bearing sequence named the Inshes Flagstone Formation. The top of the Nairnside Formation can only be recognised by the presence of its topmost Clava Mudstone Member, so that in the area south-west of Allt Lugie, where no Clava outcrop is delineated, the Leanach Sandstone merges imperceptibly into an undivided Inverness Sandstone Group outcrop incorporating strata laterally continuous with both the Nairnside and Leanach formations. Similarly, where the base of the Inshes Formation cannot be proven, northeast of Culloden Muir, the lateral equivalents of the Leanach and Inshes formations merge into an undivided outcrop of Inverness Sandstone Group.

Exposures are sparse, but suggest that the formation is about 190 m thick, the lowest beds being interbedded greenish grey flaggy sandstones and thin mudstones, and younger beds being predominantly thick reddish coarse-grained sandstones. The best sections of the basal beds lie along the River Nairn, and form the left-bank cliff [7338 4200] and pavements beneath Clava Viaduct [7642 4498]; other riverside exposures occur [at 7460 4332 and 7465 4374]. Similar greenish beds were formerly noted at the base of the deeply excavated Leanach Quarry [7480 4455], although these lie some distance above the forma-

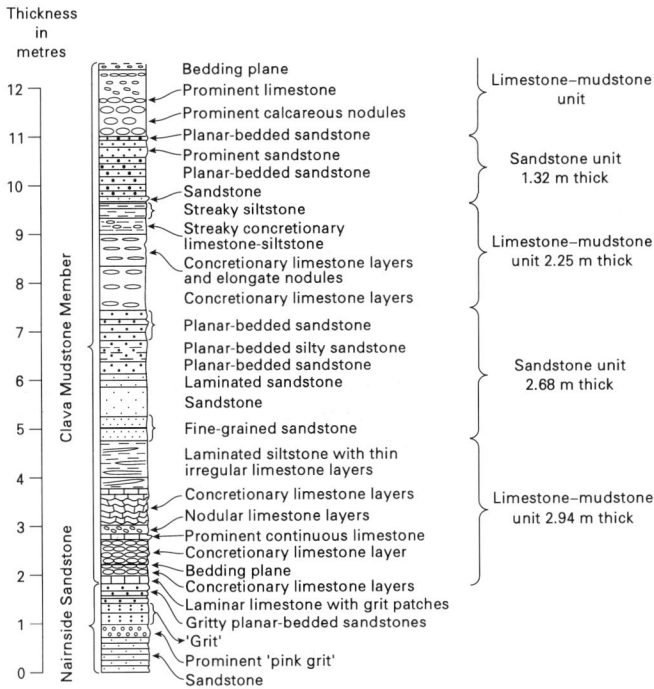

Figure 13 Type section of the Clava Mudstone Member in the River Nairn [7645 4505] opposite the mouth of Dalroy Burn.

tional base. Reid (1899, p.260 and plate opposite) stated that 6 m of such strata were overlain by 11 m of red sandstones; the latter formed the bulk of the building stones used in the Clava Viaduct and lay beneath 4 m of red cross-bedded sandstone (see also Horne, 1923 p.69). Regrettably, at the time of the resurvey this quarry was largely filled in and flooded, with little of the faces accessible. However, the wind-etched building stones in the viaduct display an impressive variety of sedimentary structures characteristic of both fluvial and aeolian regimes (Plate 10a–c). In another quarry [7615 4585] along the railway line, one metre of grey flaggy sandstone was formerly seen to be overlain by 3.66 m of red ripple-bedded sandstone, 1 cm of shaly mudstone charged with fish and plant fragments and 6.01 m of red sandstone closely resembling that in the Leanach Quarry (Horne, 1923, p.69).

The available exposed lithologies suggest a marginal lake environment subject to both fluvial and aeolian depositional processes, probably in river mouth and beach settings. Fish fragments identified as *Coccosteus* cf. *cuspidatus*, *Glyptolepis* sp., *Homosteus* sp., and *Pterichthyodes milleri* (Miller) indicate the influence of lacustrine waters, as well as a stratigraphical age in the upper range of the Achanarras Fauna.

Inshes Flagstone Formation

The relationship between the Leanach Formation and the overlying Inshes Formation is not clearly defined, because no exposed junction has been detected. In this

account, the boundary is assumed to be conformable, and probably marked by a transition from red sandstones to a predominantly darker grey-green flaggy sequence of bituminous calcareous mudstones, siltstones and sandstones forming the lowest part of the younger formation. The Inshes sequence has been likened to the flagstones of Caithness, but in this district flagstone cycles of the Caithness type are not developed; there are neither comparable laminites nor coarse-grained sandstones at cycle tops. Lithological alternations seem to reflect random sand and silt influxes, rather than chronological changes in the depth of the Orcadian Lake or changes of climate. Most of the middle section of the Inshes succession is a red-bed fluvial sequence.

In many respects, the lithologies of the greater part of the Inshes Formation are identical with the exposed beds assigned to the overlying Hillhead Formation, and the original definition (Horne, 1923, p.65) appears to have been based mainly on differences in the fossil content. In order to justify two separate lithostratigraphical units, a clearer definition is required, and since no contrasting lithological sequences can be recognised, the junction needs to be drawn at a marker bed. In keeping with the earlier concepts, the boundary should be drawn at the change from grey calcareous flagstones to the red sandstones associated with a younger faunal assemblage characterised by *Homosteus milleri* Traquair and *Millerosteus minor* (Miller) (Plate 9 and c). Regrettably, there are no exposures where this situation is manifest and the boundary line on the geological map is therefore drawn at the extrapolated base of a bright red, micaceous, flaggy sandstone sequence closely overlying dark grey, calcareous, muddy siltstones containing fossiliferous limestone nodules in the disused Leys–Culduthel Quarry [6832 4278].

General exposure of the Inshes Flagstone is poor and the overall sequence must be gauged from intermittent exposures in shallow streams flowing towards the south-eastern suburbs of Inverness; the thickness in this area is estimated at 270 m. The oldest exposed beds are grey flaggy sandstones near Newton of Leys [at 6740 4005] and at Milton of Leys [6964 4183], which are assumed to occur some little way up from the base of the formation. Mill Burn provides the most complete section, in which three units of predominantly grey-green shaly flagstones with limestones or calcareous nodules are separated by two red-bed sandstone intervals.

The lowest grey unit is overlain in Mill Burn by red coarse-grained sandstones, the boundary lying within a gap in exposure [between 6952 4193 and 6945 4208]. To the north-east, tabular cross-bedded sandstones at Muckovie Quarry [7703 4380] form part of this red sandstone sequence. The lower red unit is succeeded by grey, thinly-bedded, shaly sandstones and blackish, shaly, silty mudstone with rare crystalline limestone lenses, the mutual boundary of the units being located within close limits [between 6944 4213 and 6937 4224]; on the evidence of blackish flags with plant fragments and dipnoan or osteolepid scales in the adjacent Druid-temple Burn [at 6861 4216], the boundary is likely to lie very close to the second locality. Equivalent dark grey

Plate 10   Leanach Sandstone:
building stones in pillars of Clava
Viaduct [7645 4480]

a) Aeolian-bedded sandstone
(MNS 5696/3).
b) Cross-bedded sandstone with
mudstone clasts and chips
(MNS 5696/1).
c) Soft-sediment deformation in
sandstone (MNS 5696/2).

a

b

c

flags are exposed in St Mary's Well Burn [at 7223 4553] and alongside [7736 4457] Burnside Cottage, where limestone nodules have yielded *Osteolepis* scales; nodules occurring in the upper levels of Muckovie Quarry (now covered) formerly yielded *Coccosteus* remains (Horne, 1923, p.71).

The upper sandstone unit in and around Mill Burn is strikingly red with most of the sequence comprising bright red to purple, planar-bedded, platy sandstones interbedded with thin, red or green, laminated, shaly mudstones or ripple-bedded siltstones. Most sandstone beds are less than 30 cm thick, but rare, thicker, cross-bedded units up to 50 cm thick occur [as at 6813 4193]. The main sections in Allt na Skiath [6824 4184 to 6803 4206], Druidtemple Burn, Mill Burn [6920 4250 to 6880 4325] and an unnamed burn [between 6929 4275 and 6905 4320] fall within this red unit. Although an Old Red Sandstone facies not generally associated with fossils, red ripple-bedded siltstones contain fish scales at Mill Burn Waterfall [6916 4266], *Dickosteus* scales nearby [at 6888 4307], and *Psilophyton* plant remains in reddish grey micaceous sandstone [at 6921 4297]. To the north-east, red, wavy-bedded, flaggy sandstones in St Mary's Well Burn [at 722 4554 and 7216 4567] and near Chapelton [at 7338 4645 and 7344 4654] fall within this unit.

The top part of the Inshes Formation is a grey and green flaggy sandstone unit containing dark grey calcareous siltstones and shaly mudstones, in places with thin beds or nodules of limestone. The key section in the Leys–Culduthel quarries is much overgrown, but fish beds there have yielded *Coccosteus* and *Glyptolepis*. The base of the unit is represented in Leys Burn, where beds exposed [downstream of 6806 4200] are generally grey, fine-grained and laminated. Farther north-east, a stream section [7387 4748] alongside an old quarry at Balmachree exhibits grey, thin-bedded sandstones with shaly partings associated with bluish grey shaly strata; similar beds occur in a nearby wood, where a dark grey siltstone sequence contains a dark silty limestone 1 m thick [7414 4784]. In Rough Burn, flowing through Morayston [7525 4885], channelled sandstones with trace fossils are disturbed by small-scale faulting. The beds yield fish remains, among which *Glyptolepis* sp. has been noted, and the strata are lithologically similar to those at Balmoachree, but reddish in colour. They are considered to represent the highest part of the Inshes Formation here affected by oxidation in the fault zone.

Hillhead Sandstone Formation

The junction between the Inshes and Hillhead formations is regarded as conformable. The base of the Hillhead Sandstone is defined in the Leys–Culduthel Quarry at the facies change from grey, calcareous, laminated, fine-grained flagstones to red, coarser-grained sandstones (see above). Only two other exposures of the sequence above that boundary are known in this district, both of which lie some distance above the base. The main one forms an old quarry [7775 4985] at Hillhead

and comprises red and grey, planar-bedded, quartzose sandstone, with micaceous siltstones and silty mudstones; concretions in mudstones at the top of the quarry contain *Homosteus milleri* (Plate 9b) and *Millerosteus minor* (Plate 9c). Some slight waviness of bedding is manifest especially at sandstone bases within the more muddy intervals. In former days, about 4 m of section were quarried (Horne, 1923, p.71). Clearly a facies of lacustrine type, the sequence at Hillhead indicates that grey laminated intervals are a feature common to both the Inshes and Hillhead formations, and that the differences between them are not so well marked as those between older formations in the Inverness Sandstone Group.

Apart from red-bed exposures in streams adjacent to Culduthel, the bright red friable sandstone in Strattan Burn [at 705 459] is the only representative of such a facies in the wider district. The formation is calculated to be about 920 m thick and is considered to be overlain unconformably by the Nairn Sandstone Formation at the base of the Forres Group.

FORRES SANDSTONE GROUP

The Forres Sandstone Group was formerly called Upper Old Red Sandstone in the southern and western parts of the Moray Firth region. Following earlier works, Westoll (1951) and Tarlo (1961) recognised five subdivisions that probably can be regarded as formations. Within this district only the oldest of these comes to crop and is here named the Nairn Sandstone Formation. Because the basal contact of the group is not exposed in this district, the interpretation of an unconformity, based upon evidence in the adjacent Nairn district (Horne, 1923, fig. 2), is followed in this account.

Nairn Sandstone Formation

The Nairn Sandstone lies at the base of the Forres Group and in other parts of the Moray region is conformably overlain by the Boghole Sandstone Formation (Westoll, 1951, table III). It largely consists of red, green, grey and ochreous yellow, calcareous, cross-bedded sandstones interbedded in part with thin flaggy siltstones and shaly mudstones; conglomerate and breccia lenses are in places developed near the base.

In this district, only two exposures attributable to the formation have been recognised. A former quarry at Delnies [8322 5595], is now largely covered, and, in the past, yielded *Asterolepis maxima* (Agassiz), *Coccosteus magnus* Traquair, *Osteolepis* and *Polyplocodus* (Taylor, 1910; Horne, 1923) indicating a Givetian age, considerably younger than the Hillhead fauna in the underlying group; another exposure occurs on the foreshore at [850 569].

The Nairn Sandstone outcrop shown on the geological map is drawn to include these exposures, the junction with the Hillhead Formation being arbitrarily located some distance up-dip. Based upon this interpretation the Nairn Sandstone sequence is about 200 m thick in the district.

FIVE

Permian, Triassic and Jurassic rocks

No post-Devonian rocks occur within the land area of this district. However, the presence of Upper Kimmeridgian dark grey shelly mudstones on the Ethie shore (Figure 1), and of Corallian and Lower Oxfordian rocks as skerries off the Black Isle less than 3 km north-east of the district boundary, indicates that Jurassic strata may be preserved offshore within the district. Arenaceous Permo-Triassic rocks crop out onshore to the east of the district, resting unconformably on Upper Devonian strata. Recent maps of the Moray Firth (Andrews et al., 1990a, figs. 21, 25 and 30) show extrapolations of Lower Permian (Rotliegend) and Triassic beds, as well as Jurassic strata, falling within the northern part of the Fortrose district. However, no rocks belonging to any of these sequences have yet been proved in the firth regions of this district.

Reference to the steeply dipping exposures at Ethie, to Underhill's interpretation of the structurally complex Great Glen Fault Zone a little farther north (1991a and b), and to the Solid geological map of the Moray Firth (Chesher and Lawson, 1983), indicates a highly irregular juxtaposing of different rock sequences beneath the sea bed in the northern part of this district, though perhaps only Triassic and Jurassic beds are involved (see Figure 1).

For interpretations and accounts of the post-Devonian geological history of the inshore areas of the Moray Firth, the reader is referred to the works of Underhill and Brodie (1993) and Andrews et al. (1990).

SIX

Devonian and later structures

The structures described in this chapter affect the Devonian sedimentary sequence, as well as the underlying basement rocks. They are therefore of Devonian and younger age. The earlier deformation and metamorphism of the basement rocks have been described in Chapter Two. An analysis of the available geophysical evidence within the district is given in Chapter Seven.

STABLE MASSES AND BASINS

Devonian

Correlation of the local Devonian sedimentary succession with those in neighbouring areas shows the Fortrose district to have lain along the south-western limits of the Orcadian Basin, with significant upland areas lying to the south and north-west. The basin boundaries probably coincided with significant wide zones of Caledonian fractures rejuvenated during Devonian sedimentation. Erosion of the uplands provided detritus for deposition in the gradually subsiding basin. Initially, numerous half-grabens characterised the basin margins, but by Middle Devonian time one major basin stretching north-eastwards towards northern Caithness received sediments transported north-eastwards from an upland in the south, and westwards from an upland in the north-west.

Post-Devonian

Remnants of post-Devonian sedimentary rock sequences have not been recognised in this district, but evidence elsewhere in the Moray Firth region indicates that the Devonian rocks have been variously covered by younger deposits. Details currently accumulating about the Inner Moray Firth region show that the overall structure of the Orcadian Basin is reflected in some of the post-Devonian depositional patterns (Underhill, 1991a and b; Underhill and Brodie, 1993).

FAULTS AND FRACTURES

The general rock relationships in this district indicate that various stages of fracturing and fault movement have affected Devonian and later sequences. The first stage essentially controlled the deposition of Devonian sediments and was largely due to a prolonged period of extensional stress and thermal subsidence. The second stage reflects a compressional regime established in late-Carboniferous to Permian time, during which dextral movements along the Great Glen Fault Zone and contractional reactivation of earlier extensional faults

developed. Subsidiary conjugate faults were associated with folding, which in places was sufficiently strong to produce overturning and imbrication of strata. A third stage of fracturing and movement is indicated by certain normal faults that cut the contractional faults. Although the timing of these latest faults cannot be gauged in this district, new evidence documented from the more northerly parts of Ross-shire (Underhill and Brodie, 1993) suggests that separate extensional phases relate to two (Permo-Triassic and Late Jurassic) rift events and a subsequent (post-Early Cretaceous to Neogene) regional uplift.

Most fractures affecting Devonian and later sequences are likely to represent reactivated adjustments along shear zones and fractures already developed in the metamorphic basement.

Devonian faulting has been cited (Stephenson, 1977; Mykura, 1983) as a major control during the deposition of a marginal alluvial-fan piedmont at the onset of local Devonian sedimentation. As presently preserved, the basal Old Red Sandstone outcrops do not contain unequivocal faults that can be identified as parts of paralleling north-facing scarps. However, within the metamorphic basement of the southern hinterland, where the former Old Red cover has been stripped away, several paralleling lineaments are clearly defined, and these may have experienced movement in Devonian times. Other faults of different trend tend to step the basal Old Red outcrop and may have been active also during the Devonian.

Only one example of intra-Devonian faulting has been recognised. This occurs in Rosemarkie Glen, and though no fault is exposed, the disposition of the Den Siltstone in relation to the disconformable Kilmuir Conglomerate suggests that considerable rotation and erosion of the siltstone sequence took place prior to the covering by conglomerate. Preservation of the Den Siltstone here may indicate the remains of a small half-graben or asymmetrical block with its main fault marked by the Rosemarkie Fault on its eastern side.

All mapped faults on the land area of the Fortrose district appear to belong to sets that affect Devonian strata. Information garnered from investigations in the Moray Firth (Underhill, 1991 a, b; Underhill and Brodie, 1993) indicate that they represent adjustments largely made before the onset of Triassic sedimentation, most likely during the known period of stress in the late Carboniferous and early Permian.

The main faults probably reflect basement controls and are rejuvenations of long-standing fractures. The predominant fractures strike at approximately 040°, 140°, 080° and between 170° and 185° (Figures 6 and 14).

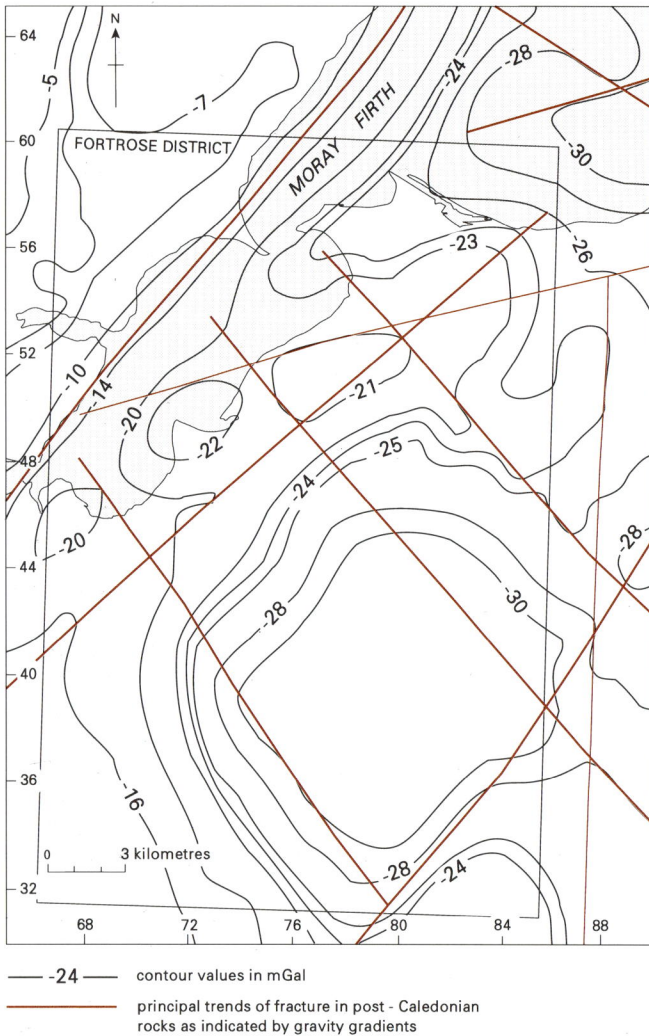

| contour values in mGal

principal trends of fracture in post - Caledonian rocks as indicated by gravity gradients

Figure 14   Bouguer gravity anomaly map derived for a reduction density of 2.72 Mg m^{-3}.

Prominent gradients in part coincide with trends of fractures mainly affecting the Old Red Sandstone south-east of the Great Glen Fault Zone.

Faults and lineaments trending 040°

BLACK ISLE

The principal fractures in this district occur along the Great Glen Fault Zone, striking approximately 040° from Inverness along the offshore region of the Black Isle (Cover photograph). At surface, the major displacement appears to be a downfaulting to the south-east, throwing younger Old Red Sandstone against basement and older Old Red Sandstone. However, correlations of the Old Red Sandstone sequences astride the fault zone indicate that significant dextral displacement took place (Rogers, 1987) prior to the main episode of downfaulting, such that the Black Isle side has moved north-eastwards about 29 km relative to the south-eastern side. Hence, the Black Isle sequence originally formed the northern extension of the Old Red cover presently exposed on the

south-eastern side of the Great Glen Fault Zone in the southern Loch Ness region of Sheet 73E (Figure 1).

Fracturing within the Rosemarkie Metamorphic Complex

The outcrop of the Rosemarkie Complex, particularly at Rosemarkie, is severely disrupted by late brittle faulting and brecciation. Most of the fault planes belong to the 040°-trending set, with an associated subset trending between north and east-north-east. These fractures, together with small lag thrusts, largely control the boundary with the Old Red Sandstone, so that only short stretches indicate the original unconformable relationship. The latest definable movement in the area is associated with minor fractures trending between west-north-west and north-west, which elsewhere (May and Highton, in preparation) are associated with fracturing during Permo-Carboniferous times.

Brittle deformation   Most of the fracture planes within the complex are vertical or steeply inclined to the east, with an apparent dextral sense of movement. The presence of drag folds [7382 5803] indicates early ductile movement. These in turn are cut by later veins of breccia and cataclasite. The extent of horizontal movement at outcrop is picked out by marker horizons of metabasic material [as at 7387 5809], and varies from no apparent displacement to several tens of metres. In many of these southern localities, however, evidence of movement is not recognisable, and the host lithologies are apparently little affected between zones of crushed rock; evidence of disturbance is largely confined to the development of fine-scale jointing. Fracture surfaces here are generally well marked, with slickensides on the hanging wall and a grey clay gouge in the footwall. The extent of any vertical movement on these fractures is uncertain, in view of the two-dimensional aspect of much of the outcrop. Vein-like masses and pods of breccia are present within most of the fracture zones. Clasts of the adjacent host rocks are comminuted and held in a carbonate-rich or cataclastic matrix. Cataclasite is common, locally forming ramifying networks of red veins up to 10 mm wide. More rarely, grey veins of ultracataclasite (as in S 72574) cut both the fracture surface and breccia [7443 5885]. Mineralisation is common, with abundant pyrite, disseminated haematite and manganese. Secondary veining is abundant in areas of severe brecciation, with the fault zone material and host invaded by carbonate and quartz-epidote veins; this is indicative of large-scale fluid movement through the fault zones. As exposed in the steep cliff-face at Cairds Cave [7455 5954] and the quarry south of Whinhill [7390 5840], the complex is cut by south-easterly dipping low-angled brittle fractures (dipping between 15° and 35°), which steepen to between 30° and 40° at higher levels. These structures define a listric geometry, with a lag or reverse sense of movement indicated by prominent slickensides principally directed east to north-east into the Great Glen Fault Zone. The timing of movement on these structures in the basement complex is uncertain, but may be early Devonian.

Fenitisation   Several areas of apparent fenitisation have been identified within the Great Glen Fault Zone (Deans et al., 1971; Garson et al., 1984). These authors documented evidence in the Rosemarkie Complex of fluidised dyke breccias with carbonate-rich matrices, albitites, fenite borders to calcite veins and metasomatic assemblages replacing the host rocks. The presence of carbonate-rich veins, with sodic amphibole-bearing margins, was cited as *prima facie* evidence of a carbonatitic origin. These phenomena were regarded as being broadly coeval and associated with the emplacement of a carbonatite complex, as yet unroofed, into the fault zone during early Devonian times (about 390 Ma).

Fenites During the original survey of this area, the presence of a dark bluish green amphibole in the metamorphic rocks was noted, but not described. The occurrence and significance of this amphibole, identified as a magnesio-riebeckite (crocidolite), was documented firstly by Deans et al. (1971) and in more detail by Garson et al. (1984).

The blue amphibole was found during the resurvey to be preferentially associated with areas of tectonic brecciation in the outcrop. The amphibole is present both as pervasive dark blue patches in the host rock, powder-blue coloured fracture infills (as in S 72572) and coatings to joint surfaces [as at 7475 5968].

A breccia exposed on the shore north of Rosemarkie [at 7438 5880] provides one of the more spectacular examples of amphibole development. Here, parts of the host rock appear to be 'soaked' with the blue amphibole (as in S 72577), producing a distinctive blue-grey colouration; this coincides with clasts of amphibolite or hornblendic gneiss in the breccia (as in S 72577). The riebeckite here is coarse grained, pleochroic from blue-green to mauve, and is clearly secondary, overprinting both as distinctive irregular, coronitic rims and replacing the pre-existing mid-green to brown pleochroic metamorphic amphibole (as in S 72577). The more common form of the amphibole is as fibrous, prismatic crocidolite exhibiting pale to mid-blue and mauve pleochroism. Birefringence colours of low to middle first order are usually masked by the intense body colour. Within the psammites adjacent to the breccia at this locality, crocidolite forms small, irregular to radially fibrous mats less than 0.1 mm long (as in S 72572). Nucleation sites lie commonly at grain boundaries (as in S 72572). Most phases adjacent to the mats are penetrated by the fibrous growth of the amphibole. However, where these masses are in contact with plagioclase, a clear corona of apparently calcic feldspar is present at the interface of the amphibole and the more sodic plagioclase.

Fracture and vein infills of carbonate, and more rarely prehnite (as in S 72573), are common. The crocidolite forms elongate, braided masses at the margins to carbonate veins. Their occurrence suggests a contemporaneous formation. Riebeckite growth, however, is nowhere seen to develop in, or cut across, the fine-grained to cataclastic matrix of the breccias.

Thin calc-silicate seams within the psammitic lithologies [at 7517 6018] contain irregular plates and aggregates of mid to dark green pyroxene, identified as aegerine-augite (Garson et al., 1984), sieved with carbonate and chlorite.

These are pseudomorphs after a pale to mid-green amphibole, which is preserved only as relict patches. The amphibole cleavage is mostly preserved mimetically by chlorite and carbonate intergrowths with anhedral crystals of sphene. This assemblage is recrystallised to eu- to subhedral sodic pyroxene with relict reaction inclusions of carbonate and sphene. The outer contacts of the pyroxene aggregates, where adjacent to plagioclase, are replaced by riebeckite. Biotite, forming primary inclusions in the regional metamorphic amphibole, is little altered, with only minor chloritisation. Where close to, or in contact with, inclusions of plagioclase or apatite, both the mica and calcium-bearing minerals are consumed by riebeckite. Sphene commonly is mimetically replaced by aggregates of carbonate, leucoxene and anatase, while magnetite is made over to haematite.

Breccias Most breccias identified at outcrop are clearly of tectonic origin, with a wide range of textural types from mylonite to ultracataclasite. However, vein-like and lensoid breccia bodies of possible igneous origin were identified at three localities [7444 5891, 7480 5972 and 7517 6018]. The largest of the veins attains a maximum thickness of 0.2 m, traceable in outcrop for some 10 m. All are slightly irregular in form (Plate 11). Contacts are everywhere sharp and cut across both the lithologies and tectonic fabrics of the metamorphic complex and granitic intrusions; the contacts are commonly slickensided. Clasts are subangular to rounded, with diameters ranging from less than one to 60 mm at outcrop. They mostly comprise wall-rock lithologies with no exotic fragments. All are matrix or carbonate supported. Sorting is generally poor, with good evidence of comminution. The matrix, where present, is mainly a brownish purple aphanitic material. Where resolvable, this appears to be essentially a micro-granite, becoming locally siliceous (as in S 72570). The presence of two feldspars is indicated by bleb-like patches of myrmekite and a graphic intergrowth of quartz and alkali feldspar. Accessory minerals include zircon, apatite, allanite and monazite. Haematite and pyrite are abundant as small, sub- to anhedral grains, with secondary chlorite, white mica and carbonate. The breccias are permeated by veins of carbonate, which also appears to fill former voids in the matrix; the veins cut both matrix and clasts respectively.

The form and internal composition of these breccia masses is uncertain. The sharpness of the contacts is not unusual in terms of classical intrusive or explosion breccia dykes and pipes (Platten and Money, 1987). However, despite the local derivation of the fragments, there is no evidence of country-rock spalling. The presence of slickensides on contact surfaces indicates some tectonic movement.

Carbonate veins White carbonate veins, patches and neocrystallisation-fracture infills are present throughout much of the Rosemarkie Complex and cut all lithologies and structures, although they are more abundant in areas of tectonic and apparent igneous brecciation. The veins are generally irregular, up to 50 mm wide and locally form ramifying networks, with several generations recognisable. The presence of crocidolite in fracture

Plate 11 Fenitisation in Rosemarkie Metamorphic Complex; fenite breccia with carbonate and cataclasite matrix [7502 5992]. Coin is 2.5 cm across (D4893).

infills has been noted above, but is not common. Whether crystallisation of this amphibole was contemporaneous with the formation of the veins or fractures is uncertain. Fractures with both amphibole and carbonate, however, may indicate the exploitation of pre-existing fractures, in which crocidolite had already formed, by carbonate-bearing fluids. On the other hand, in specimen S 72572 crystallites of chlorite and (?) sodic pyroxene in the carbonate appear to have a co-precipitative relationship, with well-formed crystal boundaries

The relationship of the carbonate veins to the late brittle structures is problematical; they both cut across, and are cut by, major and minor brittle structures.

Fracturing of the Old Red Sandstone

On the north-western side of the Great Glen Fault Zone, only two major examples of 040°-trending fractures have been detected outside the metamorphic outcrop. The main one is a small north-west-directed thrust fault with prominent shear planes cutting the Craigiehowe cliffs [6875 5198]; the other is now drift covered, but was noted by Horne in the Avoch Harbour cliff [at 8042 5512].

On the south-eastern side of the Great Glen Fault Zone, the metamorphic and igneous outcrops contain a suite of lineaments trending approximately 040°. In places brecciation has been noted without the evidence of displacement, as at Daviot [7185 3840], but faulting is conspicuous along the Conagleann Fault, which enters this district on the southern side of Strath Nairn [at 6875 3140] (Figure 14). The 040° trend is the predominant direction in this region and most faulting probably postdates Devonian sedimentation there, for fractures like the Findhorn Fault and its complementary fault cutting Allt Breac form a graben [852 388] of Devonian sediments. Bedding in the Devonian sequence is nearly vertical close to the south-eastern bounding fault, although the angle of dip decreases progressively over several hundred metres to the west and possibly indicates a syndepositional half-graben. Another outlier of shattered Devonian rocks in Meur Bheòil [around 849 402] testifies to the presence of similar parallel structures. Broad shatter zones associated with the Findhorn Fault and related structures are recognisable on Creag Ruadh [854 382], and more discrete zones of shattering occur immediately north-east, in Allt Breac. Zones of broken and sheared chloritic gneiss around Torr an Eas [843 362] dip moderately to the east.

To the south-west, outcrop evidence for the Findhorn Fault is less pronounced, although the structure controls the south-eastern boundary of the Moy Granite, and metasedimentary lithologies cannot be matched across the structure.

Between the Findhorn Fault and the edge of the main Old Red Sandstone outcrop, several subparallel faults affect various contacts between Old Red Sandstone, metamorphic and igneous outcrops. Due to the masking drift cover, possible extensions of the mapped lines remain unproven. One of these faults passes through Lochan Dubha [818 398] and is marked by breccia blocks cemented by quartz [831 412]. Some may be the splay continuations of the Conagleann Fault, affecting Devonian contacts with the metamorphic and igneous basement between Meall Mór [745 408] and Creagan Glas [767 425]. Within the Devonian outcrop, a prominent straight lineament near Daviot Castle [at 7245 4100] is of this trend.

Faults and joints trending 140°

A trend of 140° is an important structural feature on the Black Isle, where it is represented by the principal joints, best seen in coastal Old Red Sandstone sections. In contrast, a fault at Morayston [753 490] is the only known exposed fracture of this general trend south-east of the Great Glen Fault Zone.

Faults trending 080°

Fractures generally trending 080° are associated with some Old Red Sandstone folding and some faults of 040° trend. On both sides of the Great Glen Fault Zone, examples of a merging of the two trends occur, notably at the mouth of Munlochy Bay, where the Ormond [6877 5295] and Sligo [670 519] faults join along fractures roughly parallel to the Great Glen Fault Zone, and between Creagan Glas and Saddle Hill [795 435]. In the southern part of the Black Isle, the Drynie Fault is deflected to a trend closer to 080° as it crosses the district boundary [5105 6675]. Satellite images show that the 080° trend is a major feature of the area between Kilmuir and Rosemarkie. Faults have been noted within the coastal metamorphic sections at two places [748 597 and 7430 5855], the latter probably continuous with the Quarry Fault [at 729 583], which appears to postdate the Rosemarkie Fault (trend 178°).

Faults and fractures trending between 170° and 185°

Contractional faults of approximate north–south trend are well developed on both sides of the Great Glen Fault Zone, especially affecting rocks at the pre-Devonian to Devonian boundary. A recent study by Underhill and Brodie (1993, p. 521) in northern Ross-shire has shown that these faults predominate there, and are associated with northerly plunging folds. Most faults are reverse faults and series of hanging-wall anticlines are commonly developed.

On the Black Isle, the Drummarkie and Rosemarkie faults are clearly members of this set; the latter apparently predates the Quarry Fault of the 080° set and marks a slight overthrusting of a disrupted Devonian cover by the metamorphic complex. Other examples are marked near Pookandraw [681 591], along the Rosemarkie foreshore and at Arkendeith [693 558].

South of the Great Glen Fault Zone, significant faults of north–south trend greatly affect the outcrop pattern, particularly in the Drummore of Clava area [771 435], where the Finglack Fault has a westerly downthrow. Mulgrew (1985) has described this fault as dipping at about 70° and marked by a 'steep purple-coloured

mylonite zone about half a metre wide'; the dip of the Daviot Conglomerate beds increases from 15° to 50° within a few metres of this zone.

Similar westerly throwing faults occur downstream [733 410] of Daviot Castle and [at 737 415] near Mains of Daltulich.

FOLDS

All post-Caledonian folds in this district are related to compressional stresses associated with faulting and are mainly restricted to the north-western side of the Great Glen Fault Zone. All affect the Devonian succession and are likely to be late Carboniferous to Permian in age.

The Old Red Sandstone on the Black Isle forms part of the south-eastern flank of the Black Isle Syncline, whose axis lies outside the district to the west and parallels the Great Glen Fault Zone on its eastern side and the Strathglass Fault (D I Smith, 1977, fig. 1) in the west. In the Fortrose district, the south-eastern flank dips north-west at about 5°, but in the coastal region, close to the Great Glen Fault Zone and the associated splay faults that cut across Rosemarkie Glen and Munlochy Bay (Figure 14), rocks are thrown into tighter subsidiary folds with dips up to 70°, as at Knockmuir Quarry [7075 5623] and Sligo Burn [6737 5210]. Some of these splays may be reverse faults, similar to the Rosemarkie Fault and those in northern Ross-shire (Underhill and Brodie, 1993).

On the south-eastern side of the Great Glen Fault Zone, the drift cover masks much of the Old Red Sandstone succession and the only evidence of a substantial fold is indicated by the overall disposition of strata at Morayston [7525 4880], where dips to the east-north-east and east, as high as 24°, are significant deflections from the predominant north-west trend elsewhere in the region. As on the Black Isle, such changes probably reflect compression alongside a reverse fault, in this case the Morayston Fault.

SEVEN

Geophysics

This chapter describes the results of quantitative modelling of gravity and aeromagnetic data for the Fortrose region.

DATA

Gravity data

Onshore gravity data for the region around Fortrose and offshore data collected along ships' tracks have been reduced to Ordnance Datum (OD) at a uniform density of 2.72 Mg m^{-3} and interpolated on to a grid of 0.5 km spacing. Bouguer gravity anomalies are negative across the entire region. The main feature of the smoothed Bouguer gravity anomaly map (Figure 14) is a large closed subrectangular anomaly over the Moy Granite (Figure 15) with a minimum value of about -35 mGal near the south-eastern margin of the main part of the Saddle Hill Granite. Other significant closed local minima occur at Inverness, Alturlie Point and near Whiteness Head. Maximum Bouguer gravity anomalies occur in the north-western part of the Fortrose district, where values approach zero. The Great Glen Fault Zone is associated with a strong gradient in the Bouguer gravity field with more positive values to the north-west and maximum gradients across Chanonry Peninsula east of Fortrose.

A regional Bouguer gravity field was defined by fitting a third order polynomial to the observed Bouguer gravity anomalies at the intersections of a 5-km grid across the Fortrose district, but excluding points over the Moy Granite. Minimum residual values (Figure 15) are about -13.2 mGal close to the south-eastern margin of the main body of Saddle Hill Granite.

Gravity data coverage (Figure 15) is poor across much of the Moy Granite. The Findhorn Granite Complex is associated with a local positive residual anomaly suggesting a more dioritic composition overall.

Aeromagnetic data

Aeromagnetic data across the region were collected in 1964 by the Hunting Group on behalf of the British Geological Survey along east–west flight lines spaced approximately 2 km apart with north–south tie-lines spaced approximately 10 km apart. These data were collected in analogue form at a mean terrain clearance of 0.3 km and have since been digitised. The main features of the smoothed aeromagnetic anomaly map after reduction to the pole (Figure 16) are a large elongate positive anomaly along the Great Glen Fault Zone with a local maximum of about 400 nT near Fortrose, and a large discontinuous annular anomaly associated with the Moy

Granite, showing local maximum anomaly values up to about 300 nT near Loch Moy.

Seismic data

The Lithospheric Seismic Profile in Britain (LISPB) crosses the Fortrose district. The original interpretation (Bamford et al., 1978) suggested a three-layer model of the crust with a Moho at about 30 km depth close to the Great Glen Fault Zone and a step down to the south-east of about 2 km in the upper layer of Caledonian belt metamorphic rocks close to the fault zone. A revised interpretation (Barton, 1992) suggests that the upper 20 km of the crust in northern Britain is not clearly divisible into two layers, although there is a general downward increase in velocity from 6.0 to 6.4 km^{-1}.

Physical properties of rock specimens

Table 2 gives unsaturated density and magnetic susceptibility measurements made on 7 specimens from the BGS registered collection. Unsaturated density is less than saturated density, but for rocks with a low porosity the difference will be small. The Moy and Saddle Hill granites have densities close to 2.65 Mg m^{-3}, whereas the Findhorn granodiorite has densities of about 2.66 Mg m^{-3} with local dioritic varieties in the range 2.72 to 2.78 M gm^{-3}. The mean density of rocks of the Central Highland Migmatite Complex is close to 2.72 Mg m^{-3}. The anticipated maximum density contrast between Saddle Hill Granite and host metamorphic rocks is considered to be -0.10 Mg m^{-3}; the mean density contrast of the granodioritic phase is probably about -0.06 Mg m^{-3}, but might approach zero for monzodioritic rocks.

Table 2 Unsaturated density and magnetic susceptibility of rock specimens

Specimen (registered no.)	density Mg m^{-3}	susceptibility SI
Moy Granite (S 83477)	2.617	0.0001
Saddle Hill Granite (S 79385)	2.603	0.0003
Appinite-diorite (S 79375)	2.905	0.0003
Hornfelsed semipelite (S 79381)	2.722	0.0002
Findhorn granite (S 83527)	2.663	0.0044
Findhorn diorite (S 83525)	2.723	0.0198
Findhorn granodiorite (S 83527a)	2.654	0.0065

Susceptibility measurements on additional specimens of the Saddle Hill and Moy granites are generally less than 0.002 SI. Other specimens of the Findhorn granodiorite have susceptibilities up to about 0.020 SI. The suscepti-

Figure 15 Residual gravity anomaly map of the district and its surrounding area based on Bouguer gravity data derived for a reduction density of 2.72 Mg m^{-3} and a third-order polynomial field.

All data refer to the National Gravity Reference Net 1973. Contour values in mGal. Locations of onshore data indicated by crosses. Sections X–X' and Y–Y' indicate the locations of the 2.5D interpretations (Figures 18, 19). MG–Moy Granite, COG–Carn Odhar Granite, SHG–Saddle Hill Granite, FD–Findhorn diorite, FG–Findhorn granodiorite. Local lows are indicated by ticks.

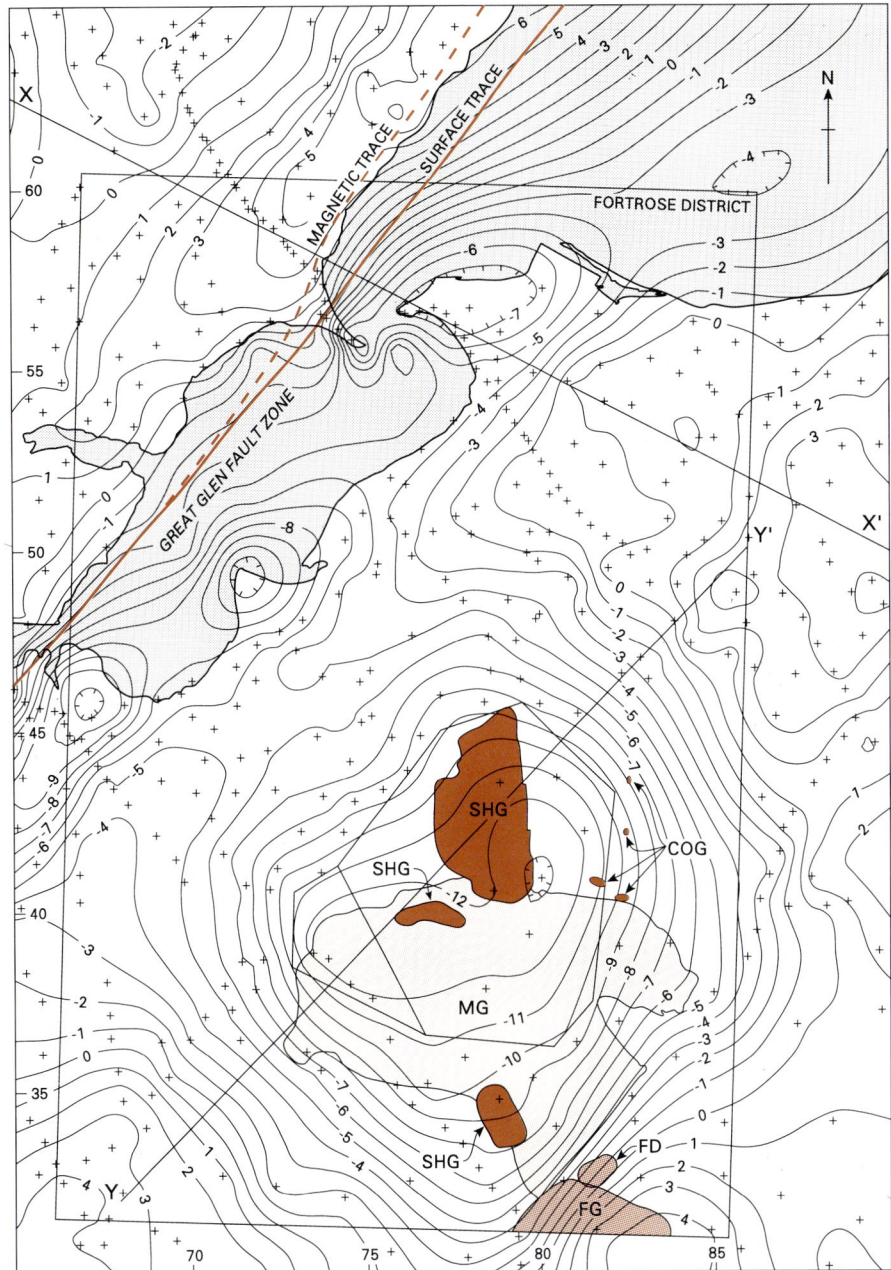

bility of a specimen of Auchnahillin Monzodiorite falls within the range recorded for the Findhorn Granite Complex. Assuming induced magnetisation in the Earth's field (approximately 40 Am^{-1}), these susceptibilities will produce rock magnetisations of about 0.80 Am^{-1}.

INTERPRETATIONS

Deconvolution

Systematic deconvolution of the aeromagnetic data across the region (Rollin, 1993) was carried out to identify the position and depth of magnetic structures in the Great Glen Fault Zone and around the Moy Granite. Euler deconvolution of the aeromagnetic data identified a magnetic boundary within the fault zone at a depth locally less than 1 km. Euler solutions do not necessarily provide a reliable depth indicator, but do indicate the plane of the magnetic discontinuity. The traces of the deconvolution solutions north-east of Fortrose suggest that the magnetic discontinuity lies north-west of the surface trace of the Great Glen Fault Zone. Several solutions occurring over the magnetic anomaly north-east of the Saddle Hill Granite indicate depths in the range 1 to 2 km below OD.

The results of Euler and Werner deconvolution for a section of flight line HG64-022 are shown in Figure 17. This line crosses the Great Glen Fault Zone near Fortrose. Euler solutions have been calculated for structural indices

Figure 16 Aeromagnetic anomaly map of the district and its surrounding area.

Total-field aeromagnetic anomalies in nT relative to a linear regional field for the UK and reduced to the pole. Data observed at a mean terrain clearance of 305 m. Thick contours are the 3D modelled depths in kilometres below OD of magnetic basement close to the Great Glen Fault Zone; shaded zones are the areas where the 3D model of the granitic rocks has susceptibilities above 10 and 20×10^{-3}SI.
MG–Moy Granite, COG–Carn Odhar Granite, SHG–Saddle Hill Granite, FD–Findhorn diorite, FG–Findhorn granodiorite.

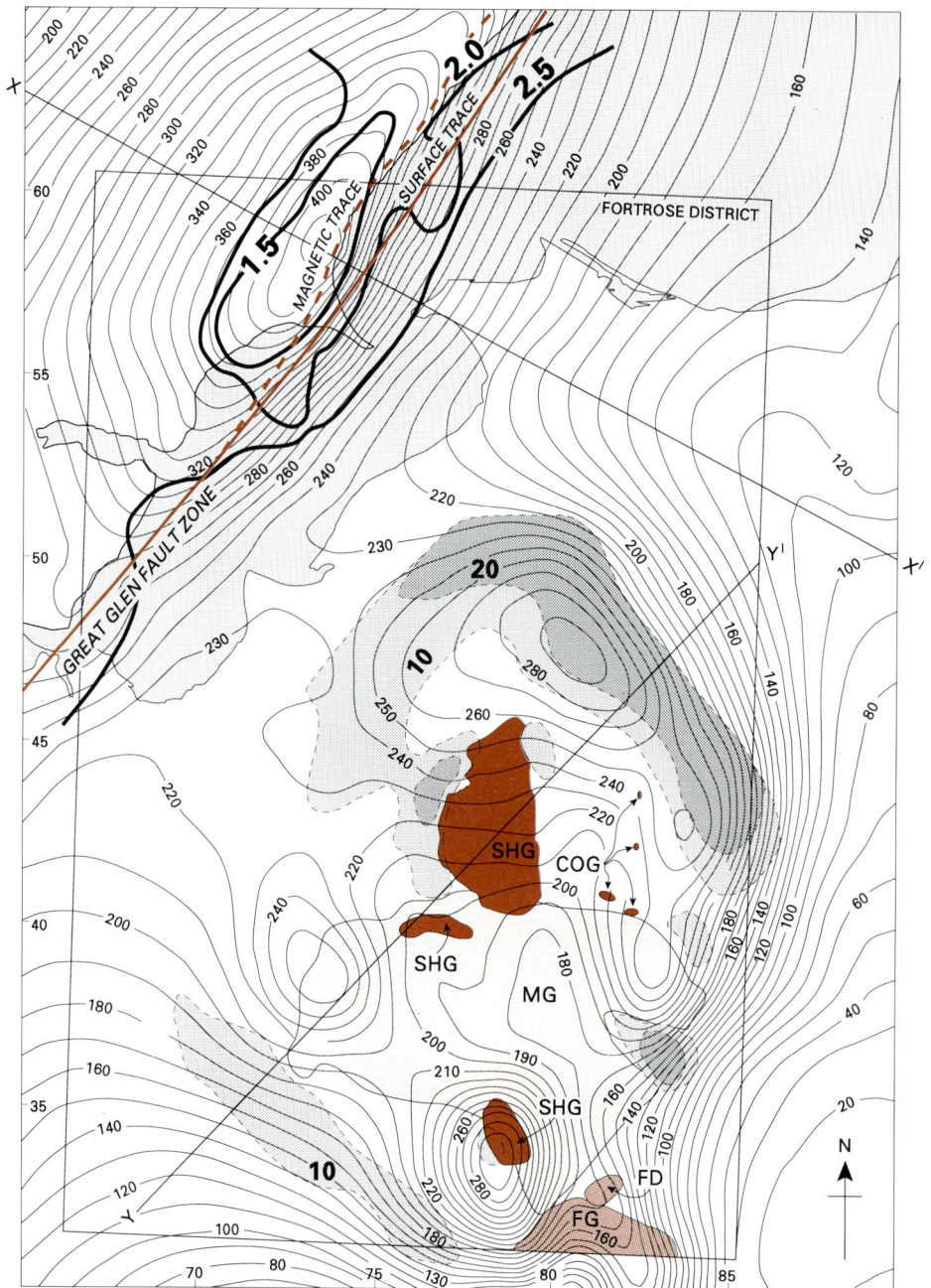

of 0.5 and 1.0, representing two simple geometrical shapes, the vertical fault and the vertical dyke. The fault zone anomaly, at about 24 km along this section of the flight line, is interpreted as a vertical discontinuity with depth solutions within 1 km of the surface. The Werner solutions, derived mostly from the total-field anomaly, are considered to give slightly better indications of depth.

2.5D modelling

Two profiles across the Fortrose district (Figure 16 X–X′, Y–Y′) were modelled using the BGS interactive GRAVMAG package, which provides integrated gravity and magnetic modelling for 2-dimensional models with a specified

strike extent. The profiles were selected to examine the main features of the district, the Great Glen Fault Zone and the Moy Granite area.

GREAT GLEN FAULT ZONE

A geophysical model of the structure across the fault zone is provided in Figure 18, taken from a longer regional profile across the Northern Highlands and Grampian Highlands, which incorporates a full crustal model to a depth of 40 km. The main features of the magnetic anomaly along the fault zone are the high amplitude, the broad extent and the local higher frequency culminations. On a regional scale, the anomaly is reasonably symmetrical about the fault zone,

Figure 17 Euler and Werner deconvolution solutions of aeromagnetic data across the Great Glen Fault (GGF).

a) Total-field aeromagnetic anomaly (T) along part of line HG64-022 (see inset).
b) First horizontal derivative (dT/dx) and first vertical derivative (dT/dz) of aeromagnetic anomaly.
c) Euler solutions for deconvolution operators of length 4 km and 8 km.
d) Werner solutions for deconvolution operators of length 4 km and 8 km.

Inset colouring as Figure 16.

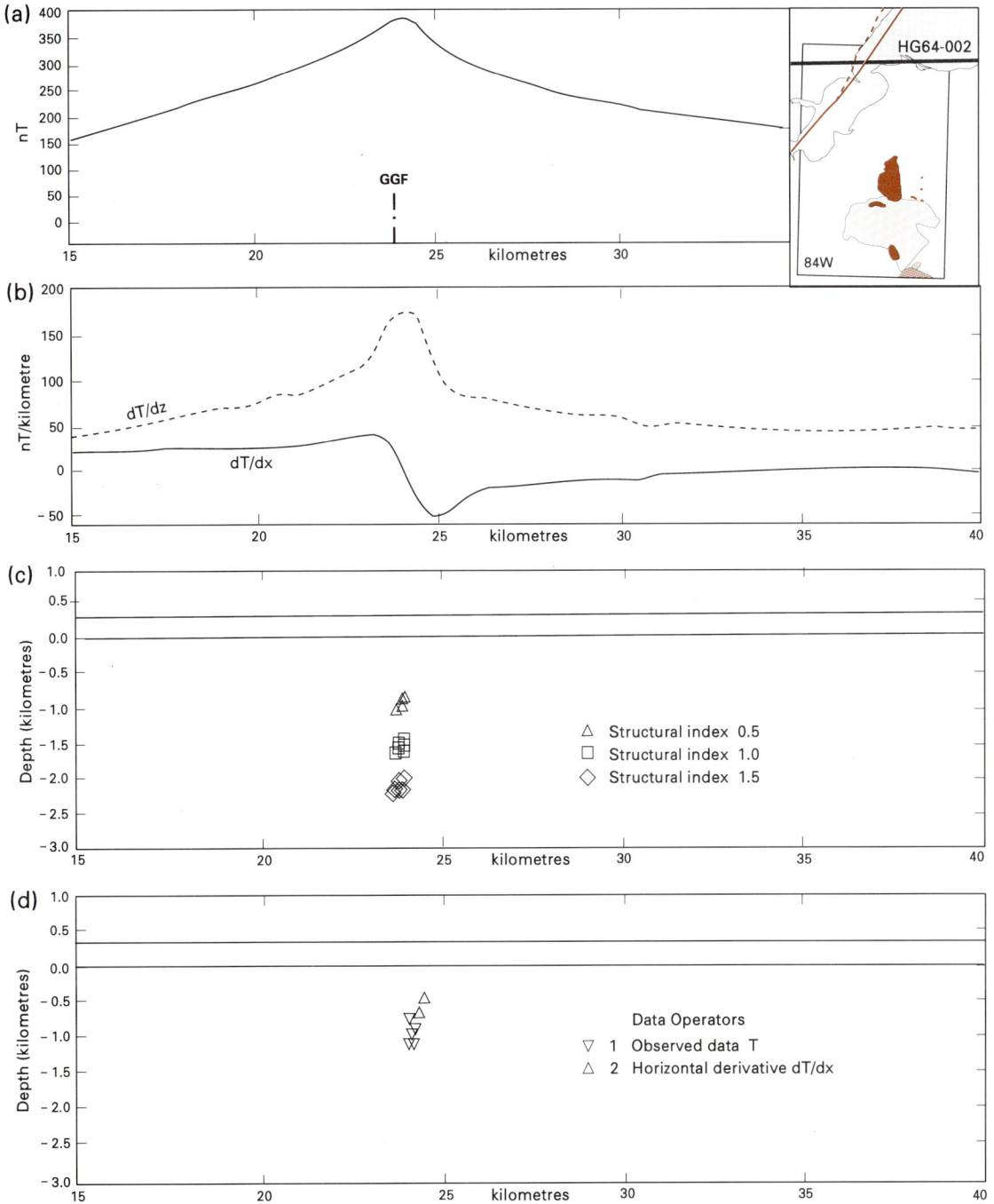

suggesting that the magnetic source occurs on both sides of the main displacement. The model shown in Figure 18 indicates a magnetic basement with a mean susceptibility of 0.040 SI at about 0.9 km below OD on the north-western side of the Great Glen Fault Zone, with a step down to the south-east of about 0.5 km. This susceptibility is higher than the mean value (0.025 SI) of the granulite-facies Lewisian gneiss of the North-western Highlands (Powell, 1970), which might suggest that Lewisian gneisses are not the source of the fault zone anomaly.

The Old Red Sandstone sedimentary rocks of the Moray Firth are modelled as up to about 2 km thick, resting directly on the magnetic basement (Figure 18).

MOY GRANITE AND ASSOCIATED ROCKS

A model profile across the granitic rocks is shown in Figure 19. A linear regional gravity anomaly has been removed. A constant regional aeromagnetic anomaly of 160 nT was also removed before modelling. These regional corrections produced a residual gravity anomaly of about -18 mGal and residual magnetic anomalies of

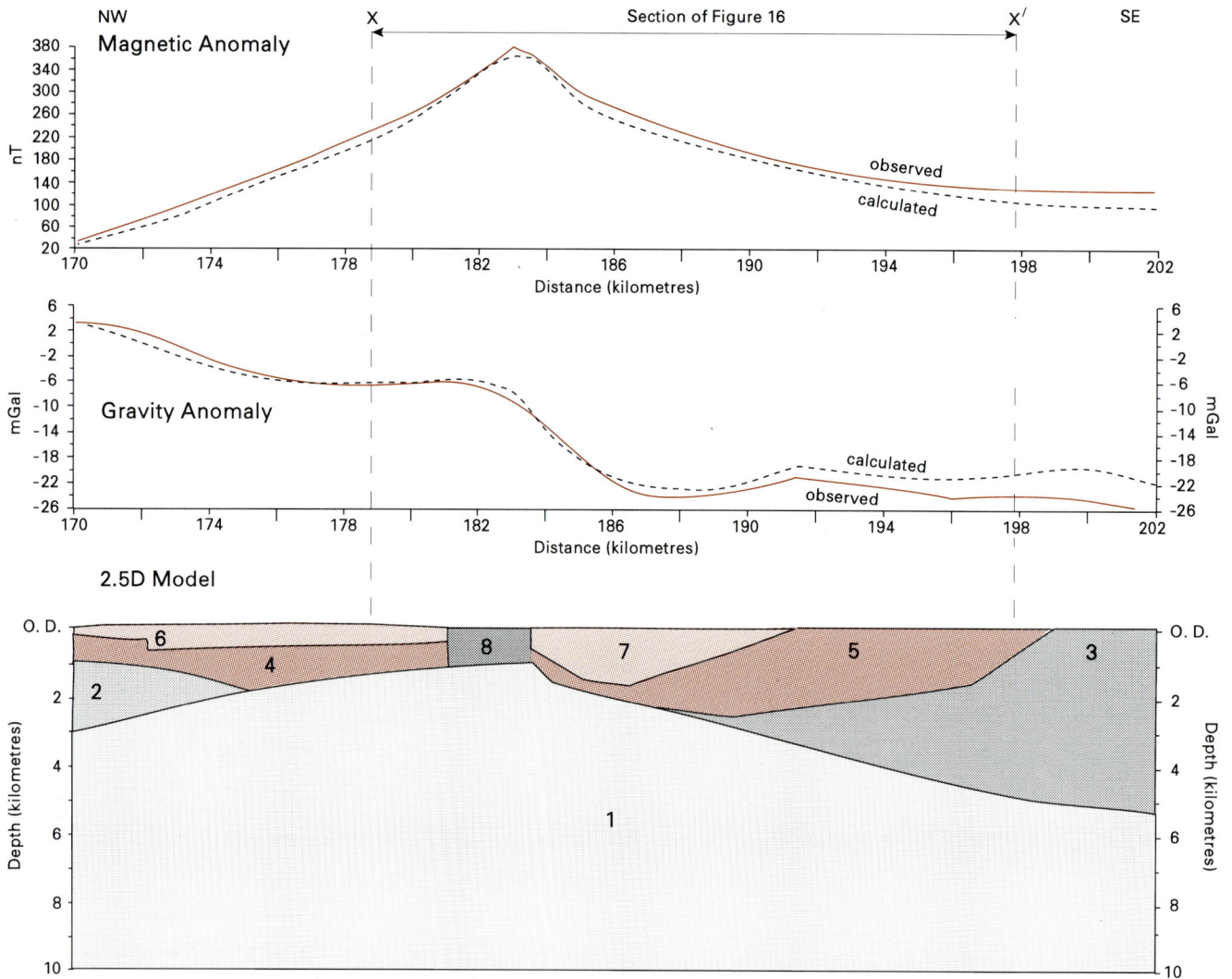

Figure 18 A 2.5D model of the upper crust along section X–X′ (Figure 16).

The model is part of a longer profile across the Grampian and Northern Highlands. The observed and calculated gravity and magnetic anomalies are shown. The model polygon properties (density Mg m^{-3}/ susceptibility SI) are:

1	2.78/0.04	Early Proterozoic basement
2	2.78/0.0	Moine
3	2.72/0.0	Central Highland Migmatite Complex
4, 5	2.60/0.0	Old Red Sandstone
6, 7	2.40/0.0	Old Red Sandstone
8	2.60/0.0	Rosemarkie Metamorphic Complex

about 140 nT. The model incorporates three distinct phases of intrusion: the Moy Granite (density 2.62 Mg m^{-3}), an unexposed phase considered to be monzodioritic (density 2.66 Mg m^{-3}, susceptibility 0.015 to 0.020 SI) and the younger Saddle Hill Granite (density 2.63 Mg m^{-3}). The depths of the magnetic phase north-east of Saddle Hill are in accord with the results of deconvolution, although ground profiles across similar annular magnetic anomalies in the Grampians generally indicate a more localised and shallower source. The granite model extends to a depth of about 7 km below OD.

3D modelling

The main features of the residual gravity anomaly and the polar magnetic anomaly have been modelled in three dimensions.

GRAVITY MODEL

The central area of the granitic rocks was modelled with a density contrast of -0.10 Mg m^{-3}, representing the Saddle Hill Granite. The outer parts were assigned a density contrast of -0.06 Mg m^{-3}, representing the more magnetic

Figure 19 A 2.5D model of the granitic rocks along section Y–Y′ (Figure 16). A linear regional gravity anomaly has been removed, with values of -10 mGal at 0 and -22 mGal at 26km along the section. A regional aero-magnetic anomaly of 160 nT has been removed. The model polygon properties (density Mg m^{-3}/ susceptibility SI) are:

1, 5   2.63/0.0 Saddle Hill Granite
2   2.66/0.015 monzodiorite
3   2.66/0.02 diorite
4   2.62/0.0 Moy Granite

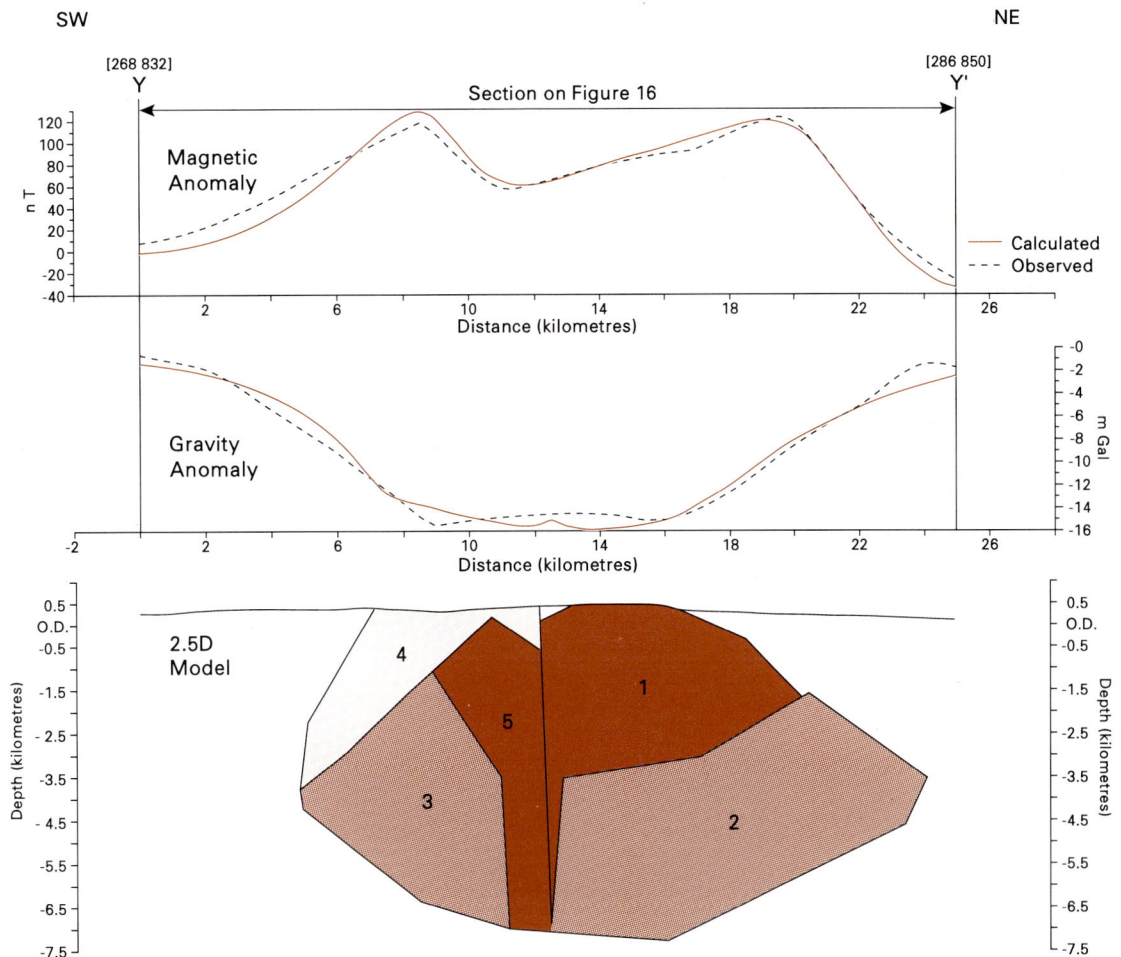

and dioritic phases. The base of the initial granite model was set at 3 km depth. The elongate trough in the residual gravity anomaly map (Figure 15) extending north-eastwards from Inverness into the Moray Firth is considered to reflect a basin of Old Red Sandstone adjacent to the Great Glen Fault Zone and was modelled using a mean density contrast of -0.15 Mg m^{-3}.

The upper and lower surfaces of the initial model were simultaneously adjusted upwards and downwards respectively, until the calculated gravity anomaly matched the observed residual anomalies. The modelled upper surface of the granitic rocks (Figure 20) is at or close to ground level over a large part of the zone of hornfelsing shown in Figure 7. Shallow granitic rock is also modelled north-west of the main Moy Granite outcrop, where hornfelsing is not recorded. The residual anomalies adjacent to the Great Glen Fault Zone have been interpreted as a basin, with Old Red Sandstone to depths of about 3 km below OD (Figure 20).

MAGNETIC MODEL

The main features of the aeromagnetic anomaly after reduction to the pole (Figure 16) were interpreted by a 3D model of the magnetic basement and the model of the Moy Granite derived from gravity modelling. The

anomalies have been interpreted in broad terms, after subtraction of a uniform regional anomaly of 100 nT across the Fortrose district, by magnetic basement modelled upwards from a depth of 7 km below OD. The main elements of the calculated model surface north-west of the Great Glen Fault Zone are shown in Figure 16. The modelled basement surface was close to 1 km below OD near Fortrose. The basement surface south-east of the Great Glen Fault Zone was then modified using the top surface of the model derived from gravity modelling, and the susceptibility was adjusted. The main zones of modelled high susceptibility are shown superimposed in Figure 16. Inferred maximum granitic rock susceptibilities are about 0.030 SI close to the margins of the intrusive rocks. The aeromagnetic anomalies along the south-eastern margin of the Moy Granite can be explained by granitic susceptibilities less than 0.010 SI close to the ground surface; the margin of the granitic rock at greater depth has been modelled with a higher susceptibility.

CONCLUSIONS

1   Residual gravity anomalies of about -13 mGal over the Moy Granite can be interpreted in 3D by a composite intrusion, which has a central zone density contrast

Figure 20 3D models of depths to granitic rocks (brown) and depths to base of the Old Red Sandstone basin (grey). The depth to the modelled upper surface of the granite is shown in kilometres below ground level. The depth to the base of the basin is in kilometres below OD. Dashed line encloses the zone modelled as granite using a density contrast of -0.06 Mg m^{-3} (-0.10 Mg m^{-3} inside pecked line).

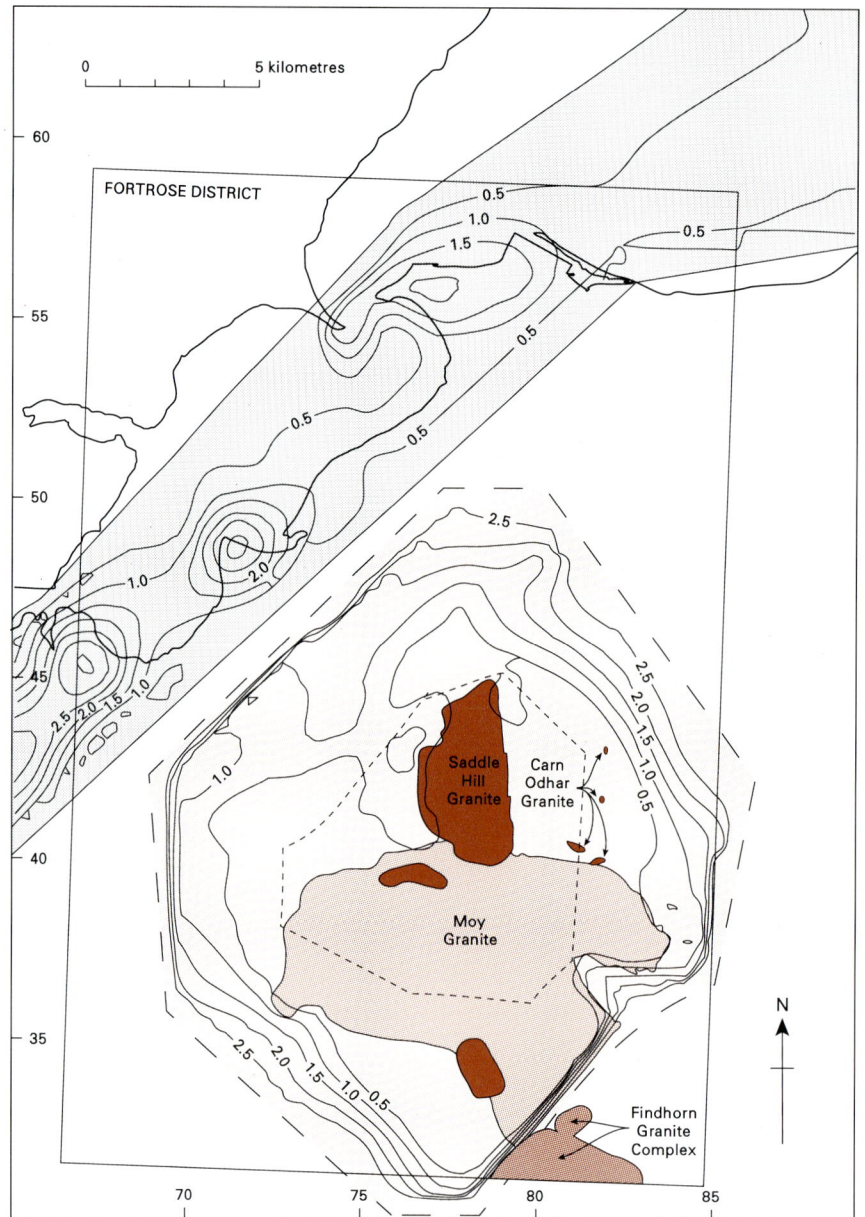

about -0.10 Mg m^{-3}, a marginal zone of density contrast -0.06 Mg m^{-3}, and a depth extent up to about 7 km. The area of hornfelsing in and around the Moy Granite is underlain by very shallow granitic rocks. The higher density phase is mostly associated with the zones of local positive aeromagnetic anomalies. These zones might represent parts of an intrusion of a more granodioritic magma formed after the Moy Granite and before the Saddle Hill Granite was intruded.

2   The higher-density intrusive phases of the Moy Granite are interpreted to be magnetic, although only a few of the measured specimens show significant susceptibility. A 2.5D interpretation of aeromagnetic anomalies in this area suggests that magnetic granodioritic rocks with susceptibilities of about 0.02 SI were intruded by low-susceptibility rocks inferred to be equivalent to the Saddle Hill Granite. The 3D model of the granitic rocks derived from gravity interpretation can be used to reproduce the main features of the aeromagnetic anomaly by modelling the marginal magnetic phases with susceptibilities locally up to about 0.030 SI.

3   Euler and Werner deconvolutions of the aeromagnetic data over the Great Glen Fault Zone suggest a fairly shallow source with an upper surface at about 0.75 km below OD. 3D modelling of the polar magnetic anomaly suggests that a magnetic basement within the Great Glen Fault Zone might lie within 1 km of OD near Rosemarkie.

4   The amplitude of the Great Glen Fault Zone magnetic anomaly requires a reasonably high bulk susceptibility. Symmetry of the aeromagnetic anomaly

suggests that the magnetic material must extend across the fault zone. This might suggest that the source of the anomaly is magnetic Early Proterozoic crust possibly similar to the basement on the Rhinns of Islay. 2.5D modelling of the observed gravity and magnetic anomalies near Fortrose using a susceptibility of 0.040 SI provides a model with a source extending up from depths of about 10 km to within 0.5 km of sea level.

EIGHT

Quaternary

INTRODUCTION

Quaternary deposits cover about 85 per cent of the district, being most extensive beneath the coastal lowland backing the southern shores of the Inverness and Moray firths. Few parts of the British Isles can rival this area in the range, quality and accessibility of glacially related features and Quaternary sediments. The district contains the enigmatic deposits of shelly clay at Clava, made famous in the 19th century, and several other sites of international importance providing evidence of both climatic and sea-level change during the last 130 000 years (Auton et al., 1990). A summary of events in the Quaternary has been given in Chapter One.

Climatic change had a dominant influence on sedimentation during the Quaternary, but, there was little evolution of lifeforms, and so there is no basis for biostratigraphical subdivision. Consequently, the period is subdivided into a series of climato-stratigraphical stages. Apart from a few isolated remnants, glacial erosion has removed onshore evidence of events predating the last glaciation of the district, which took place during the Dimlington Stadial period (Rose, 1985) of the Late Devensian Stage (Figure 21). Sediments formed during some of the earlier stages are found offshore, however (Andrews et al., 1990; Cameron et al., 1987). Deposits of the last glaciation include sheets of till, hummocky glacial deposits and glaciofluvial silt, sand and gravel. Although ice covered the entire district, some of the highest ground is almost free of glacial drift, and shows only the effects of weathering and erosion. The following outline of geological history summarises the considerable body of research that has been undertaken on the numerous and complex Quaternary deposits and associated landforms in the district. Much of the Quaternary stratigraphy is based on successions at Dalcharn, Allt Odhar and Clava, which are described separately in Appendix 3a, 3b and 3c respectively. Detailed sedimentological logs of natural sections and other sites in the district are contained in a separate report (Merritt and Auton, 1993).

OUTLINE OF QUATERNARY HISTORY

Events predating the maximum development of the Main Late Devensian Ice Sheet

Until recently, few deposits or landforms in the district had been attributed with certainty to glacial events before the Dimlington Stadial. Several isolated pockets of deeply weathered bedrock occur, notably in the Findhorn Valley, near Drynachan, and on the western slopes of Carn nan Tri-tighearnan, where the Moy

Granite is decomposed locally to a gruss over 20 m deep. Gruss weathering occurred under temperate climatic conditions postdating the Miocene (Hall, 1986). More-severe chemical weathering commonly affects metamorphic and igneous rocks close to the unconformable base of the Old Red Sandstone. The survival of this weathered material, which probably formed during the early Devonian, indicates that parts of the area are now close to the level of erosion established at that time (Hall, 1991). Indeed, the major river valleys such as those of the Nairn and Findhorn are thought to have been initiated during the Devonian (Hall, 1991; Mykura, 1991), re-occupied in the Neogene (Bremner, 1939b, 1942; Linton, 1951) and then modified during the Quaternary.

One of the earliest descriptions of a possible pre-Late Devensian Quaternary deposit was made by Jamieson (1874). He discovered a mass of fossiliferous clay containing shells of 'Arctic' affinity at Kirkton, near Ardersier, and claimed that it had been transported by a glacier and redeposited within glacial sediments. However, the Kirkton clay is now recognised as being part of a glacitectonised sequence of glaciomarine sediments that formed during the initial deglaciation of the Inverness Firth in the Late Devensian.

A better known deposit of shelly clay occurring beneath till at Clava has long been regarded as pre-Late Devensian in age (Gordon, 1990). The Clava Shelly Clay (Fraser, 1882a; Horne et al., 1894) lies at an altitude of 150 m above OD and contains a well-preserved, shallow-marine fauna of a high-Boreal to low-Arctic character (Appendix 3c and 4). The deposit provoked considerable debate during the 19th century because, if it is in place, it would indicate either substantial preglacial submergence of the country, or transport by ice from offshore. The recent resurvey has confirmed that this clay, together with masses of shelly diamicton (Clava Shelly Till) occurring nearby, are indeed glacially transported allochthons, probably derived from the Loch Ness Basin during the build-up of the Late Devensian Ice Sheet (Merritt, 1992). Amino-acid dates obtained from shells collected from the main raft of shelly clay confirm that the deposit accumulated during a mid-Devensian interstadial period.

In the 30 m Red Craig cliff section [7360 5785] behind the village of Rosemarkie on the Black Isle, till is underlain by glaciolacustrine silts containing dropstones. These deposits overlie clast-supported diamictons and an older till; both contain angular, mostly locally derived clasts. The clast-supported diamicton formed as a subaerial alluvial fan within Rosemarkie Glen—a deep gorge probably excavated by glacial meltwaters. J S Smith (1968) implied that the entire drift sequence within the

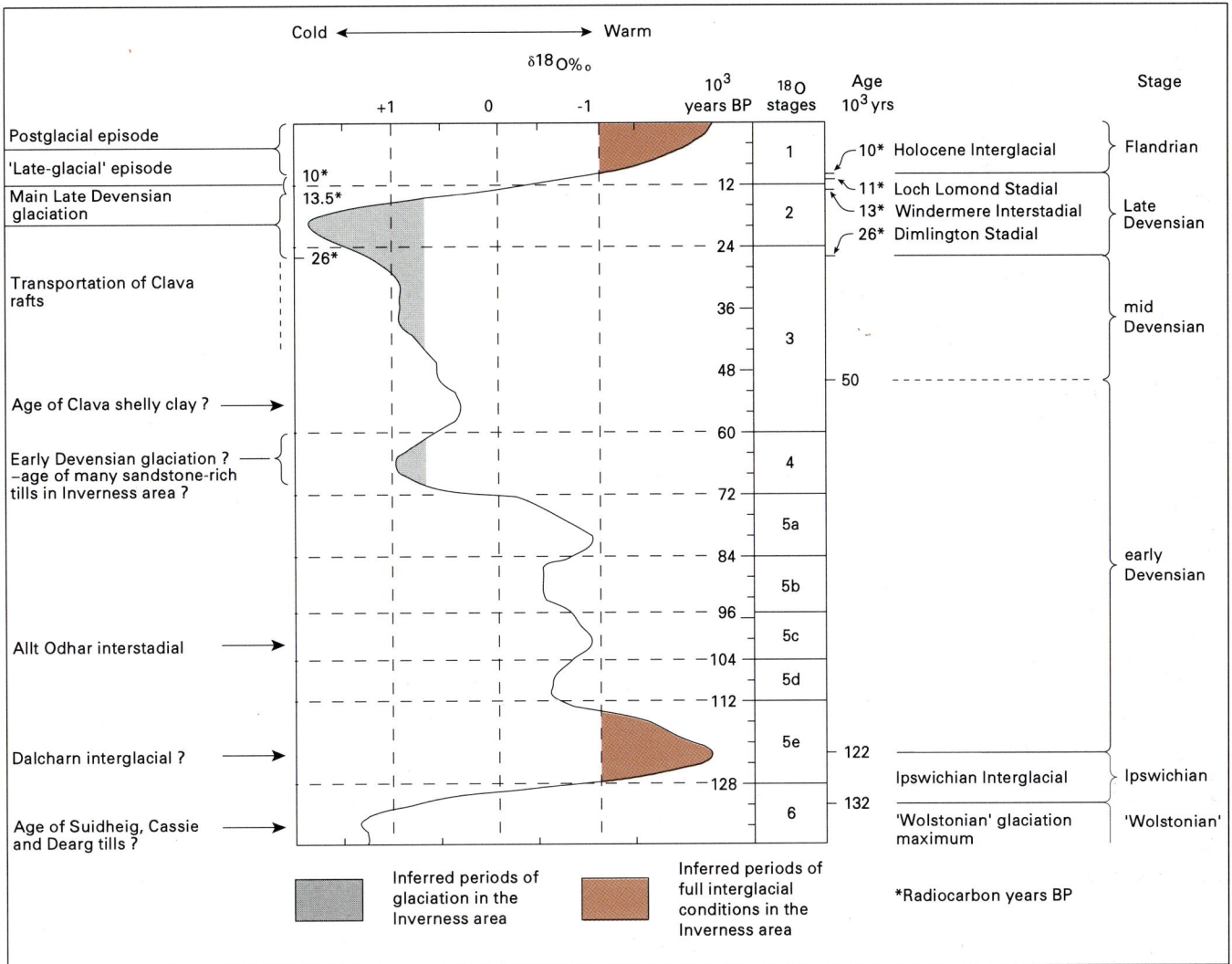

Figure 21   Climato-stratigraphical stages and a proxy climate curve for the last glacial-interglacial cycle. The climate curve is based on oxygen-isotope variations in deep oceanic faunas (after Imbrie et al., 1984).

glen formed during the Ardersier Readvance (see below), whereas Firth (1989b) argued that it formed earlier in the deglaciation of the district. It is more likely, however, that the gorge formed during a pre-Devensian glacial episode, that the basal diamictons are early Devensian or older, and that the ponding responsible for the glaciolacustrine deposits developed during the build-up of the Late Devensian Ice Sheet, which laid down the uppermost till.

At Dalcharn [8146 4521], 6 km south-west of Cawdor, disturbed biogenic sediments containing pollen of inter-glacial affinity rest on deeply weathered gravel and are overlain by two distinct formations of lodgement till (Appendix 3a); an older till underlies the gravel. The biogenic sediments could be assigned to the Ipswichian Interglacial on the basis of experimental optically stimu-lated luminescence dating (H McKerrell, personal com-munication), but the pollen evidence suggests that they are older (Walker et al., 1992). Pre-Late Devensian

biogenic deposits also occur at the Allt Odhar site [7978 3678], which lies within Moy Estate, 16 km south-east of Inverness (Appendix 3b). Here, compressed peat of interstadial affinity lies between two till formations. On the basis of its included pollen and insect remains, experimental Uranium Series Disequilibrium dating and an infinite radiocarbon age of 62 300 ^{14}C years, the peat has been assigned to an interstadial event in the early Devensian, perhaps equivalent to Oxygen Isotope Stage 5c (Walker et al., 1992) (Figure 21).

Ice-moulded landforms and glacial striae are generally aligned south-west to north-east on the lower ground and along the major valleys, indicating that ice moved north-eastwards during the final phases of glaciation (Figure 22). On higher ground in the southern part of the district, however, some striae are reported by Horne (1923) to be aligned north or north-north-west. He concluded that these striae indicate earlier ice movement towards the north, but it is possible that they

Figure 22
Glacial striae and
ice-moulded
features in the
Inverness area
(after Auton et
al., 1990, fig. 1).

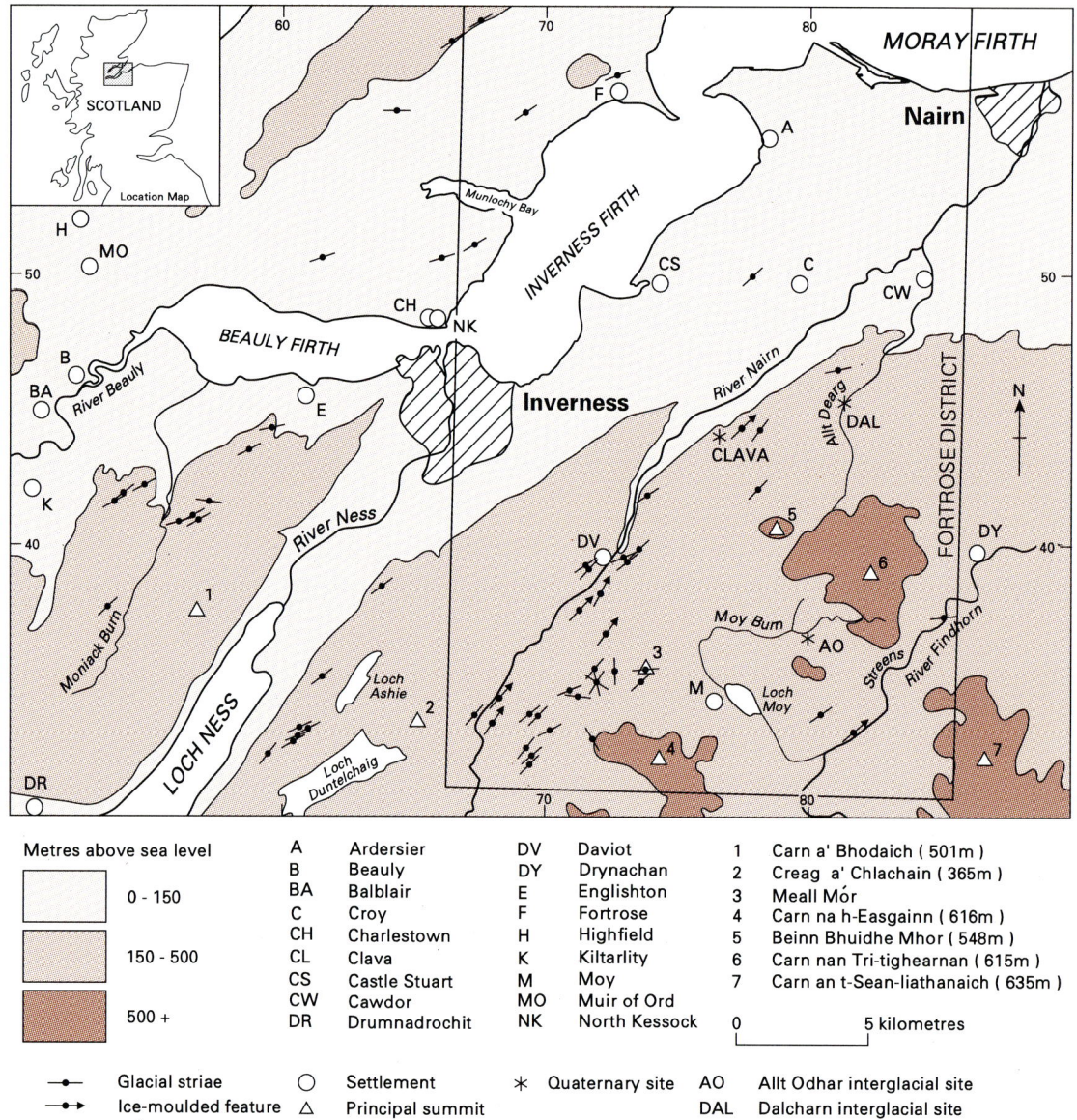

were developed during a phase of southward ice movement. Numerous erratics of sandstone, flagstone and breccio-conglomerate have been transported from Old Red Sandstone outcrops bordering Loch Ness and the Moray Firth on to one of the highest point in the area (Carn nan Tri-tighearnan, 615 m above OD) and beyond into the Findhorn Valley, where they are incorporated within distinctive sandstone-rich tills. Although some of these deposits may indeed be testimony to a phase of southward ice movement, it is clear from the broader distribution of erratics (Figure 23) that the dominant direction of carry was towards the east or east-south-east. A similar movement of ice occurred at Clava, during the build-up of the Late Devensian Ice Sheet (Merritt, 1992).

The age of the sandstone-rich till on the high ground south of the firths is not known. Indeed, it seems that tills of similar composition formed during more than one glacial event. Horne (1923) found so-called 'inter-

glacial beds' lying above a sandstone-rich till between Cawdor and Moy, implying that the till was deposited prior to the last glaciation. Angular head gravel resting on bedrock and overlain by sandstone-rich till occurs within the Allt Breac valley, west of Daless (Auton, 1990a) and indicates that periglacial conditions prevailed before the glaciation which laid down the till. Weathered sandstone-rich tills are also found at depth in Moy Gap, a broad, north-west-oriented depression that links the valley of the River Findhorn, near Tomatin, with that of the River Nairn, at Daviot. If the gap is a glacial breach, it also formed prior to the Late Devensian.

Main Late Devensian glaciation and deglaciation of the upland areas

The first comprehensive model of glaciation and deglaciation of the district was put forward in the

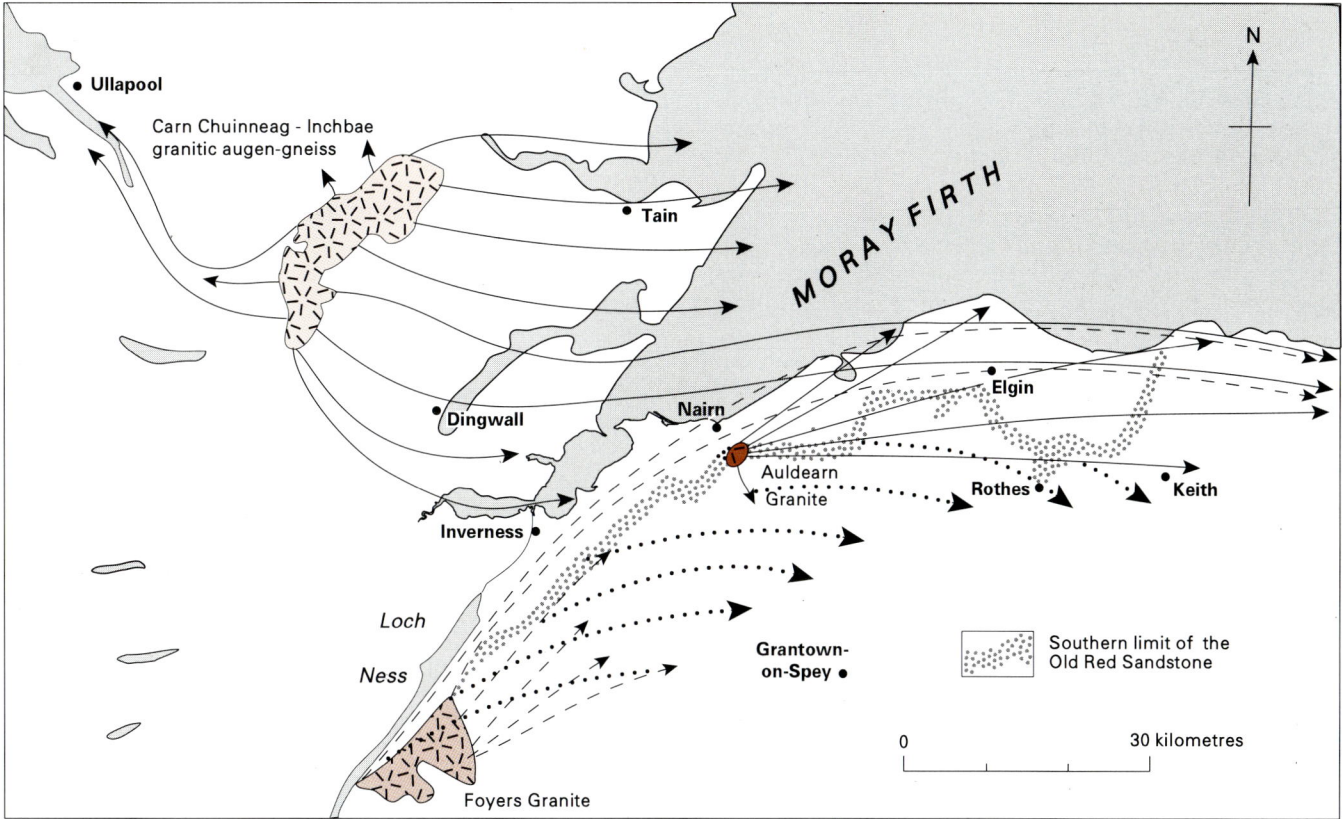

Figure 23 Map showing the transport paths of some erratics across the southern part of the Moray Firth.

Compiled mainly from information in Mackie (1905) and Geological Survey publications. The outcrop of the Carn Chuinneag–Inchbae granitic augen-gneiss has been extended to include the Old Red Sandstone conglomerate in which clasts of this gneiss occur (modified after Sissons, 1967, fig. 31).

explanatory memoirs for One-inch sheets 83, 84 and part of 94 (Horne and Hinxman, 1914; Horne, 1923). The Quaternary geology of the area had already excited the interest of pioneer workers, notably Fraser (1877, 1879, 1882a, b), Jamieson (1874, 1882, 1906), Jolly (1876), Macdonald (1881), Horne (1880), Wallace (1883, 1898) and Bell (1891, 1893a, b, 1895a, b, 1896, 1897a, b), who recorded evidence of glaciation in the form of striae, moraines and erratics. They also recognised evidence of deglaciation, in the form of terraces and outwash fans, and postulated the former existence of glacial lakes at many localities.

Horne (1923) recognised three distinct episodes of glaciation in the southern hinterland of the Moray Firth. The first was a period of 'maximum glaciation' in which an ice sheet built up from the south and west to cover even the highest ground. This was followed by a 'confluent glacier period' when ice retreated from the high ground to form individual glaciers that occupied the main valleys and coalesced on the lower-lying ground. The final 'valley glacier period' witnessed valley glaciers retreating up the glens.

The early workers thought that the valley glaciers retreated slowly whilst ice-dammed lakes were ponded in the valleys by ice, or by moraines left by the ice as it

decayed. This idea of 'active recession' was developed subsequently by Bremner (1939a, 1939b) and Charlesworth (1956). The former identified an ice-dammed lake in the Findhorn Valley, between Dulsie Bridge [932 414] and Dounduff [993 494]; the latter postulated ice-dammed lakes within most of the valleys in the region (Auton, 1990b). Young (1977, 1978, 1980), however, proposed that the valley glaciers wasted down in situ rather than by active retreat. He dismissed most of the evidence put forward earlier in favour of ice-dammed lakes, particularly in respect of the upper reaches of the Findhorn catchment. However, the recent discovery of glaciofluvial fan-delta deposits and glaciolacustrine deposits with dropstones, notably at various altitudes on the flanks of the Findhorn Valley and near Moy, indicate that some small, temporary ice-dammed lakes did indeed exist (Auton, 1990a), as envisaged by the early workers.

Horne's contention that ice overrode the highest ground of the district is supported by the presence of Old Red Sandstone erratics within till close to the present Nairn–Findhorn watershed at more than 600 m above OD. It seems that most of the till mantling the watershed was laid down by ice during the Main Late Devensian glaciation, although some of the sandstone erratics in the till may have been derived from earlier glacial deposits.

Much of the evidence cited by Horne as indicating active retreat following the period of maximum glaciation is questionable. The moraines and 'moraine terraces' supposed by him (Horne, 1923, fig. 3) to mark the former margins of confluent glaciers are now recognised to include non-morainic features such as ridges of till or bedrock, kames, kame-terraces and eskers; few of them are true moraines. However, recessional moraines associated intimately with ice-marginal drainage channels and kame-terraces do occur along the southeastern flanks of Strath Nairn between Daviot and Cawdor, and low ridges of gravelly diamicton found in the vicinity of Dalcross [777 483] may be recessional winter push-moraines.

Deglaciation of the Inverness and Moray firths and accompanying changes in relative sea level

On deglaciation, isostatic rebound, due to unloading of the crust, failed at first to keep pace with the worldwide (eustatic) rise in sea level that resulted from the melting of ice sheets. Consequently, raised beaches and associated marine deposits are now found up to at least 35 m above OD around the Inverness and Moray firths. Isostatic uplift was greatest where the ice was thickest, with the result that raised beaches are tilted northeastwards, away from a centre of uplift in the western Grampians. The raised beaches are divisible into a Late Devensian (late-glacial) series, which began to form during retreat of the Main Late Devensian Ice Sheet, and a Flandrian (postglacial) series. As the rate of isostatic rebound has decreased with time, the older, higher beaches are tilted more steeply than the lower, younger ones (Figure 24).

The first comprehensive models of deglaciation and sea-level change around Inverness were based on the work of Kirk et al. (1966), J S Smith (1966, 1968, 1977), Small and Smith (1971), Synge (1977a, b) and Synge and Smith (1980). Smith and Synge both suggested that the Late Devensian Ice Sheet initially retreated as far west as Inverness, where a marine shoreline at a height of 42 m above OD was formed, and subsequently readvanced north-eastwards to terminate at Ardersier. They named this later event the 'Ardersier Readvance', correlating it with the Perth Readvance of Sissons (1967). They also suggested that the arcuate asymmetrical ridge forming the coast between Castle Stuart [738 498] and Ardersier was a push moraine and that much of Chanonry Peninsula on the Black Isle was essentially a 'kame moraine'. The Ardersier Readvance was followed by retreat to Alturlie, where a minor still-stand occurred (Figure 28). It was proposed that the ice sheet subsequently retreated westwards again to Inverness and Kessock, where meltwaters formed extensive marine deltas when relative sea level stood between 33 and 34 m above OD. As the ice front then retreated farther westwards along the Beauly Firth and south-westwards up the Great Glen, relative sea level dropped to an altitude close to that of the present day. The retreat was thought to have been followed by a transgression during which the sea invaded Loch Ness, forming shorelines at heights of between 28 and 31 m above OD around the loch and in the valley of the River Ness.

Following extensive instrumental levelling work on Late Devensian shorelines in the region, Firth (1984, 1986, 1989b) concluded that there is no evidence to substantiate a major Ardersier Readvance and that there was no subsequent marine transgression into Loch Ness. He proposed that at Inverness there was a progressive, uninterrupted fall in relative sea level from at least 35 m above OD to about the present-day OD, while the Late Devensian Ice Sheet receded both westwards and southwestwards (Firth, 1989a). Ten raised shorelines of Late Devensian age were recognised by Firth, each sloping down towards the north-east at a gradient of between 0.57 and 0.15 m per km (Figure 24). He proposed that the shorelines formed in close proximity to the margin of the retreating ice sheet and that rates of retreat were rapid where the ice terminated in the sea, as a result of iceberg calving, but slower where the ice sheet was landbased.

Between Inverness and Forres, some 40 km to the north-east, there are large tracts of moundy, ice-contact deposits lying below the elevation of the highest Late Devensian marine shoreline (the Late Devensian 'marine limit'). Parts of this area appear to have been modified by marine processes up to, and possibly above an altitude of about 27 m above OD (Synge, 1977a; Firth, 1989a and b, 1990); there is, however, little unambiguous evidence of any marine incursion. Firth considered that much of the ground remained covered either by ice or by glaciofluvial sediment containing buried ice masses until relative sea level had fallen to below 13 m above OD.

The absence of marine fossils from sediments within several large kettleholes occurring within these tracts, such as those at Alturlie [720 487], certainly indicates that some ice remained buried seaward of the marine limit while relative sea level fell. On the other hand, the widespread occurrence of deltaic sequences (not glaciofluvial as proposed by Firth) suggests that the ice occupying the Inverness and Moray firths must have retreated substantially (by iceberg calving) in order to create the deep water into which the deltas prograded. The ice within the firths probably became uncoupled from surrounding land-based ice, to form a tidewater glacier. Unlike the land-based ice, which either stagnated or slowly retreated westwards or south-westwards, the tidewater glacier was affected by minor changes in sea level and hence became oscillatory, because its mass balance would have been determined mainly by the rate of iceberg calving (cf. Boulton and Deynoux, 1981). The sea probably stood at 35 m or more above OD at this time. The bottomset beds of marine deltas and subaqueous fans that prograded in front of the tidewater glacier during still stands and minor readvances form the Ardersier Silts Formation; the foreset and topset beds of such bodies are included in the Alturlie Gravels Formation.

Following a critical re-examination of key sites on Ardersier Peninsula (namely Jamieson's Pit, the 'Contorted Silts' exposure north of Ardersier village, and

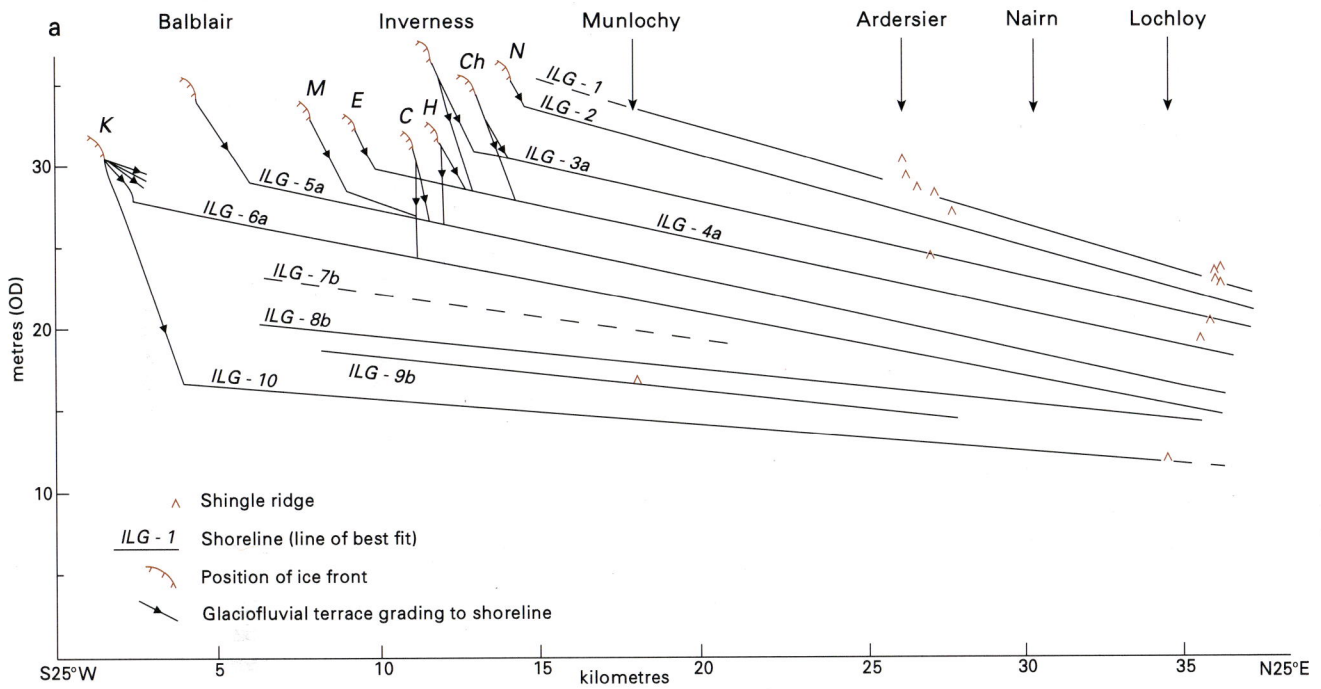

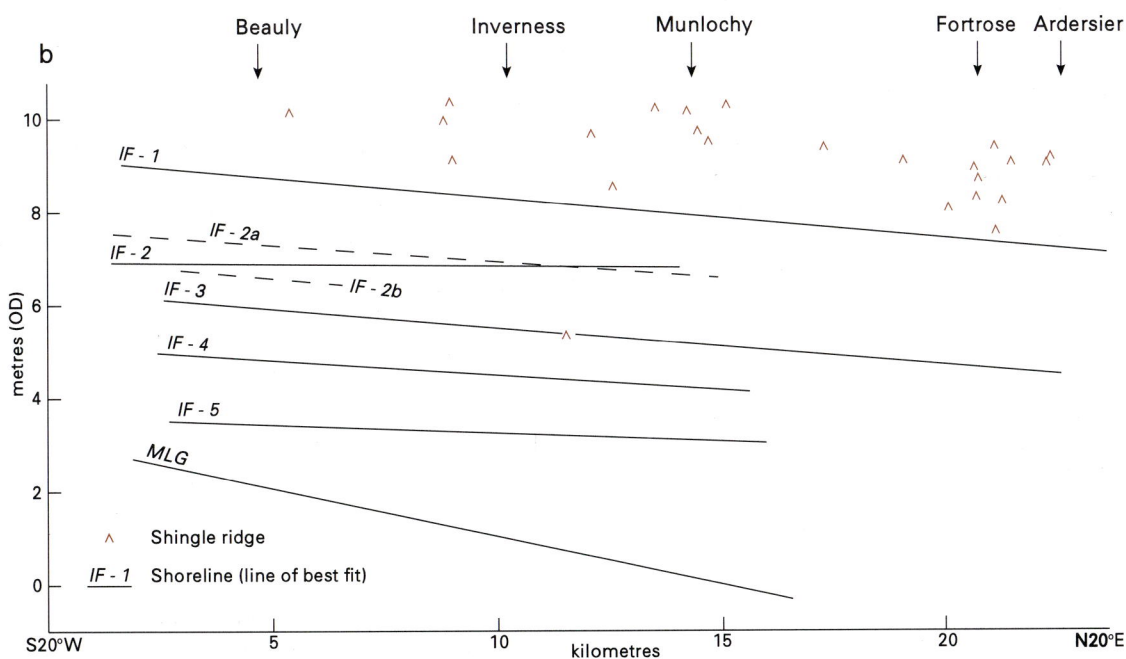

Figure 24   Height–distance diagrams of shorelines in the Inverness area.

a) Late Devensian (Late-glacial shorelines ILG-1 to 10, modified after Firth, 1989, fig. 7). C–Contin, Ch–Charlestown, E–Englishtown, H–Highfield, K–Kiltarlity, M–Muir of Ord, N–North Kessock.
b) Main Late-glacial Shoreline (MLG), probably of Loch Lomond Stadial age, and Flandrian shorelines IF-1 to 5 (modified after Firth and Haggart, 1989, fig. 5).

sections at Kirkton and Hillhead), the recent resurvey has established that there was indeed a minor readvance at Ardersier during deglaciation, but that it occurred earlier than envisaged by Smith and Synge (Gordon and Merritt, 1993). In order to avoid possible confusion, it is suggested that the readvance at Ardersier established here should be known as the Ardersier Oscillation, in the fashion of the earlier Elgin Oscillation described by Peacock (1968). Evidence obtained during the resurvey has confirmed that there was indeed a subsequent still-stand at Alturlie.

Relative sea-level changes during Loch Lomond Stadial and Flandrian times

Sissons (1981a) proposed that the prominent Main Flandrian (Postglacial) Cliffline bordering much of the Inverness and Moray firths was largely formed by marine erosion during the Loch Lomond Stadial and that the feature was later trimmed in mid Flandrian times. He suggested that the most extensive shoreline associated with this period (Figure 24, MLG) slopes north-eastwards from a maximum height of about 2 m above OD at the head of Beauly Firth and that it is equivalent to the 'Main Lateglacial Shoreline of the Forth Valley' (Sissons, 1969, 1974, 1976). Peacock (1977) suggested that the cobble gravels underlying the low-lying ground at Longman Industrial Estate [670 465], north-east of Inverness, are the topset beds of a delta that formed by successive floods during the Loch Lomond Stadial. Sissons (1981a) maintained that such coarse material was more likely to have formed in a single event, proposing that it resulted from the catastrophic drainage of the 260 m OD, ice-dammed lake in Glen Roy/Glen Spean towards the end of the Loch Lomond Stadial. He argued that the lake drained subglacially to Fort Augustus and then via Loch Ness and the Ness Valley to the sea at Inverness.

Flandrian relative sea-level movements and shoreline displacements around the Inverness and Moray firths have been deduced from detailed morphological, strati-graphical and palaeobiological investigations, mostly at the head of Beauly Firth (Haggart, 1986, 1987, 1988; Firth and Haggart, 1989). Haggart (1982, 1986) suggested that at about 9600 BP, estuarine flats (Barnyards Beds) lying at an altitude of 6 m above OD at the head of Beauly Firth, became dry land as relative sea level fell. He correlated the Barnyards Beds with the 'Main Buried Beach of the Forth Valley' (Sissons, 1966), which formed at about 9600 BP. Relative sea level continued to fall until it reached a lowstand at about 8800 BP, but just how far it fell is unknown. Subse-quently, relative sea level rose to 9 m above OD at Beauly (c.8 m at Inverness), the rise being associated with the formation of the Beauly Beds—grey sands, silts and clays with intercalations of peat (Firth and Haggart, 1989, p.43). This estuarine sequence is correlated with the highest (Main) Flandrian shoreline in the district (Figure 24, IF-1), which slopes towards 015° at a gradient of 0.076 m per km. This prominent shoreline is backed by the Main Flandrian Cliffline, formed between 7100

and 5775 BP (Firth and Haggart, 1989), and is correlated with the 'Main Postglacial Shoreline of eastern Scotland' (Sissons et al., 1966). The Main Flandrian Shoreline was formed at the culmination of the postglacial transgres-sion in the area, when the isostatic recovery of the land was outstripped by a rapid phase of eustatic rise in sea level. Its abandonment resulted from a slackening in the rate of eustatic rise such that the continued isostatic recovery, though itself diminishing in rate, lifted the shoreline and its associated estuarine flats clear of the sea (Cullingford et al., 1991).

The late Flandrian witnessed generally falling relative sea levels, with minor transgressive events that produced four, and possibly five, shorelines (Figure 24, IF-2 to 5) (Firth and Haggart, 1989). At the present time, relative sea-level movement is minimal in north-eastern Scotland (Woodworth, 1987).

Climate, vegetation and landscape change since deglaciation

Windermere Interstadial and Loch Lomond Stadial periods

Analysis of pollen and insect remains from lake sediments and peat bogs outwith the district shows that there was a rapid amelioration of the climate in the Highlands at about 13 000 BP (Walker, 1975). As the Late Devensian Ice Sheet wasted, the newly deglaciated areas were colonised by a pioneer vegetation of open-habitat species followed by the immigration of crowberry heath, juniper and birch. In some areas there was a short-lived reversion to open habitats at about 12 000 BP, and during the Loch Lomond Stadial (10 000 to 11 000 BP) there was a general return to sparse tundra vegetation.

Glaciers reappeared in the western and central Highlands during the Loch Lomond Stadial, but there is no direct evidence of this glaciation in the district, nor are there any known deposits firmly attributed to the preceding Windermere 'Lateglacial' Interstadial, although they may exist. Many of the higher river terraces, together with the older alluvial fans, head deposits and scree, probably formed during the Loch Lomond Stadial. Soils would have been destroyed at that time, in a severe periglacial regime where vegetation was insufficient to retard the runoff of surface water, particu-larly during rapid snow melt.

Holocene (Flandrian) times

The abrupt increase in temperature following the Loch Lomond Stadial led to rapid re-colonisation by crowberry heath and juniper scrub, and subsequent expansion of mixed deciduous woodland, in which birch and hazel were followed by oak, elm and pine, and then by alder. The inward migration of Scots Pine reached the adjacent Cairngorm/Speyside region by about 8000 BP (O'Sullivan, 1976; Bennett, 1984) and pine forest is thought to have expanded rapidly between 7500 and 4000 BP, before it underwent a dramatic decline. The decline was the result partly of the cooler, wetter climate which favoured the growth of blanket peat, and partly of human activities. Pears (1968, 1970) published detailed studies of the upper altitudinal limit of pine growth in the Cairngorms and recognised a Flandrian limit at 793 m above OD, some 178 m higher than the highest summit in this district. Stumps and trunks of pine are exposed beneath peat up to 5 m thick at 530 m above OD, east of Carn Torr Mheadhoin [at 856 343]; stumps are seen at similar elevations in cliff sections near the head of the Allt Odhar valley [7978 3678] and at about 560 m above OD at the head of Caochan Odhar.

Although the imprint of glaciation remains dominant, postglacial processes have superimposed subtle, but distinctive, modifications on the landscape. Steep mountain sides have been modified by rockfalls, soil creep and debris flows, whereas valley floors have been modified by rivers forming spreads of alluvium and river-terrace deposits and by the accumulation of alluvial fans at the mouths of tributary streams. Some of these depositional landforms are relict features, either paraglacial deposits (Church and Ryder, 1972) that accumulated immediately after deglaciation, or periglacial deposits formed at the end of the Dimlington Stadial and during the Loch Lomond Stadial (Ballantyne, 1991). Nevertheless, there is good evidence that accelerated erosion and mass transport have occurred within the past few centuries and that much of the increase in debris flow may be related to the destabilisation of mountain soils by overgrazing.

A veneer of blown sand is widespread along the coastal lowlands south of the firths, and thicker deposits blanket the raised Flandrian beaches that extend eastwards from Fort George. Although dunes are mainly stabilised by vegetation, large amounts of sand and dust are blown around in dry weather, particularly during the spring when ploughed fields are bare of crops. Spits of sand and shingle are accreting at Whiteness Head [805 587] and Chanonry Point, and intertidal mudflats immediately east of Inverness are being reclaimed progressively by the tipping of domestic waste and rubble.

GLACIAL DEPOSITS

Tills

Till, the most widely distributed of the glacial deposits, crops out over much of the district and also occurs beneath younger superficial deposits. It consists of ice-transported material laid down subglacially at the base of active ice sheets, or formed supraglacially and paraglacially at the margins of retreating ice sheets. Deposits formed in the last two environments commonly occur at the surface and comprise a metre or so of heterogeneous, very poorly sorted, crudely stratified, gravelly diamicton intercalated with gravel, silty sand, silt and clay. These sediments accumulated at the ice front, mainly as debris flows that were modified and redeposited by ephemeral meltwater streams and sheet wash. They are generally permeable and include large boulders up to several metres in diameter. In contrast,

tills formed in the subglacial environment are generally much thicker (up to 30 m locally) and are relatively homogeneous. They are typically overconsolidated, more clayey, fissile and relatively impermeable.

In general, it is not practical to map the supraglacial/paraglacial and basal tills separately, even though the boundary between them is invariably unconformable. Locally, however, the supraglacial and paraglacial deposits are several metres or more thick, form constructional mounds and can be mapped out. In such areas, they are distinguished on the 1:50 000 geological map as 'Hummocky Glacial Deposits' (see below).

The subglacially formed diamictons generally take the form of lodgement till. This type was deposited beneath actively moving glaciers as a result of frictional retardation of debris particles and debris-rich ice masses against the glacier bed (Boulton and Deynoux, 1981). Such tills are typically very stiff, stony, sandy, clayey diamictons with matrix support and little stratification. Boulders are generally not as large as those occurring in associated supraglacial and paraglacial deposits, but they typically have bevelled and striated surfaces. Subhorizontal fissures resulting from pressure release are common towards the upper surface of lodgement tills and impart a platy structure; concavo-convex discontinuities and subglacially formed shear planes are common throughout. All types of discontinuity may be lined with silt, clay or indurated silty fine-grained sand, the latter commonly ferruginous.

Many overconsolidated tills have the diagnostic features of lodgement till, but are also weakly stratified on a scale of a few millimetres to a few centimetres. The stratification, which is most common in sandstone-rich till, is formed of laterally impersistent laminae and wisps of pale, very fine-grained sand and silt. The presence of these laminae probably indicates that some resedimentation in running water occurred at the ice to till interface during the lodgement process, although such stratification may also result partially, if not wholly, from subglacial shearing and associated comminution of granular material (Boulton, 1979; Boulton et al., 1974). These stratified lodgement tills usually include boulder pavements and clusters of imbricated cobbles.

Pronounced planar stratification is most commonly associated with sandstone-rich tills. These deposits mainly comprise friable, sandy, matrix-supported diamictons interbedded with diamictic pebbly silty sand, clast-supported diamicton and thinly laminated silt and clay. Individual beds are generally less than 0.5 m thick (typically 5 to 20 cm), laterally impersistent, and grade one into another. Thicker beds of water-sorted sand and gravel also occur. There is commonly evidence of penecontemporaneous slumping and mass flowage, scour-and-fill, soft-sediment deformation and 'pull-apart' structures. Many of these stratified diamictons are similar to supraglacial flow-till complexes described elsewhere (Boulton, 1968, 1971, 1972; Boulton and Paul, 1976), but lenses of clast-supported gravel with sharp, horizontal bases and arched tops have been observed in some, as near Clava [at 7646 4386] (Merritt, 1992a). Lenses such as these are indicative of fluvial deposition within cavities

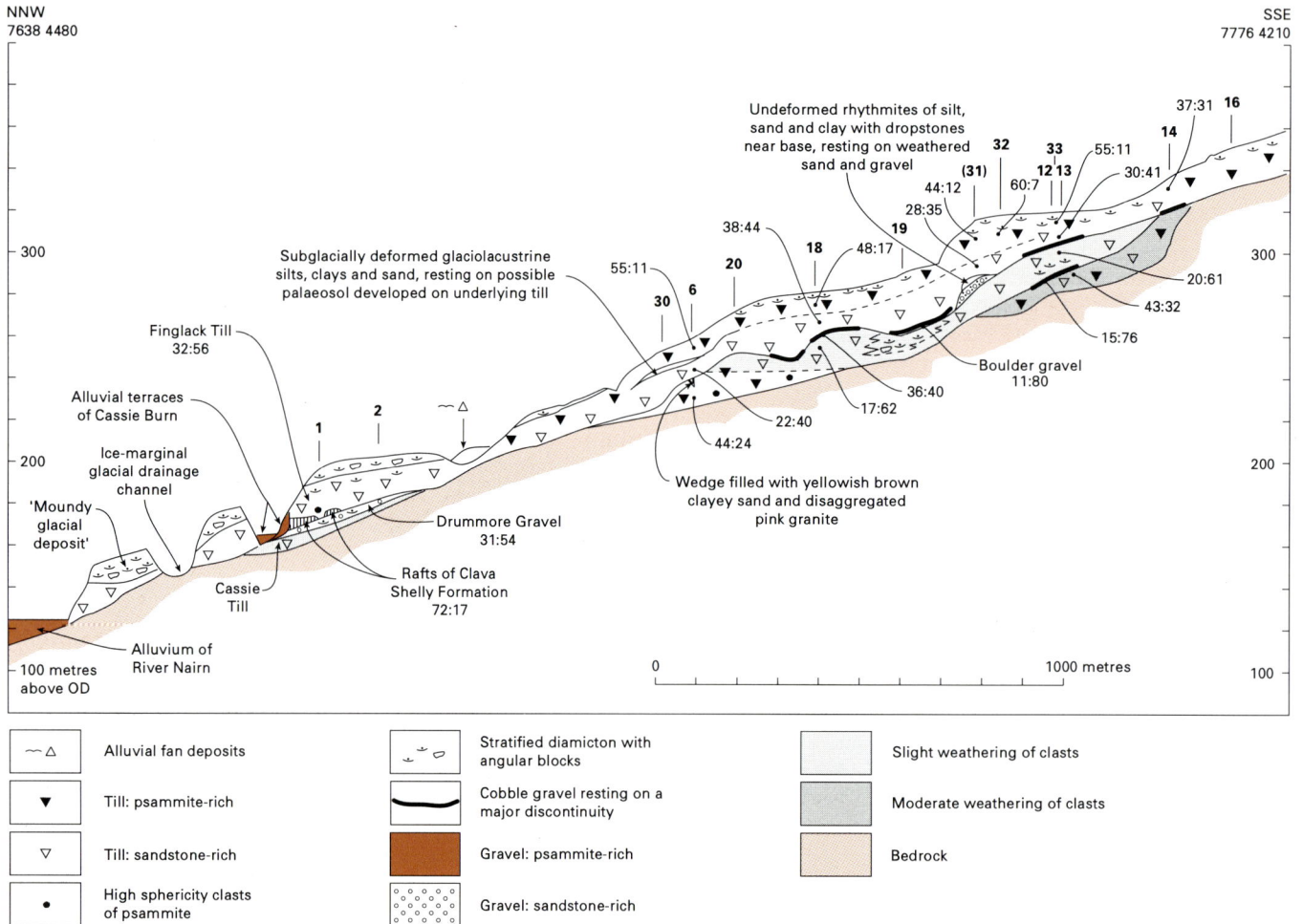

Figure 25   Schematic longitudinal profile of the Allt Carn a' Ghranndaich valley at Clava.

Stone-count data (e.g. 20:60) are percentages of metamorphic clasts and Old Red Sandstone clasts. Bold numbers refer to recorded sections on 1:10 000 sheet NH 74 SE. Note: details of Section (**31**) have been extrapolated to the line of section.

and channels in the soles of melting ice sheets and are diagnostic features of basal melt-out tills (Boulton, 1970; Shaw, 1979; Dreimanis, 1989). Thus, the stratified diamictons may have been formed by basal melt-out and/or sediment gravity-flow processes, though sediments formed by the former process have poor preservation potential (Paul and Eyles, 1990).

LATE DEVENSIAN TILLS

It is difficult to date tills unless they include, or overlie, organic remains. Exceptionally, at Clava (Appendix 3c), the surficial **Finglack Till** overlies a raft of supposed mid Devensian shelly clay (Figure 25) and thus is almost certainly of Late Devensian age. The Finglack Till (Merritt, 1990a) is correlated with the Dalcharn Upper Till Formation (Merritt and Auton, 1990) at Dalcharn (Appendix 3a), with the upper member of the Moy Till Formation (Merritt, 1990b) at the Allt Odhar site (Appendix 3b), and with surficial tills throughout the district. It follows that almost all the till shown at

outcrop on the 1:50 000 geological map was laid down during the Late Devensian glaciation, but older tills and associated deposits, which cannot be delineated for cartographical reasons, occur at depth in many localities (see below). A diamicton interpreted to be either a flow till or a basal till overlies glacitectonically disturbed sands of the Ardersier Silts Formation at Jamieson's Pit [7939 5616], near Ardersier. This is here named the **Baddock Till**. It formed at the ice front during a minor readvance of the ice sheet and is the youngest known till in the district.

Over much of the district, till is between 2 and 10 m thick. On the Black Isle, there are numerous scattered exposures of bedrock that suggest that the till cover is generally thin, but it thickens to as much as 15 m in subdrift depressions [as at 7130 5579]. Boreholes drilled along the line of the A9 road across Drummossie Muir, south of Inverness, were unbottomed in diamicton at depths of 5 m and 10 m, reaffirming that this ice-smoothed, convex feature is largely till covered. Some of the thickest deposits of till in the district form tabular sheets beneath the north-westerly facing slopes of the Nairn–Findhorn divide, where streams, such as Allt Carn a'Ghranndaich, Riereach Burn and Allt Dearg and its tributaries, have formed many splendid river cliff sections exposing up to 40 m of glacigenic material.

Apart from a separate older till (not mapped) within the Quaternary sequence exposed at the mouth of Rosemarkie Glen (see below), only one formation of till is recognised on the Black Isle. Largely covering Old Red Sandstone bedrock, the diamictons here are stony and have matrices that are predominantly very silty or sandy. In places, for example in a quarry [6744 5080] near Pitlundie, the matrix is almost wholly of silt, as is also the case south of Munlochy Bay, although patches with a more sandy matrix occur around Sligo [6773 5184], East Lundie [6751 5121] and Taindore [6785 5110]. Colours of the till matrix reflect those of the underlying bedrock and are predominantly red, but oxidation has produced a variety of shades of reddish pink and browny purple, with sporadic goethitic yellow-ochre.

Most tills cropping out south of the Inverness and Moray firths also have compositions that reflect the nature of local bedrock. On Drummossie Muir and across the coastal lowland where bedrock is mostly Old Red sandstone and siltstone, these rock types form the predominant clasts in the tills and contribute to their sandy and silty matrices. Colours vary from yellowish brown to moderate brown and reddish brown.

On the higher ground backing the coastal lowland, tills are generally olive-grey, brownish grey or dark yellowish brown, and typically contain a preponderance of clasts of igneous and metamorphic rocks derived from the Proterozoic basement and granitic bedrock there. Clasts of Old Red Sandstone rocks are also present, but they occur in subordinate amounts, especially in the south-west of the district, where they probably represent reworkings from older glacigenic deposits.

Although there are few exposures of till to study on the highest ground, it is clear that above about 400 m above OD, mature lodgement till is not commonly present at surface. For example, stiff, stony, psammite-rich lodgement till exposed in river cliffs of Allt Odhar and its tributaries can be traced upstream, where it becomes increasingly gravelly, stratified, less cohesive and less consolidated. This immature basal till is typically pale olive-grey and has a distinct planar stratification, with beds 5 to 15 cm thick. Beds of friable sandy diamicton are intercalated with very compact, water-winnowed sand and gravel. Similar deposits occur at the head of the Meur Garbh valley [8190 4115] (515 m above OD) and in the Meur Bheòil valley [8431 4012] where 10.8 m of interstratified immature till and clay-bound gravel rest on bedrock at about 440 m above OD. At one place [8185 3732], a deposit of this type rests unconformably on a stiff lodgement till, but as the latter is distinctly weathered, it probably was laid down during a pre-Late Devensian glaciation. This juxtaposition of immature basal till on an older lodgement till suggests that, although the Late Devensian Ice Sheet was insufficiently thick for mature lodgement till to have been deposited

above 500 m above OD, the ice sheet that laid down the older till was thicker and more extensive.

SANDY TILLS, UNDIFFERENTIATED

A suite of sandy pale yellowish brown or reddish brown tills containing a high proportion of clasts of Old Red Sandstone (principally brown, grey, mauve and green sandstones, carbonaceous siltstones, flagstones, conglomerates and breccias) can be traced across much of the northern side of the Nairn–Findhorn interfluve up to at least 500 m above OD, where typically they underlie dark olive-grey till containing clasts mainly of psammite and grey and red granite.

Such psammite-rich and sandstone-rich tills are exposed within the catchment areas of Dalriach Burn and Moy Burn and its tributaries, where they are recognised as the upper and lower members, respectively, of the **Moy Till Formation** (Merritt, 1990b). Contact between the till members is generally represented by a sharp colour change, but it is rarely demonstrably unconformable. These tills may interdigitate locally [as at 8022 3696], and the lower member appears to be more weathered than the upper one, though this may reflect its more sandy matrix and greater permeability. If the two members are conformable, it would seem that during the initial phase of the Late Devensian glaciation, ice moved eastwards into the area, carrying with it Old Red erratics from outcrops to the west and south-west of Daviot (Figure 23). By the end of the glaciation, ice had begun moving north-eastwards, as evidenced by striae and streamlined glacial landforms (Figure 22), and carried clasts mainly of psammite into the area.

DETAILS

Two of the best places to study sandy tills are in the valleys of Allt Carn a'Ghranndaich and Allt Cromachan, which both drain north-westwards towards the Clava site (Appendix 3c). Results of detailed sedimentological studies (Merritt and Auton, 1993) in the former valley are summarised in Figure 25, from which three important observations can be made. Firstly, the presence within the glacigenic sequence of several major erosion surfaces overlain by gravels, sands and silts, demonstrates that sandstone-rich tills were deposited in the area during more than one glaciation. Secondly, the Late Devensian Finglack Till, sandstone-rich at the Clava site, passes up-slope into a bipartite sequence of psammite-rich till overlying a sandstone-rich till. As in the Moy area, the two tills are not always separated by a discontinuity, leaving the age of the sandstone-rich members somewhat uncertain. However, this evidence again suggests that, during the Late Devensian glaciation, ice carried Old Red erratics eastwards on to high ground before the final north-easterly flow became established. Thirdly, at Section 30 of Figure 25 [7714 4337], the sandstone-rich member of the Finglack Till Formation appears to have a slightly weathered top, being separated from the upper, psammite-rich member by a sequence of subglacially deformed glaciolacustrine deposits. This section provides the only known evidence to suggest that deglaciation may have occurred between the times of accumulation of these two till units.

Sandstone-rich tills, some possibly predating the Late Devensian, are also found in the catchment area of Allt Dearg and its tributaries, and Riereach Burn south of Cawdor (Figure 26). The

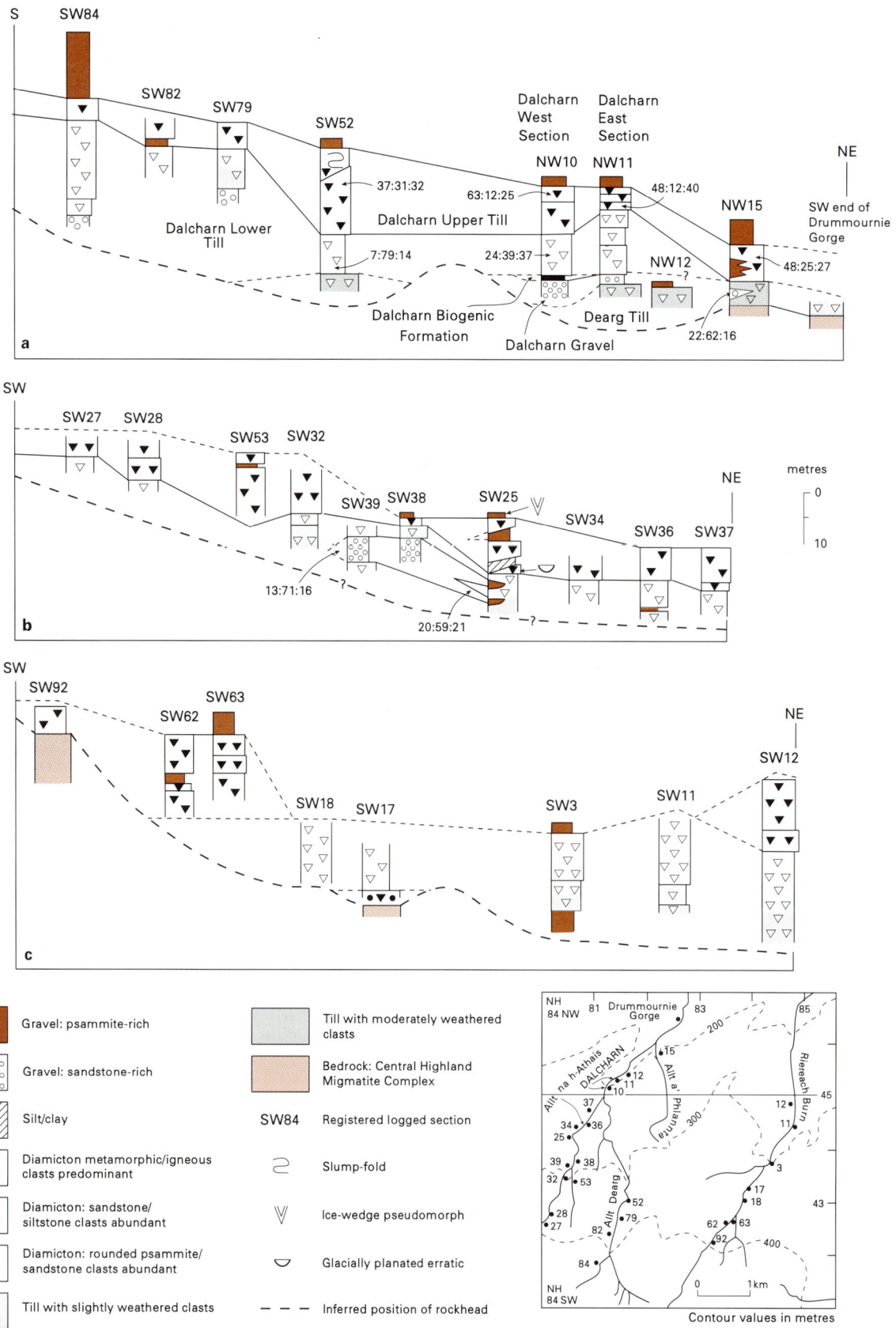

Figure 26   Stratigraphical relationships of Quaternary deposits in the valleys of a) Allt Dearg,
b) Allt na h-Athais, c) Riereach Burn. Stone-count data (e.g. 24: 39: 37) are percentages of
metamorphic, Old Red Sandstone and granitic clasts respectively.

contact between sandstone-rich tills and overlying psammite-rich glacigenic sediments is generally well defined in this area; for example, a sharp subhorizontal contact associated with small-scale tension fractures, including pull-apart structures filled with sand and gravel, is exposed in the Dalcharn East section (Figure 30). At Section SW25 of Figure 26b [8059 4409] in the Allt na h-Athais valley, sandstone-rich till is separated from the overlying psammite-rich till by a bed of very compact, thinly interlaminated silt, clay and sand. This bed, up to one metre thick, may be analogous to the subglacially deformed glaciolacustrine deposits in the Allt Carn a'Ghranndaich valley described above. An erratic of gneissose semipelite, 2 m long, with its upper surface planed off by glacial abrasion, lies at the top of the sandstone-rich till. The lateral persistence of this unconformable boundary demonstrates that a considerable hiatus occurred between the deposition of the two tills, but whether the tills are products of distinct glaciations is not resolved.

Reddish brown tills containing a notable proportion of sandstone clasts are also present locally on parts of the upland flanking the Findhorn Valley. Sandy reddish brown till is exposed [at 8480 3859 and 8487 3864] in the Allt Breac valley, where it overlies clast-supported pre-Late Devensian head gravel.

The red-brown colouration of these sandstone-rich tills is probably largely due to inclusion of locally derived, reddened psammite debris rather than comminuted Old Red Sandstone, although rounded brown sandstone clasts are numerous in the tills at both sites. Inclusions of reddish brown sandy diamicton occur within till containing a preponderance of clasts of igneous and metamorphic rocks, exposed in river cliffs of Allt Breac [8542 3866 and 8543 3876]. An absence of sharp discontinuities separating the two till types suggests that in the Allt Breac area, at least, both are products of a single glaciation, presumably in the Late Devensian.

PRE-DEVENSIAN TILLS

Conclusive evidence of pre-Late Devensian glaciation is only recorded from three sites in the district: Dalcharn, Allt Odhar and Clava (Appendix 3a, b and c). Datable organic material is interstratified within the glacigenic sequences at these localities, but its absence elsewhere does not preclude the possibility that many of the thick drift sequences in the upland areas also include sediments formed during earlier glacial and interglacial periods.

DETAILS

The oldest known till, the **Dearg Till** (Merritt and Auton, 1990; Auton, 1993) occurs at the bottom of the river cliffs of Allt Dearg at Dalcharn East [8157 4537], where it is overlain by deeply weathered gravel (Dalcharn Gravel) capped by sediments containing compressed and disseminated biogenic matter of interglacial affinity (Dalcharn Biogenic Formation). The gravel and the organic sediments are exposed beneath a sequence of tills up to 19.5 m thick. The Dearg Till, which is pre-Ipswichian, if not pre-Hoxnian in age, is a very stiff, poorly stratified, matrix-supported diamicton (lodgement till) containing a high proportion (46 per cent) of Old Red Sandstone clasts. The deposit is moderate brown to moderate yellowish brown; many of the igneous and metamorphic clasts have been decomposed, but are much less altered than those in the overlying gravel.

The stratigraphical relationship between the Dearg Till and the overlying glacigenic sequence can only be established with certainty at the Dalcharn East section, but similar grossly overconsolidated tills occur for about 800 m downstream (Figure 26a, NW 12 and 15). The tills are typically moderate brown to reddish brown and contain a high proportion of sandstone clasts and

many rotten pebbles and cobbles of granitic and psammitic rocks. At the base of a river cliff of Allt a'Phlannta [8226 4553] (Figure 26a, NW15), a reddish brown clayey till, at least 3 m thick, passes laterally into matrix-rich gravel and both the till and the gravel contain many clasts with orange-weathering outer skins. Stiff till with numerous decomposed pebbles is also exposed at the base of the succession in a river cliff [8153 4268] 2.5 km upstream of the Dalcharn East section (Figure 26a, SW52).

Extremely consolidated lodgement tills containing many decomposed clasts also occur at the bases of several river cliffs along Allt Carn a'Ghranndaich, upstream of Clava (between sections 32 and 14 of Figure 25). Like the Dearg Till, these diamictons contain much sandstone and siltstone but, perhaps significantly, in smaller proportions than in overlying glacigenic deposits. Deeply weathered, not unduly sandstone-rich, tills also occur at the bases of river cliffs of Dalriach Burn [7793 3937], north of Moy, and in the catchment of Allt Odhar, north-east of Moy [7984 3671, 8134 3731, 8185 3732, 8238 3757 and 8236 3761]. West of Moy, a weathered till containing little, if any, sandstone is exposed at the base of a river cliff [7409 3343] of Allt na Slanaich, whereas in the Allt na-h-Airigh Samhraich valley, south of Moy, a weathered sandstone-rich till is exposed [7820 3178] in an area where Old Red Sandstone rock types are otherwise rare in tills. Most of the above-mentioned weathered tills, like the Dearg Till, almost certainly predate the Ipswichian Interglacial period.

Another pre-Late Devensian diamicton, the **Suidheig Till** (Merritt, 1990b) occurs at the base of the Allt Odhar section (Appendix 3b), where it lies beneath a moderately weathered gravel (Odhar Gravel), which in turn underlies a lens of compressed peat (Odhar Peat). The latter has been assigned to an interstadial early in the Devensian (Walker et al., 1992). The Suidheig Till is almost certainly pre-Ipswichian in age, because it is most unlikely that the area was glaciated between the Ipswichian and the early Devensian interstadial. It is a very consolidated, sandstone-rich, stony sandy clayey diamicton of pale brown to moderate yellowish brown hue.

In the south of the district [at 7820 3178], a weathered lodgement till is overlain by weathered gravel, which is in turn overlain by fresh psammite-rich lodgement till. Both weathered deposits contain abundant clasts of pale reddish brown sandstone and conglomerate together with pebbly cross-bedded psammite, porphyritic granite and felsite. This assemblage of rock types is known to occur only in Stratherrick and in the Great Glen, both to the west-south-west, indicating that sandstone-rich till was laid down by ice that moved east-north-eastwards into the district. The weathered state of the clasts suggests that in this instance the glaciation was a pre-Late Devensian event.

At sections 6 and 20 of Figure 25 [7708 4329 and 7710 4320], a dark olive-grey, psammite-rich lodgement till containing many well-rounded clasts appears to pass upwards into a vivid moderate brown till dominated by Old Red Sandstone rock types. This sequence may indicate that during a pre-Late Devensian glaciation, ice moved from high ground in the south and west towards the coast before it began carrying Old Red erratics eastwards on to the high ground; this is the reverse of what happened during the Late Devensian event.

Hummocky Glacial Deposits

These deposits are typically hummocky and make up a distinctive sediment-landform assemblage, shown on older maps as 'morainic drift'. Lithologically, the deposits are highly variable and include complex interdigitations of matrix- and clast-supported diamicton, stratified and unstratified silty boulder gravel, and beds of sand, silt and clay. Most of the deposits are moraines, i.e. constructional features formed at ice margins, and most, but not all, were formed primarily by debris-flow and sheet-wash processes (see above). On the 1:50 000 geological map, most areas of hummocky glacial deposits are fully delineated (pale green), but some are marked by blue form-lines in areas of till, and locally they are too thin and patchy to be shown. Ten geographically distinct suites of hummocky glacial deposit are shown on the detailed 1:10 000 geological maps of the district, and are described below.

DETAILS

Dalcross Suite This suite takes the form of subparallel ridges, typically some 2 to 5 m high, spaced 75 to 150 m apart and trending north-north-west. The most prominent ridges, shown on the 1:50 000 geological map as blue lines, occur in an area of till to the south and south-east of Dalcross Castle [7786 4828], where they have traditionally provided the best free-draining sites on which to erect houses and farm buildings. The ridges run perpendicular to the final direction of ice flow (Figure 22), suggesting that they may have formed at the ice-front as winter push-moraines during the local south-westward retreat of the Late Devensian Ice Sheet. Modern examples of such 'De Geer' or 'wash-board' moraines (Mawdsley, 1936; Hoppe, 1959) are typically asymmetrical in profile, with steeper faces on the side away from the former glacier; they commonly coalesce and are generally lobate in plan, with intervening re-entrants (Boulton, 1986). The Dalcross Suite shows some of these diagnostic features, but the ridges were more probably formed as a result of till squeezing up into, and water flowing within, transverse crevasses in the downwasting Late Devensian Ice Sheet, possibly following a local surge (cf. Sharp, 1984; Beaudry and Prichonnet, 1991).

There are few exposures, but south-west of Croy [at 7919 4922], about 3 m of stratified diamicton were exposed in an excavation in the side of a 5 m-high ridge. The deposit comprises thinly interstratified, stiff, sandy clayey diamictons and more friable, diamictic clayey pebbly sand. The former lithology is most prevalent towards the top of the exposure, where it includes laminae of fine-grained sand that are probably the water-winnowed tops of discrete cohesive debris flows. Clasts are well dispersed in the deposit as a whole; they are chiefly of local sandstone, siltstone and conglomerate, and include some large ice-bevelled boulders. Similar material is exposed in small disused gravel pits [8065 4865] 600 m south-west of Holme Rose.

Drummossie Suite Ridges of similar height and disposition to those described above are ubiquitous on Drummossie Muir to the south-west of the A9 road, being especially well preserved in the vicinity of Leys Castle [680 410]. Unlike the Dalcross Suite, however, these ridges are oriented parallel to the north-eastward direction of ice movement. The whole outcrop lying to the north-east of Carr Bàn [670 365] has a corrugated form, with heather-clad ridges and peat-filled furrows of 2 to 4 m amplitude spaced 10 to 20 m apart; the larger ridges are shown as blue lines on the 1:50 000 geological map.

The ridges are formed mainly of clast-supported, unstratified, angular rubble of local sandstone and flagstone with a matrix of silty fine-grained sand. Many of the ridges are probably rock cored, but it is difficult to distinguish frost-shattered bedrock from glacially transported material. The entire sediment-landform assemblage is most likely to have formed as fluted moraine by subglacial erosion and deformation (cf. Sissons, 1967; Boulton, 1976; Hodgson, 1986) with slight modification caused by glacial meltwaters during deglaciation.

Craobh Sgitheach Suite The fluted moraine described above peters out along a line linking Mains of Gask [6805 3605] and the northern end of Loch Bunachton. It merges southwards into an area of more-irregularly shaped and chaotically distributed mounds and peat-filled kettleholes, the Craobh Sgitheach Suite. On the lower-lying ground east of the loch, the mounds are strewn with boulders of sandstone, flagstone and conglomerate; in a gravel pit [6807 3488] near Beachan, up to 7.5 m of very poorly sorted, crudely stratified silty sand and gravel are capped by up to 1.5 m of clast-supported, angular rubble formed exclusively of conglomerate. Although partly waterlaid, the sand and gravel were mostly deposited as debris flows, and contain scattered boulders up to 1 m in diameter, mainly of sandstone. The sequence exposed here was probably formed in close proximity to ice, as supraglacial flow till.

The disorganised hummocks that characterise this suite of moraines pass northwards at Mains of Gask into a discontinuous, steep-sided ridge bordering the eastern shore of Loch Caulan [686 367]. This boulder-strewn ridge, some 5 m high, might have formed in an ice-walled tunnel cut by subglacial meltwaters and thus be regarded as an esker; a subparallel ridge 150 m north-west of the loch is definitely an esker and is mapped as such.

The high ground of Craobh Sgitheach [670 336] (not named on the 1:50 000 geological map) is strewn with very large boulders, as are the scattered outcrops of hummocky glacial deposits occurring to the south and south-east. On the northern slopes of Craobh Sgitheach, the blocks are mostly of psammite and conglomerate, whereas to the south they are almost wholly of gneissose semipelite. Excavations into hummocks [6725 3265 and 6726 3270], near Ballone, revealed angular blocks of gneissose rock up to 2.5 m long set in a matrix of compact, micaceous, silty, fine- to coarse-grained pebbly sand. The deposit, which is probably typical of those in the area, is crudely bedded, being composed mainly of debris flows with little water-sorted material.

Craggie Suite Boulder-strewn mounds also occur in the lower reaches of the Craggie Burn valley, east of Daviot. The mounds are most widespread on the southern side, where they extend uphill towards distinct lateral moraines (formed of gravel) associated with a kame-terrace and several short glacial drainage channels at about the 300 m contour. Deposits of glaciolacustrine silt and fine-grained sand wrap around, and partially blanket the mounds towards the centre of the valley. Exposures are few, but a temporary section in a mound [7325 3889] revealed 1.5 m of angular rubble composed of gneissose rock and grey granite in a matrix of silty fine-grained sand. The deposit rests with an erosional base on cross-laminated sands and silts, suggesting that it originated as a debris flow that slid into an ice-dammed lake.

Stairsneach nan Gaidheal Suite Rounded elongate mounds, 15 to 25 m high, are scattered for about 4 km along Stairsneach nan Gaidheal, the glacial breach to the south of Meall Mór [737 355], which contains the re-aligned A9 road. Unlike the deposits described above, those of this suite contain relatively few boulders. A section in a borrow pit at Uaigh an Duine-bheo [7274 3482] revealed 1 m of stiff sandy clayey diamicton overlying crudely interbedded, very poorly sorted, diamictic clayey gravel and thinly stratified gravelly sand. The deposits clearly were formed close to ice, mainly by debris-flow processes with limited water sorting. Isolated mounds of a similar nature occur to the south and east [7260 3410, 7425 3375 and 7665 3385].

Balvraid Suite A distinct suite of hummocky glacial deposits lies in Gleann an Tairbhidh and Gleann Seileach, north-east of Balvraid [829 314]. A good river cliff section [8360 3251], typical of many in the area, occurs in the latter valley, where 7 m of very compact, stratified, silty, sandy diamicton rest on stiff lodgement till. The stratified deposit forming the mounds includes laminae of silt and silty clay, lenses of poorly sorted gravel, and clusters of angular boulders of gneissose semipelite up to 0.5 m in diameter. Sparse lenses of cross-laminated sand indicate a palaeocurrent flow towards the north-east. Bedding is generally thin and laterally discontinuous. Discrete beds of diamicton with flow noses and water-winnowed tops indicate that deposition mainly involved debris-flow and sheet-wash processes. Other sections in the moundy deposits occur upstream [at 8486 3344], and in Gleann an Tairbhidh [at 8374 3192].

Cantray Suite   Unlike the hummocky suites described above, in which mounds are primarily constructional features, those of the Cantray Suite are intimately associated with ice-marginal drainage channels. Hence, the moundy topography is due mainly to fluvial erosion. Nevertheless, these hummocks comprise deposits that were laid down primarily by debris-flow and sheet-wash processes, like most other hummocky glacial deposits.

This suite occurs on the south-eastern side of Strath Nairn and is best examined in the vicinity of the Clava site (Appendix 3c). There is a stepped profile to this side of the valley, each step rising south-eastwards; the steps take the form of prominent north-easterly-trending benches up to about 265 m above OD (see 'ice-sculptured benches', below). The benches are typically bound on their uphill sides by arcuate, steep-sided, glacial meltwater channels. In contrast, escarpments facing the main valley are essentially ice-contact slopes, although many have been trimmed by ice-marginal drainage and are now bordered by terraced spreads of glaciofluvial sand and gravel. The whole landform-sediment assemblage was formed during the final stages of deglaciation, as ice became restricted to progressively lower levels in the valley of the River Nairn, and meltwaters were constrained to flow between the ice and the valley side.

A good section [7646 4386] in one of the most extensive benched deposits occurs 350 m upstream of the Clava site (Figure 31, Section VI), where 3 to 4 m of poorly sorted, silty cobble gravel cap a sequence of stratified, matrix-supported diamicton interstratified with thin, laterally discontinuous beds and laminae of finely laminated silt, sand and fine gravel. The matrix of the diamictons varies from very compact, silty, fine-grained sand to friable, sandy silty clay. There is ample evidence of sedimentation by sheet-wash and cohesive debris-flows, with associated penecontemporaneous slumping, 'pull-apart' structures and shrinkage cracks. The diamictons typically include subhorizontal wisps of white silt and fine-grained sand and are more massive in the lower half of the section, where beds are up to 0.5 m thick and interbedded seams of sand (up to 10 cm thick) have scoured bases. Peacock (1975a) interpreted the sequence as comprising supraglacial flow-tills similar to those described by Boulton (1968), but towards the centre and base of the section there is a lens of clast-supported gravel with a sharp horizontal base and an arched top (Merritt, 1990a). Lenses such as these are thought to indicate deposition within cavities and channels in the soles of melting ice sheets and are diagnostic features of basal melt-out tills (Shaw, 1979). Thus the sequence exposed here may have been formed by a range of basal meltout and sediment gravity-flow processes, the latter becoming more important in the latest stages.

A ridge, formed substantially of stratified diamicton similar to that beneath the cobble-gravel at the section described above, is exposed [7912 4599] near Cantraydoune. Here, the diamicton is overlain by about 1 m of very compact, interlaminated silt, silty clay and fine-grained sand with lenses of gravel

and diamicton and sparse dropstones. These glaciolacustrine deposits are in turn overlain by about 3 m of very poorly sorted, silty, pebbly sand that coarsens upwards. Boulder gravel caps this ridge, and similar glaciofluvial gravels cap many contiguous ridges farther down-valley where stratified diamictons occur, but are too thin to distinguish from till on the 1:50 000 geological map. Likewise, the stratified diamictons underlying the higher benches shown on that map and depicted in Figure 25 have been included in the till.

Balloch Suite   This suite of deposits, which stretches from Dell of Inshes [6903 4415] north-eastwards for about 7 km, generally forms subdued, irregularly shaped mounds. Better-organised, linear features around Balmachree [7383 4757] are shown on the 1:50 000 geological map by blue lines. Although the landforms are quite distinct from the smooth convex slopes underlain by till to the south-east, the boundary with the till is gradational. The hummocky deposits are dissected by numerous arcuate, shallow glacial drainage channels that probably formed at the ice front as the glacier occupying the Inverness Firth retreated south-westwards. All the features become subdued seaward of the conjectural marine limit shown on the geological map.

The deposits of this suite have been revealed in many shallow excavations for new housing. At surface, the deposits comprise yellowish brown to orange, crudely stratified, very poorly sorted, mainly matrix-supported clayey gravel. They contain angular to well-rounded clasts up to boulder size of sandstone, psammite, quartzite, schistose and gneissose semipelite, and conglomerate. A roadside section [7155 4615] in Smithton revealed about 2.5 m of thin, laterally discontinuous beds of silty, sandy diamicton interstratified with thinner beds and laminae of pebbly sand. Fold noses in the more massive beds of diamicton indicate that the lower part of the sequence formed mainly as debris flows. However, the deposit becomes increasingly gravelly and clast-supported upwards and is capped by up to 1.5 m of glaciofluvial, clast-supported gravel.

Shenachie Suite   Discontinuous hummocky spreads of glacial debris flank the Findhorn Valley between Ruthven [813 331] and Quilichan [854 378], and are particularly well developed upstream of Shenachie [827 349]. Similar spreads occur on the sides of the valley of Allt nam Meannan, a misfit stream occupying a glacial trough [816 340] west of Tom na Slaite, and also along the eastern boundary of the district, north-east of the Meur Bhcòil valley. At the latter locality [856 404], the deposits form a series of benches (in some ways analogous to those of the Cantray Suite described above) flanking the lower parts of the Caochan na Féithe Seilach valley.

Exposures in these deposits are sparse, but pale brown sandy diamicton with abundant boulders of psammite, gneissose semipelite and granite caps the mounds on the eastern side of the Findhorn Valley upstream of Ruthven; in a borehole sited 850 m north-east of Ruthven, 19.5 m of clayey gravel and sand overlie bedrock. The records of two boreholes sited on the moundy deposits on the western bank show 6.9 m of coarse gravel and sand overlying broken bedrock, and bouldery sand and gravel (unbottomed at 24.2 m depth) respectively.

Small exposures [8190 3426 and 8188 3429] on the eastern side of the Allt nam Meannan valley show greyish to light yellowish brown thin gravelly diamicton resting on broken granite; mapping indicates that many of the mounds on the western side of the valley have cores of granitic bedrock. Diamictic deposits, up to 12 m thick, are exposed in the sides of steep gullies west of Kincraig [839 365]. In some exposures [as at 8372 3674], scree blankets loosely consolidated diamicton composed of very angular blocks of psammite and granite in a sandy and gravelly matrix. Up to 12 m of matrix-supported diamicton and clast-supported claybound gravel are visible in a nearby exposure [8373 3672]; the lower two thirds of the section show well-developed, high-angle stratification dipping towards the valley floor, the upper third shows evidence of slumping, with the axes of large gently inclined slump folds plunging both towards and away from the valley floor. These structures suggest that the deposits were laid down as lateral moraines against stagnating ice towards the end of the Late Devensian glaciation of the valley (Auton, 1990a).

A triangular spread of poorly sorted, glacial debris extending southwards from Shenachie into the Findhorn and Allt na Seanalaich valleys, has the appearance of a medial moraine that formed where glaciers from both valleys coalesced during the late stages of glaciation.

Riereach Suite   Ridges, in the catchment of Riereach Burn [840 424], aligned east–west and up to 10 m high, are thought to be lateral moraines laid down during minor still-stands as the ice sheet retreated on to lower ground. The ridges, which are largely blanketed by peat and associated with ice-marginal glacial drainage channels, occur in two groups; one at about 400 m above OD, and a second between 365 and 385 m above OD. Large angular boulders are scattered along the crests, and the ridges are composed of sandy bouldery diamicton up to at least 5 m thick. A similar ridge lies within the col [7835 4305] separating Saddle Hill and Beinn Bhuidhe Mhór.

GLACIOLACUSTRINE DEPOSITS

Fine-grained sand, silt and clay, laid down in standing water, form part of many glaciofluvial sequences in the district, but generally only those seen to have lamination quite clearly disrupted by dropstones have been distinguished as glaciolacustrine deposits. These deposits have a wide altitudinal range, but are best developed within valleys such as Rosemarkie Glen and in those of the Findhorn and Craggie Burn. They were laid down in temporary ice-dammed lakes that mostly formed at the margins of the Late Devensian Ice Sheet as it decayed and retreated from the higher ground to the south of the Inverness and Moray firths.

DETAILS

Rosemarkie Glen

A sequence 6.5 m thick, comprising thinly interbedded fine-grained sand and thinly laminated silt, is exposed in the prominent Red Craig cliff [7360 5785] overlooking Rosemarkie. The deposit, named here the **Kincurdy Silts**, is a coarsening-upward sequence with the basal metre or so comprising thinly interlaminated silt, sandy silt and plastic clay. The laminae are graded and may be varves. Similar rhythmites occur higher in the sequence as discrete packages of between 5 and 10 silt–clay couplets. The sequence includes dropstones and lenses of diamicton, all of which were probably deposited from floating ice. Soft-sediment deformation has occurred and flame structures affecting beds of sand and silt are invariably overturned towards the east. The Kincurdy Silts are overlain by 5.5 m of stony diamicton, presumed to be a Late Devensian basal till. They overlie an older lodgement till that caps a distinctive gravel (Red Craig Gravels) exposed at the base of the cliff. The silts were probably laid down in an ice-dammed lake during the build-up of the Late Devensian Ice Sheet when ice advancing into the Moray Firth impeded drainage down Rose-

markie Glen. The uppermost lodgement till has been eroded into excellent examples of earth pillars (Plate 13) in Fairy Glen, on the southern side of Rosemarkie Glen.

Craggie

Micaceous fine-grained sands and silts are widespread at surface within the Craggie Burn valley, east of Daviot. The deposits both overlie, and interdigitate with, hummocky glacial deposits and glaciofluvial sands and gravels. In a river cliff section [7368 3911], a 7 m-thick sequence of thinly interlaminated clayey silt, silty clay and very fine-grained sand incorporates several splendid dropstones up to 1 m in diameter. Upstream [7406 3891], 3 to 4 m of thinly bedded, horizontally laminated sandy silts and silty clays overlie about 6 m of micaceous fine-grained sand with stringers of gravel, lenses of coarser-grained sand and thin seams of silt. Climbing ripple-drift cross-lamination indicates a palaeocurrent flow towards the north-west. The sands are affected by normal faults, presumably the result of consolidational collapse, but the overlying silts are not.

Laminated silts beneath glaciofluvial sand and gravel on the western side of Strath Nairn crop out immediately downstream of the confluence of this river and Craggie Burn. Boreholes drilled in the vicinity show that the silts occur at the base of a coarsening-upward sequence, as in a typical prograding delta.

It would seem that during deglaciation, ice continued to block Strath Nairn downstream of the gorge at Daviot, while upstream, ice retreated from the main valley and from the Craggie Burn valley leaving a proglacial lake into which the ice calved, debris collapsed and outwash streams formed deltas.

Ballachrochin

A flat-topped spread of sand and gravel overlying glaciolacustrine deposits occurs near to the confluence of Allt a'Choire Bhuidhe and the River Findhorn, 150 m south of Ballachrochin [847 368]. Exposure is limited, but a small cliff [8484 3655] cut in the lower part of the sequence shows 1.5 m of silty fine-grained sand with thin clay drapes and silt partings. The deposit shows horizontal lamination and low-angle cross-bedding. The sand passes down into 2 m of finely interlaminated sandy silt and clay, with draped and convolute-draped lamination, rounded dropstone cobbles and sparse interbeds of diamicton. This coarsening-upward sequence was laid down as a glaciolacustrine delta within a small glacial lake, ponded by ice in the main valley during the final stages of deglaciation (Auton, 1990a; Werritty and McEwen, 1993).

Other areas

Rhythmites with dropstones similar to those at Ballachrochin are exposed in a river cliff of Allt Carn a' Mhàis Leathain [8521 4031]. Well-developed cryoturbation structures are present in the upper part of the sequence, which is penetrated to a depth of 60 cm by a frost-wedge cast infilled with silty fine sand. The dropstones, up to 20 cm in diameter, occur within interlaminated beds of silty clay and fine sand, beneath a bed of clast-supported sandy gravel containing angular boulders up to 50 cm long. The fine-grained sediments overlie 2 m of poorly stratified gravel that passes gradationally down into silty diamicton; both gravel units were probably laid down as subaqueous debris flows within a temporary ice-marginal lake.

Deposits not delineated on the map

Thinly laminated deposits, possibly varves, have been located within thick glacigenic sequences at several localities on the high ground south-east of Strath Nairn. At Cnoc a' Chinn Leith, laminated silts and fine-grained sands with sparse dropstones are exposed near the base of a thick coarsening-upward sequence of sands and gravels [8371 3850] that formed a fan-delta prograding into a deep ice-marginal lake (Auton, 1990a). At the foot of a river cliff [7749 4280] of Allt Tarsuinn, about 2 m of greyish olive interlaminated silt, clay and micaceous fine-grained sand underlie two lithologically distinct formations of lodgement till and overlie a weathered deposit of sand and gravel. The laminated deposit, which is very compact and gently folded, includes dropstones and is probably early Devensian or older. The sequence at this site is depicted in Figure 25, Section (31), but it should be noted that the exact relationship between the weathered sand and gravel deposit and the underlying sandstone-rich till here is not known. Glaciolacustrine silts and clays occupy a broadly similar stratigraphical position within the glacigenic sequence in the Moy Burn valley [7859 3664] (Merritt, 1990b, fig. 23, section 73 NE 10).

Crudely laminated deposits of possible glaciolacustrine origin separate psammite-rich and sandstone-rich tills in the Allt Carn a'Ghranndaich valley [7714 4337] (Figure 25, Section 30) and at in the Allt na h-Athais valley (Figure 26b, Section SW 25). These deposits, both of which have been sheared and folded subglacially, may indicate that the high ground at least was deglaciated between the times of formation of the two till units, possibly in the mid Devensian. It is also possible, however, that the ponding occurred subglacially during one uninterrupted glaciation.

Micaceous fine-grained sand with clay and silt drapes, exposed beneath diamicton and sand and gravel on the northern flank of the Flemington esker system near Bemuchlye [827 531], was probably deposited within the esker conduit (Auton, 1992; Gordon and Auton, 1993). A subglacial origin has also been proposed for faulted and contorted interstratified beds of silt, sand and pebbly diamicton, exposed beneath sand and gravel in a river cliff [8424 3742] along Ballaggan Burn (Auton, 1990a).

GLACIOFLUVIAL DEPOSITS

Glaciofluvial deposits are sediments laid down by waters issuing from ice masses. Three categories are recognised within this district, namely Ice-contact, Fan-delta and Sheet deposits.

Glaciofluvial Ice-contact Deposits

Ice-contact deposits mostly comprise sand and gravel, but include subsidiary beds of diamicton, silt and clay; on older maps they are shown as 'Glacial sand and gravel'. The sediments were laid down on, under or against glaciers or ice sheets; hummocky topography is characteristic, but flat-topped spreads also occur. Steep-sided ridges of gravel (eskers) are shown individually as red lines on the 1:50 000 geological map, whereas rounded hillocks of sand and gravel (kames) and flat-topped mounds (kame-plateaux) are not delineated from other, more irregularly shaped mounds and undulating spreads. Such deposits are typically associated with kettle-holes, and former ice-contact slopes are commonly recognisable. The kames and kame-plateaux are typically composed of coarsening-upward sandy deposits that formed as fan-deltas in emphemeral ice-marginal lakes,

but as most are difficult to delineate, few have been mapped out as fan-delta deposits *per se*.

The district includes some of the best examples of eskers in Scotland, the most famous being the braided Flemington Eskers and the Littlemill Eskers, a system of high, parallel ridges. There are also fine examples of ribbon eskers associated with subglacial meltwater channels.

DETAILS

Esker systems

Flemington Eskers The Flemington Eskers probably form the best-preserved braided esker system in Britain, since it remains untouched by large-scale quarrying. The system comprises an almost continuous series of north-east-trending ridges, 5 to 10 m high, between Balnabual [775 490] and Meikle Kildrummie [856 539] (Plate 12). The ridges are principally composed of poorly stratified coarse-grained sand and well-rounded cobble gravel, although beds of diamicton and silty fine-grained sand occur locally. The eskers and the ground surrounding them have been the subject of geological controversy for more than a century, and a variety of hypotheses has been put forward to relate the esker system to the Late Devensian history of glaciation, deglaciation and sea-level change (Auton, 1992; Gordon and Auton, 1993).

Braided eskers occur between Meikle Kildrummie and Loch of the Clans [834 531], between Bemuchlye [827 531] and the south-western end of Loch Flemington, and between Balblair [804 515] and Culaird [782 500]. The system is aligned with the final direction of ice-movement across the area, as deduced from the orientation of glacial striae and ice-moulded landforms on the flanks of Strath Nairn (Figure 22). A flat-topped kamiform ridge [860 538], south and east of Meikle Kildrummie, has been interpreted as a glaciomarine delta laid down at the mouth of a subglacial channel (Synge, 1977a) or as a large subaerial crevasse-fill (Firth, 1990b). Both authors infer that the ridge was initially contiguous with the esker system to the west, and was separated from it by subsequent erosion (of either glaciomarine or glaciofluvial character) in the intervening ground. The main development of eskers extends for about 10 km, and isolated eskers on the same alignment to the south of Auldearn [918 555], beyond the eastern margin of the district, suggest that the system formerly extended for more than 15 km.

Exposed sections in the eskers are sparse and degraded. The upper parts of the features are generally composed of clast-supported cobble gravel with a preponderance of clasts of psammitic rocks. In places, the gravel includes up to 30 per cent of cobbles and pebbles of mauve and brown sandstone, many containing mudstone intraclasts. Small numbers of pelitic and granitic clasts are also present. No distinctive far-travelled erratics have been recognised within the esker gravel.

Littlemill Eskers This system of eskers (Gordon, 1993a) lies on the south-eastern side of Strath Nairn between Inverarnie [689 336] and Lairgandour [721 376]. A single esker leads north-eastwards into a system of eskers, kamiform mounds and kettle-holes, which then converge into a broad, flat-topped mound at Lairgandour, now bisected by the re-aligned A9 road. There are four major steep-sided subparallel ridges, 2 km long. These are generally about 15 m high, but in Mid Lairgs Pit [710 365] reached as much as 40 m in height before they were dug away for aggregate. Sections in the eskers [7080 3665, 7015 3535 and 6905 3420] reveal up to 25 m of gravel. Their 'cores' typically show crude horizontal stratification and are composed of cobble-gravel with boulders up to 0.5 m in diameter. Better-sorted sand and gravel, with planar bedding inclined parallel to

Plate 12 The Flemington esker system. Aerial view looking west from above Meikle Kildrummie (Cambridge University Collection, RAF aerial photograph RD48. Crown copyright/MOD, reproduced with the permission of the Controller of HMSO).

the sides of the esker, form a drape. Normal faults parallel to the axes of the ridges are common and were formed when the lateral ice support melted. The gravel comprises mainly sub-rounded to well-rounded clasts of hard, sound rock types; clasts of micaceous psammite and quartzite predominate, with granite, diorite, fine-grained sandstone, porphyry, amphibolite and gneissose semipelite also present. There are sparse, less durable, clasts of Old Red Sandstone conglomerate and garnetiferous schistose semipelite.

Peat-filled kettleholes and terraced deposits of poorly sorted silty sands and gravels occur between the Littlemill ridges, and an undulating spread of glaciofluvial deposits lies to the east of them. The latter deposits are exposed in the eastern part of Mid-Lairgs Pit [7140 3673], where a layer of clast-supported gravel up to 2.5 m thick overlies a unit of cross-laminated fine- to medium-grained sand, micaceous silt and sandy diamicton. Fold noses within the diamictons suggest that they formed as cohesive debris flows. Beds of silt with dropstones intercalated between the diamictons indicate that deposition probably took place within a shallow proglacial lake ponded while the ice

front retreated south-westwards, leaving the freshly formed eskers as islands. North-east of Lairgandour, glaciofluvial deposits are mostly sandy and appear to grade laterally into glaciolacustrine silts mapped in the Craggie Burn valley (see above). These sands are thus probably deltaic in origin.

Allt a'Chuil Esker   Unlike the eskers described above, which formed relatively late in the deglaciation of the district, after meltwaters had become restricted to lower-lying ground, the splendid Allt a'Chuil Esker [7680 3160] relates to earlier, ice-directed drainage that was constrained to flow north-eastwards across the topographical grain (Young, 1980). This esker links with a subglacial drainage channel lying immediately south of the district in a suite of interconnected channels and eskers traceable across the north-central Grampian Highlands (Young, 1977, 1978). A section in the esker [7685 3169] reveals about 15 m of loose gravel on 5 m of stratified sandy diamicton, resting on highly weathered semipelitic gneiss.

Saddle Hill Eskers   A group of ribbon eskers, with subglacial meltwater channels lying parallel to them, emerges from the col between Saddle Hill [788 435] and Beinn Bhuidhe Bheag [792 423] and trends east-north-east towards Creag an Daimh [834 443]. Rounded boulders of granite and psammite are scattered along the esker crests, which rise up to 18 m above the surrounding moorland. The present drainage bisects the eskers and degraded exposures [8069 4415, 8065 4389, 8113 4410 and 8235 4466] show 15 to 20 m of poorly sorted cobble gravel with a preponderance of clasts of psammitic and granitic rocks; sandstone clasts comprise 15 to 30 per cent of the gravel.

Findhorn Eskers   Two groups of eskers occur on the north-western side of the Findhorn Valley, one group on the southern side of the Ballaggan Burn valley, the other to the south of Kincraig (Auton, 1990a). Up to 5 m of poorly sorted gravel, overlying planar cross-bedded sand, are exposed in river cliff sections cut into the sides of several of the Ballaggan eskers. The gravel contains subangular boulders of psammite and granite up to 60 cm in diameter. Three metres of coarse gravel, predominantly composed of rounded clasts of psammite, pink granite and grey granite, are exposed in a section [8378 3603] in one of the Kincraig eskers. The eskers were all deposited by subglacial or englacial streams draining through a mass of ice that remained within the Findhorn Valley, and which blocked fluvial drainage along the valley after the Nairn–Findhorn watershed had become ice free.

Moundy deposits, kames and kame-plateaux

Farr   The Littlemill esker system passes south-westwards into a moundy, kettled terrace formed of very poorly sorted cobble gravel. The terrace is strewn with large blocks of gneissose rock up to several metres in diameter, forming a remarkable train crossing the valley; blocks become more numerous south-westwards towards their source. South-west of Ballone [6715 3277], an outwash fan merges north-westwards into a steep-sided ridge in which a small gravel pit reveals 5 m of moderately well-sorted sand and gravel.

Moy Gap   Moundy, kettled deposits of sand, gravel and silt abound in Moy Gap, a misfit valley now occupied in the north by Craggie Burn and in the south by the Moy and Funtack burns. The gap lies at right angles to the former direction of ice movement, and in contrast to valleys aligned parallel to the ice-movement direction, where there was a relatively organised up-valley retreat during deglaciation, this valley witnessed stagnation and in-situ downwasting of the ice. Loch Moy probably occupies part of a large kettlehole that formerly extended farther north-westwards.

River cliff sections [7539 3790, 7531 3808 and 7523 3820] along Craggie Burn downstream of Lochan a' Chaorainn reveal up to 15 m of well-stratified sand and gravel. The deposits hereabouts are predominantly sandy, with gravel restricted to lenses and channel lags. Bedding is mainly planar horizontal, although large-scale tabular cross-sets locally occur towards the base of the sequence. Units of fine-grained sand reveal climbing ripple-drift cross-lamination, suggesting that these deposits accumulated in the distal part of an outwash fan delta.

Trackside sections [7569 3752 and 7573 3737] east of Lochan a' Chaorainn reveal up to 8 m of unstratified, coarse, clast-supported gravel locally overlying deeply weathered granite. Clast-supported gravels also underlie undulating spreads south-south-east of the lochan, towards Moymore [7638 3623], and isolated mounds towards Moy. At Moy, there is a complex of mounds and ridges of gravel separated by terraces and deep kettleholes (Young, 1980). These features form a belt of ground that leads westwards towards hummocky glacial deposits in

Stairsneach nan Gaidheal. Over 10 m of very poorly sorted gravel are exposed in a small pit [7760 3382] near Moy School, and a borehole sited nearby proved 21 m of gravel and gravelly diamicton resting on bedrock. Gravels hereabouts are composed mainly of micaceous psammite, gneissose semipelite, granite and granodiorite with some amphibolite, porphyry, felsite, quartzite, conglomerate and brown, fine-grained sandstone.

The belt of gravel mounds described above stretches to the south-east of Loch Moy and abuts the floodplain of the River Findhorn. South-west of Dalmagarry [7878 3230], the gravel thins rapidly against till, but south-east of the farm, where the moundy deposits have been trimmed by glacial meltwaters [as at 7955 3150 and 7920 3265], the deposits probably reach 40 m or more in thickness. A face 6 m high in a small gravel pit [7942 3179] revealed coarse, clast-supported gravel broadly similar in composition to that described from the pit near Moy School. The gravel includes thin seams of stony clay diamicton (flow till), clayey sand and gravel (debris flow), but is mainly glaciofluvial in origin. Cross-stratification and cross-lamination indicate a general north-eastward palaeocurrent flow.

Findhorn Valley   A steep-sided conical hill [847 378], 60 m high and 250 to 300 m wide, formed mainly of poorly sorted gravel, lies on the western side of the Findhorn Valley near Ballaggan. Clast-supported boulder gravel is exposed in the track running eastwards along the southern side of the hill, and pebbly fine-grained sand interlaminated with sandy silt is exposed 30 m above the base of the hill and in a degraded face 20 m high [8493 3762]. These deposits probably constitute a large kame, laid down in contact with stagnant ice. Similar moundy spreads are present 300 m south-east and 500 m north-east of Quilichan [854 379]; a much smaller deposit occurs nearby [835 356].

Kildrummie and Cawdor   Extensively kettled spreads of sand and gravel occur south and south-west of Cawdor and, inland of the Late Devensian marine limit, between Easter Glackton [824 534] and Tornagrain [767 499]. In places, notably between Meikle Kildrummie and Drumdivan [844 547], individual flat-topped and conical kames (composed mainly of cobble gravel) are clearly delineated. In other areas, however, as around Kirkton of Barevan [837 472] and Newlands of Inchnacaorach [856 492], there is no clear-cut boundary between the moundy and terraced spreads; the mapped line is merely an approximation of where terraced landforms with few kettle-holes give way to moundy topography.

Blackcastle–Delnies Ridge This feature, which stands 5 to 10 m above the level of the surrounding glaciofluvial sheet deposits, is a kettled, elongate kame-plateau that stretches north-eastwards from the vicinity of Blackcastle [829 541] to beyond the boundary of the district. Exposures in Drumdivan Gravel Pit [840 547] show clast-supported gravel with tabular cross-bedding, pronounced bimodal sorting and imbrication of clasts; coarsening-upward sequences are commonly developed, although fining-upward successions occur locally. Evidence of cryoturbation, in the form of incipient ice-wedge pseudo-morphs and erect pebbles, occurs towards the top of the sequence in one exposure [8391 5470]. In a gravel pit [863 552] at Moss-side (just beyond the eastern margin of the district) cryoturbation structures, soft-sediment deformation structures and small-scale normal faults affect laminated silt and fine-grained sand infilling kettleholes and drainage channels. Synge (1977a) described the Moss-side exposures as 'horizontally bedded glacial outwash gravels punctured by kettleholes filled to the brim with gravel'. He attributed the 'gravel fill' as being due to wave action and the planated ridge as being a raised marine shoreline (at 30.5 m above OD) equivalent to the 'Culcabock Shoreline' at Ardersier (Synge 1977a, fig. 4–2). Although kettleholes are developed in the gravel, most are infilled with silt and fine sand; undisturbed exposures in the gravel show planar beds dipping 5° towards the south-east. The gravel overlies planar and trough cross-bedded fine-grained sand. The generally coarsening-upward nature of the sequence suggests that the sands and gravels were laid down as deltaic deposits, the kettled and irregular upper surface indicating deposition in contact with decaying ice, without modification by subsequent marine activity.

Riereach Burn   Mounds and kame-plateaux of sand and gravel occur in association with east–west-trending ice-marginal drainage channels and hummocky glacial deposits of the Riereach Suite. Two conical mounds lie on the western side of Allt Creag a' Chait, a headwater tributary of Riereach Burn. A degraded river cliff [8309 4215] near the base of the larger mound, which rises about 25 m above the surrounding moorland, displays 10 to 15 m of silty, matrix-supported boulder gravel interbedded with ferruginous fine-grained quartz sand.

Spreads of sand and gravel cropping out around the confluence of Riereach Burn and Caochan nan Seangan [843 437] form mounds that rise 5 m above the surrounding till surface. Many have a terraced form, but rounded hummocks are also common. The sand and gravel deposit is exposed in several places in the Caochan nan Seangan valley; it overlies lodgement till [at 8431 4336] and gneissose bedrock [at 8430 4358]. An exposure in a small working [8448 4323] shows cryoturbated, cross-stratified sand 60 cm thick, overlying 2.4 m (unbottomed) of gravel and silty sand interbedded as fining-upward couplets; ripple lamination and small-scale trough cross-bedding are developed in the sandier beds.

Black Isle   Few deposits of sand and gravel of glacial origin occur in this part of the district. East of Janefield, a prominent ridge of gravel is developed [7307 5924] and fine-grained sand crops out [between 7265 5982 and 7300 6002]. The latter deposit possibly merges with a stony sandy diamicton that occurs close to the base of the drift sequence locally and is exposed in the burn near Lower Whitebog [7251 5999 to 7258 6023]. Along the northern side of the Suddie–Rosehaugh Gap, moundy deposits of densely packed clean-washed coarse gravel are conspicuous. Higher up slope [671 549], such deposits form a perched terrace remnant and dense cobble gravel is exposed in a small circular mound [at 6737 5505].

Deposits of sand and gravel buried beneath till *(not delineated on the 1:50 000 geological map)*

Rosemarkie Glen   A distinctive deposit of moderate reddish brown very compact gravel and diamicton, named here the **Red Craig Gravels**, forms the base of the high Red Craig cliff [7360 5785] at the mouth of Rosemarkie Glen. Glaciolacustrine silts and till (Plate 13) occurring higher in the cliffs in this area have been described above. The gravel is predominantly clast-supported with a muddy sand matrix, but beds with matrix support also occur, especially towards the top of the unit, where there is a distinctive bed of probable lodgement till 3 m thick. The gravel comprises angular to subrounded clasts of red sandstone and psammite up to 40 cm in diameter. Bedding is well defined, thick to very thick, generally planar and almost horizontal, although there are some shallow channels infilled with pebbly sand and more rarely with thinly interlaminated yellowish brown silt and clay. Individual beds of gravel exhibit either normal or reverse grading. Firth (1989b) has stated that cross-bedding indicates a palaeoflow 'up' Rosemarkie Glen, but such bedding is rare and its origin is not clear. Judging by their composition, it is likely that the gravels accumulated as a subaerial fan that spread down the glen towards the sea, possibly during a periglacial climatic episode early in the Devensian.

South of the Inverness Firth   Fragmentary deposits of sand and gravel beneath, or within, sequences of till occur widely in the south-eastern half of the district (Merritt and Auton, 1993). The oldest known deposit, the Dalcharn Gravel (Merritt and Auton, 1990), occurs at the base of the Dalcharn East and West sections (Appendix 3a, Figure 30). It has a severely weathered top (Bloodworth, 1990) and underlies biogenic material ascribed tentatively to the Hoxnian Interglacial period, though it may represent the Ipswichian episode (Walker et al., 1992). At the Allt Odhar site (Appendix 3b), the Odhar Gravel

Plate 13   Earth pillars capped by boulders in Fairy Glen, The Dens, Rosemarkie [731 577]. The whole drift sequence, mostly till, has been gullied by erosion. Photographed about 1910 before the present-day tree cover was established (C1881).

(Merritt, 1990b) underlies peat that accumulated in the early Devensian (Walker et al., 1992). Both gravel deposits are fluvial, if not glaciofluvial, in origin.

Weathered, probably pre-Late Devensian, fluvial or glacio-fluvial gravels also occur downstream of the Allt Odhar site [7847 3661 and 7859 3664] (Merritt, 1990b, fig. 23, section 73NE10), in the nearby Allt na Beinne valley [7988 3811 and 8016 3833], and in the valley of the Dalriach Burn [7988 3811 and 7834 3956]. They are also exposed in many cliff sections along the Riereach Burn [notably at 8412 4350, 8459 4407 and 8461 4405], along Allt Carn a' Ghranndaich (Figure 25), south-east of Clava, and in the adjacent valley of Allt Cromachan [at 7691 4290, 7695 4275 and 7707 4238]. All these gravels contain a relatively large proportion of sandstone. Weathered gravels occur at more than one stratigraphical level within the glacigenic sequence exposed along Allt Carn a' Ghranndaich (Figure 25) and it is not clear which deposits, if any, correlate with the Drummore Gravel (Cassie Gravel of Merritt, 1990a. tab. 4) underlying the Clava Shelly Clay (Appendix 3c). Weathered gravels, mostly sandstone rich, also occur beneath supposed Late Devensian tills at scattered localities south of the Moy Gap; these lie in the Allt na Slanaich valley [at 7409 3343], in the Allt na Loinne Moire valley [at 7565 3267], and in river cliffs [7823 3166 and 7820 3178] of Allt na h-Airigh Samhraich.

In the Allt Breac valley, a section [8487 3864] reveals reddish brown sandy till overlying clast-supported head gravel up to 3 m thick, composed of angular fragments of reddened psammite. The gravel mantles a small knoll of highly shattered psammitic bedrock close to the Findhorn Fault and was probably formed by weathering and mechanical breakdown of the bedrock under periglacial conditions preceding the glaciation responsi-ble for depositing the overlying till. The gravel is possibly of early Devensian age.

Relatively fresh glaciofluvial gravels are present within the till sequence in many sections in the Allt na h-Athais valley (Figure 26b). Sandstone clasts are abundant within the gravels (over 60 per cent), and most are fresh. Clast-supported open-framework beds are common and beds of planar-stratified sand also occur. Similar sandstone-rich gravel is exposed close to the base of a sequence of reddish brown tills in the Allt Breac valley [8542 3867].

Glaciofluvial Fan and Fan-delta Deposits

These deposits form fan-shaped landforms with smooth, gently sloping upper surfaces and are made up of gravel, sand, silt and some clay. The material at the surface of a fan is generally bouldery and poorly bedded, whereas the underlying component characteristically comprises coarsening-upward, cross-bedded successions of deltaic origin. Two altitudinally separate groups have been mapped in the Findhorn Valley area; 'high-level' deposits occur east of Carn nan Tri-tighearnan, at elevations of between 450 and 560 m above OD, and 'low-level' deposits are present between 310 and 370 m above OD. A single glaciofluvial fan has been mapped on the Black Isle near Sligo [6705 5155].

DETAILS

High-level deposits

A thick coarsening-upward sequence of sand and gravel is exposed in sections cut by the headwater tributaries of Allt Breac to the south of Carn an Uillt Bhric. The deposits underlie a fan [836 384] 300 m wide, termed the Cnoc a'

Chinn Leith Fan-delta by Auton (1990a), the upper surface of which slopes at about 10° towards the east. At its western end, the upper surface of the fan-delta stands at 550 m above OD. This corresponds to the 'upper limit of the Findhorn Glacier' as shown by Horne (1923, fig. 3), who suggested that the sand and gravel formed a 'moraine terrace' at the glacier margin. The proximal deposits of the fan-delta are poorly stratified clast-supported coarse gravel (topset beds), generally 2 to 3 m thick. These rest with a sharp erosional contact on large-scale, planar cross-bedded, coarse-grained sands (foresets). The topsets locally reach 12 m in thickness and trough cross-bedding is developed towards the top of the sequence [8345 3827]. The foresets thicken eastwards and crop out in a cliff section 20 m high [8371 3850], where they pass down into tabular cosets of fine- to medium-grained sand, with ripple-drift lamination and silt drapes, dipping at 10 to 12° eastwards. The sand with silt drapes passes down into horizontally laminated beds of sandy and clayey silt (bottomset beds). The sedimentary sequence as a whole indicates deltaic deposition within a deep lake that stood at about 540 m above OD, ponded by ice to the east. Similar ponding, between 550 and 560 m above OD, is indicated by a coarsening-upward deltaic sequence exposed in the upper reaches of Meur Bheòil [8375 3947].

Low-level deposits

A glaciofluvial fan [855 373] is also present on the south-eastern side of the Findhorn Valley, standing between 74 and 100 m above the level of the river floodplain. Close to its apex (at 330 m above OD), the upper surface of the feature slopes 5 to 8° towards the north-west. Discontinuous exposures in its frontal bluff show that the fan deposit is about 30 m thick and rests on bedrock; the steeply sloping bluff is probably an ice-contact slope. A section about 4 m high [8541 3747] shows that the upper part of the deposit is a clast-supported cobble gravel. Lesser sections lower in the sequence also expose gravel, indi-cating that the deposit does not conform to the classic Gilbert-type deltaic model (Fairbridge, 1978, p.240). Gently sloping terraced features [851 385] and the **Kincraig Terrace** [837 363] at an elevation of about 320 m above OD (Auton, 1990a) have also been interpreted as glaciofluvial fans, and a small terraced spread of sand and gravel [846 364] at about 365 m above OD, on the interfluve between the Allt a' Choire Bhuide and Findhorn valleys, was possibly laid down as a fan adjacent to decaying ice in the two valleys.

Black Isle

The upper surface of the glaciofluvial fan [670 517] at Sligo on the Black Isle rises to about 50 m above OD. The feature, which was trimmed by the sea in late-glacial times, stands approxi-mately 23 m above the adjacent raised beach deposits. The apex of the fan lies just south of the farmstead where coarse- and fine-grained gravel and sand appear to pass downslope into whitish very fine-grained sand [6699 5172] and gravelly silt [6668 5177].

Glaciofluvial Sheet Deposits

Sheet deposits, classed as 'Fluvio-glacial sand and gravel' on older maps of adjacent districts, are found only south of the Inverness Firth, where they accumulated in four con-trasting depositional environments during the deglaciation of the district. The oldest deposits formed as ice-marginal spreads on the high ground after hilltops were free of ice. In the Nairn and Findhorn valleys, a second group of sheet deposits forms an extensive series of kame-terraces at lower altitudes. These are best developed on the south-eastern flanks of the valleys where they interlink, and are cut by, numerous ice-marginal glacial drainage channels. A third suite of sheet deposits is associated with eskers, particularly the Flemington Eskers; they accumulated proglacially after decay of the ice within which the eskers had formed. The fourth, and youngest, suite of sheet deposits includes the widespread terraces upon which much of Inverness is built. These features formed as outwash plains (sandar) and deltas that prograded into the late-glacial sea.

DETAILS

High-level spreads

Once the high ground around Carn nan Tri-tighearnan [823 390] was free of ice, sheets of sand, gravel and flow till were laid down in the upper catchments of the streams that now drain that area. The material accumulated as outwash fans against ice that still occupied the lower ground, and as fan-deltas within temporary ice-marginal lakes. These deposits are largely concealed beneath blanket peat.

A large spread of compact, pale olive-grey, silty, sandy gravel with distinct planar, subhorizontal stratification occurs in the catchment area of Allt Odhar [81 37], upstream of its confluence with Allt na Beinne. This laterally variable deposit, the Carn Monadh Gravel of Merritt (1990b), includes medium- to coarse-grained pebbly sand, clast-supported gravel and matrix-supported, silty, sandy, gravelly diamicton. Individual beds rarely exceed 0.5 m in thickness and the deposit as a whole is generally less than 10 m thick. Similar gravels, largely concealed beneath peat, are dissected by the headwaters of Allt Dearg [81 41] and by Allt Creag a'Chait [83 43].

Kame-terraces

Flights of narrow terraces occur on the south-eastern side of Strath Nairn downstream of Daviot. The lowest features are sometimes difficult to distinguish from fluvial terraces, but the higher ones interconnect with meltwater channels, include kettleholes, and are quite clearly glaciofluvial in origin. The highest terraces were laid down by meltwaters draining a temporary ice-dammed lake (or lakes) in the vicinity of Daviot and in Moy Gap, whilst ice blocked Strath Nairn downstream of Daviot Gorge. A remnant of the highest of these terraces occurs at Easter Craggie [7343 4031], and a small gravel pit [7371 4142] in a slightly lower-lying terrace near Mains of Daltulich revealed 3 m of coarse, clast-supported gravel. This deposit, which is probably typical of most terraces within this part of Strath Nairn, comprises subrounded to rounded clasts of gneissose psammite with granite, gneissose semipelite, quartzite, porphyry and sparse sandstone and conglomerate. As the ice wasted, meltwaters laid down kame-terrace gravels and cut drainage channels at progressively lower levels. Most of these

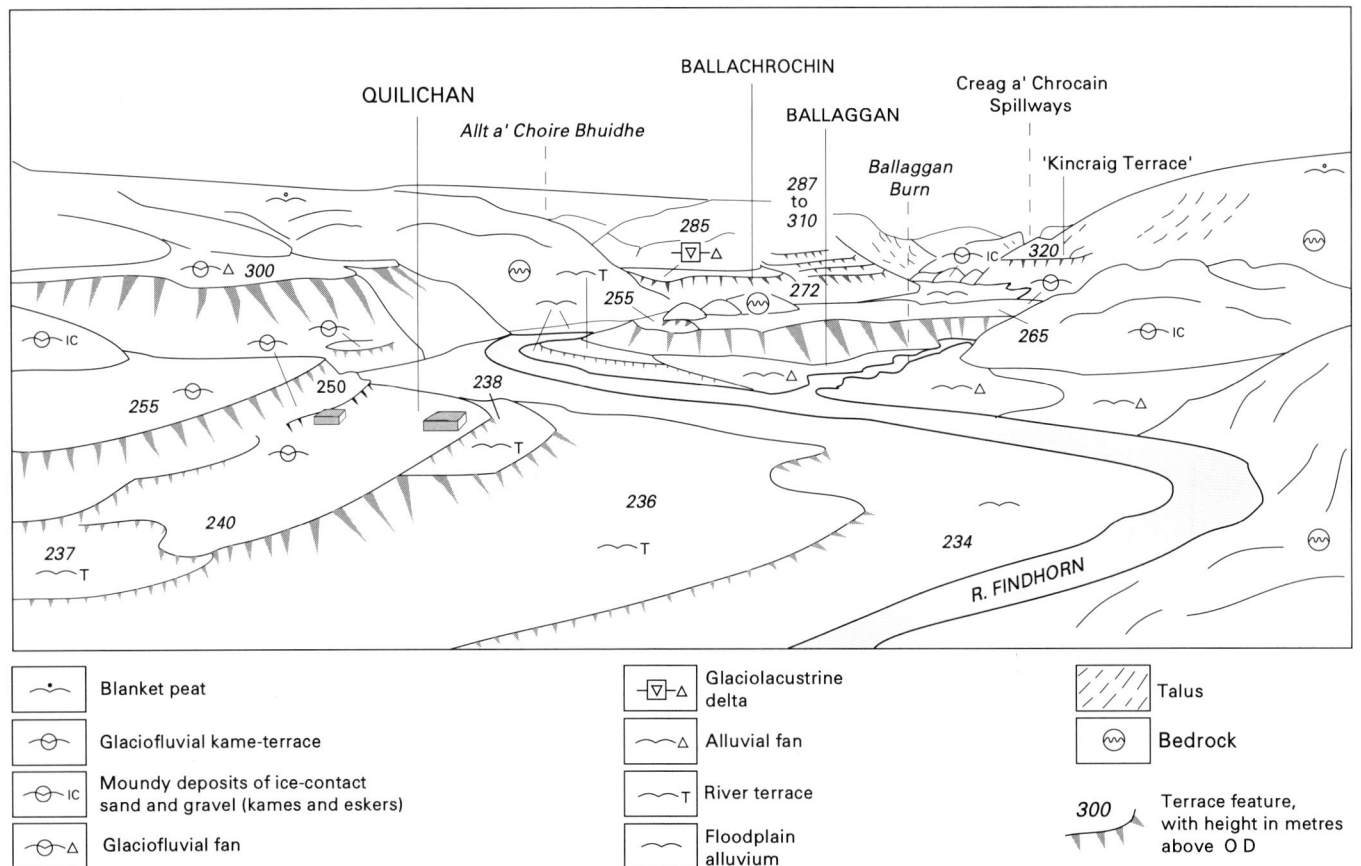

Figure 27 Landforms and sediments in the Findhorn Valley between Daless and Creag a' Chròcain (see also Plate 14).

Plate 14 The Findhorn Valley looking south-west from Daless (D4271). See also Figure 27.

terraces are underlain by cobble gravel; near Croygorston, over 2.5 m of boulder gravel with clasts up to 1 m long were exposed in a temporary excavation [7806 4523]. Sedimentary structures typical of braided outwash streams can be examined in a good trackside section [7654 4363], near Drummore of Clava.

Similar terrace features are developed near Cawdor [845 500], where they are also associated with ice-marginal drainage channels. Clast-supported cobble gravel, at least 2.5 m thick, is exposed in a small pit [8416 4968] in one of the higher spreads and up to 15 m of well-rounded cobble gravel overlying till are exposed in degraded sections in a terrace bluff [8308 4958]. Elsewhere, cobble gravel, resting on till and bedrock, forms extensive terraced spreads on the eastern side of the Allt Dearg valley between Dalcharn [811 452] and Kirkton of Barevan [838 472], and isolated terraces are preserved in the higher parts of the Dearg and Riereach catchments. Fragmentary kame-terraces also occur on both flanks of Strath Nairn upstream of Daviot, the most extensive and thickest deposits lying south of Milton of Farr [683 326].

A superb flight of kame-terraces occurs on the south-eastern side of Streens Gorge [835 355] (Plate 14 and Figure 27), at elevations between 287 and 310 m above OD . The features were largely interpreted by Horne (1923) as river terraces, although he realised that some of the highest were of glaciofluvial origin. The kame-terraces have almost planar upper surfaces strewn with angular blocks of psammite, commonly up to 0.5 m long; a small

kettlehole [8431 3642] occurs in one of the terraces. These terraces were formed by meltwaters draining along the south-eastern side of a mass of ice that temporarily blocked drainage along the valley during deglaciation, and also created a temporary lake upstream of Streens Gorge (Auton, 1990a). Thick deposits of sand and gravel also underlie a flight of three kame-terraces on the north-western side of the Findhorn Valley near Ballaggan [851 377] (Plate 14 and Figure 27). These terraces, which have upper surfaces at altitudes of 272, 265 and 255 m above OD respectively, occur downstream of a ridge of ice-abraded bedrock. Similar terraces occur at lower elevations on the south-eastern side of the valley at Quilichan. Kame-terraces, locally kettled, occur on both sides of the Findhorn Valley upstream of the Streens Gorge, and also flank Allt nam Meannan west of Ruthven [815 331], and Allt na Seanalaich west of Shenachie [827 349].

Sections [8249 3482 and 8240 3469] in the lower reaches of Allt na Seanalaich expose up to 4.1 m of clast-supported coarse gravel overlying dark yellowish brown lodgement till. Thicker gravels occur farther upstream; 9 m of ferruginous gravel overlie decomposed Moy Granite [at 8120 3476] and up to 13.2 m of gravel overlie a till that rests on decomposed granite [at 8112 3475].

Terraces flanking the Flemington Eskers

Extensive terraced spreads of sand and gravel occur on the southern side of the Flemington esker system between Tirfogrean

[805 504] and Meikle Kildrummie, and also extend farther north to reach the Late Devensian marine limit near Drumdivan. Terraces south-west of Wester Lochend [819 523] are clearly of glaciofluvial origin, being pitted with small kettleholes and dissected by south-west-trending glacial drainage channels.

The origin of the terraces between Meikle Kildrummie and Drumdivan is less certain (Auton, 1992; Gordon and Auton, 1993). Many workers, notably Ogilvie (1923), Gregory (1926), Small and Smith (1971), Smith J S (1977) and Synge (1977a) have argued that the features are marine terraces or raised beaches formed during periods of relatively high sea level soon after deglaciation. Marine erosion was also said to have trimmed the esker ridges near Meikle Kildrummie. In particular, Gregory recorded 'raised beach material', comprising sand, sandy shingle and coarse boulder gravel resting against a degraded face of esker gravel in a small pit (no longer visible) near Meikle Kildrummie. He noted that boulders within the 'beach' were coated in places with calcium carbonate, which he attributed to the dissolution of included shells.

The recent mapping, however, supports the views of Firth (1989a; 1990a) that the terraces in the Meikle Kildrummie area are of glaciofluvial origin, laid down from meltwaters that drained across the ground after the ice in which the eskers formed had melted. The stratification of the 'beach material' illustrated by Gregory more closely resembles that in nearby river terraces and kame-terraces, rather than that displayed in sections through undoubted raised marine and beach deposits exposed seaward of the Late Devensian marine limit. The presence of calcium carbonate may be due to the dissolution of calcareous Old Red Sandstone clasts.

Glaciofluvial outwash terraces at Inverness

The low-lying ground of Inverness city centre is backed by an abandoned cliffline that truncates glaciofluvial terraces upon which much of the southern part of the city has been built. These terraces developed during the deglaciation of the area, when meltwaters flowed from a glacier in the Great Glen and the glacier front lay a short distance south-west of the city. A remnant of the highest terrace at about 45 m above OD underlies Hilton (**Hilton–Lochardil Terrace**). This feature slopes northwards and formed while relative sea level in the Inverness area stood at about 34 m above OD (Figure 24a, ILG-3A) and possibly as high as 37 m above OD (Firth, 1990a). The much more extensive **Raigmore Terrace** grades towards shoreline remnants underlain by a veneer of raised marine deposits capping the abandoned cliffline at about 31 m above OD (Smith J S, 1968, 1977; Synge 1977a; Synge and Smith, 1980; Firth, 1984, 1989a). The Raigmore Terrace, which is cut by lower terraces locally, emerges from a large meltwater channel near Torbreck [648 409] immediately beyond the western margin of the district, and descends north-eastwards (Figure 24a, ILG-4A). Large-scale foresets were formerly exposed in a sand and gravel pit [6822 4562] near Raigmore, indicating that the terrace deposit is partly of deltaic origin. Boreholes drilled along the line of the re-aligned A9 road provided evidence to show that these deltaic deposits rest on dense, matrix-rich gravels (see cross-section B–B' on the 1:50 000 geological map) that formed earlier, possibly as a sub-aqueous marine ice-contact fan. These older gravels (Alturlie Gravels Formation), unbottomed at a depth of 48 m in a borehole [at 6838 4506], appear to be banked against a concealed escarpment or channel margin lying beneath Raigmore Hospital; the feature is cut in bedrock and slopes steeply towards the north-west. Gravels forming the terrace hereabouts are 9 to 10 m thick, but thin rapidly south-eastwards.

Good sections in terrace deposits are scarce, but the material appears to be mainly sand and subrounded to rounded gravel. The deposits fine towards the north-east, where they grade laterally into Late Devensian raised tidal flat deposits comprising very fine-grained sand and silt (see below).

GLACIOMARINE DEPOSITS

Two formations of raised glaciomarine deposits have been recognised in the district (Figure 28). They occur along the southern shores of the firths, between Inverness and Nairn, and extend inland towards the Late Devensian marine limit. The Ardersier Silts Formation mainly comprises silt, very fine-grained sand and diamictic mud, whereas the Alturlie Gravels Formation is mostly sand and gravel. Where the two formations are juxtaposed, they generally form a single coarsening-upward sequence and the boundary between them is gradational. Both formations are locally overlain by nonglacial marine deposits of Late Devensian age. The successions were probably deposited during the deglaciation of the district, when ice in the Inverness Firth had retreated rapidly south-westwards by calving, then re-advanced as a tidewater glacier. In general, the Ardersier Silts represent the bottomset beds of marine deltas and subaqueous fans that prograded in front of the tidewater glacier during still-stands and minor readvances. The Alturlie Gravels are the foreset and topset beds of such bodies.

Ardersier Silts Formation

Sediments of this formation form much of the abandoned Main Flandrian Cliffline between Inverness and Nairn, and thin southwards. The type area on Ardersier Peninsula has been studied for more than a century and its landforms and deposits have featured in most reconstructions of the Quaternary history of the area (Jamieson, 1874; Wallace, 1883; Ogilvie, 1914; Horne, 1923; J S Smith, 1968, 1977; Small and Smith, 1971; Synge, 1977a; Synge and Smith, 1980; Firth, 1984, 1989b, 1990b). Smith and Synge both cited geomorphological and sedimentological evidence for a major readvance of ice between Inverness and Ardersier (the 'Ardersier Readvance'), but their evidence, together with their correlation of the event with the Perth Readvance (Sissons, 1967) was refuted by Firth (1989b). However, the recent resurvey has established that a minor readvance did occur, but that the date of the event was earlier than proposed by Smith and Synge (Gordon and Merritt, 1993); here it is referred to as the Ardersier Oscillation.

DETAILS

There are few permanent exposures at which the Ardersier Silts can be examined. A type section through the lower levels of the formation has been established in an old clay pit [7849 5647] 450 m south-east of Kirkton, which was dug out and investigated in 1990. The section displayed a sequence 8.5 m thick of thinly interbedded and interlaminated clay, silty clay and silty fine-grained sand. As is typical of other exposures in this formation, couplets (cyclopels of Cowan and Powell, 1990) of

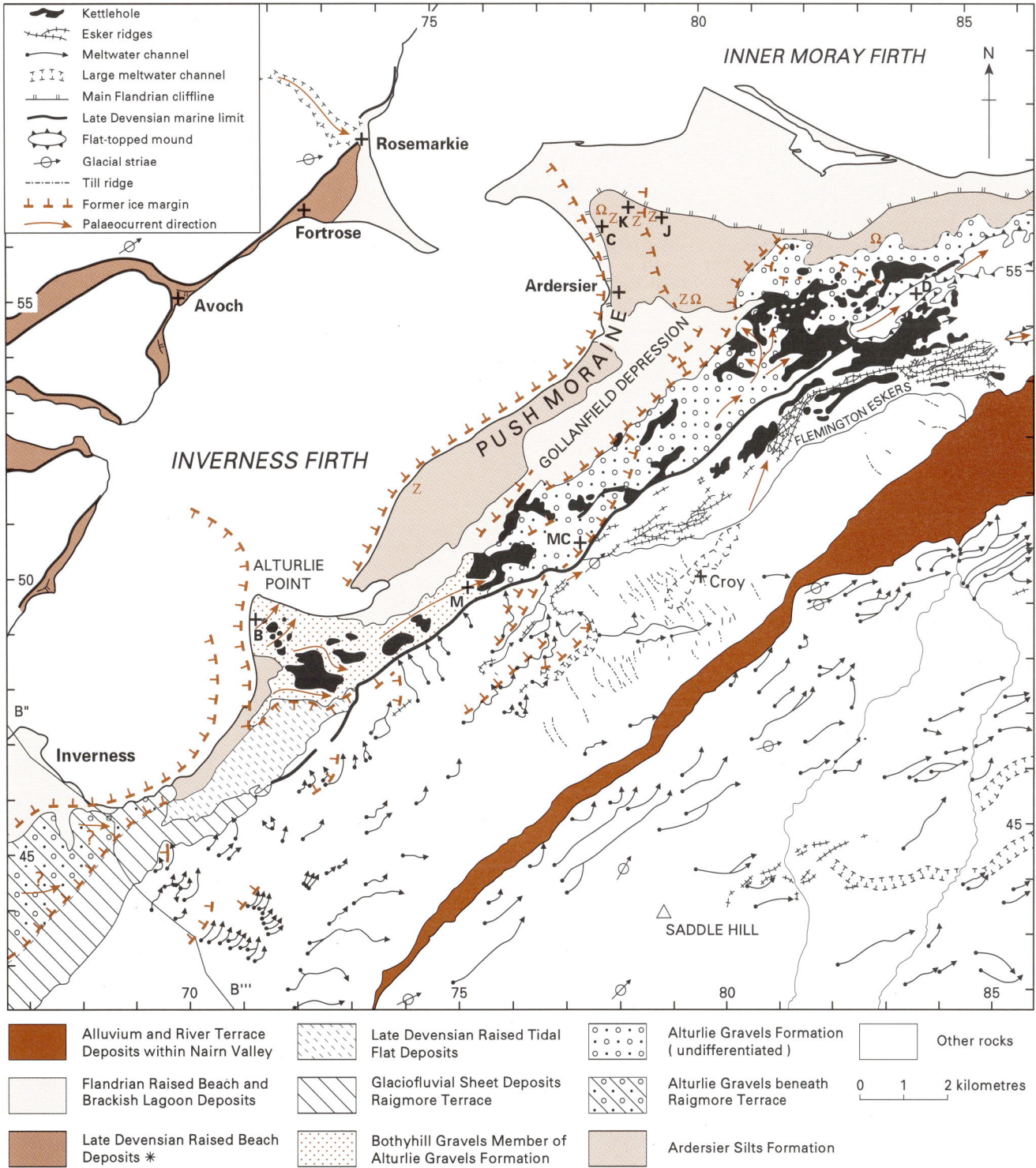

Figure 28   Features and deposits associated with the retreat of the Main Late Devensian Ice Sheet and interpreted former ice margins in the northern part of the district.

Legend

- Kettlehole
- Esker ridges
- Meltwater channel
- Large meltwater channel
- Main Flandrian cliffline
- Late Devensian marine limit
- Flat-topped mound
- Glacial striae
- Till ridge
- Former ice margin
- Palaeocurrent direction

Map labels: INNER MORAY FIRTH, INVERNESS FIRTH, Rosemarkie, Fortrose, Avoch, Ardersier, PUSH MORAINE, GOLLANFIELD DEPRESSION, FLEMINGTON ESKERS, ALTURLIE POINT, Croy, Inverness, SADDLE HILL

Key:
- Alluvium and River Terrace Deposits within Nairn Valley
- Flandrian Raised Beach and Brackish Lagoon Deposits
- Late Devensian Raised Beach Deposits *
- Late Devensian Raised Tidal Flat Deposits
- Glaciofluvial Sheet Deposits Raigmore Terrace
- Bothyhill Gravels Member of Alturlie Gravels Formation
- Alturlie Gravels Formation (undifferentiated)
- Alturlie Gravels beneath Raigmore Terrace
- Ardersier Silts Formation
- Other rocks

0   1   2 kilometres

B Bothyhill Pit, **C** The contorted silts of Ardersier, **D** Drumdivan Pit, **J** Jamieson's Pit, **K** Kirkton Section, **M** Morayhill Pit, **MC** Mid Coul Pit, Ω Glacitectonic structures, Z Water-escape structures

* Most deposits occurring to the south of the Inverness Firth are not shown for cartographical reasons

B" ___ B'''   NW part of line of cross section B - B' on 1:50 000 geological map silt grading upwards into clay are stacked into discrete packages (0.3 to 0.8 m thick); the latter may be varves. Such bedding is typical of ice-proximal glaciomarine sediments (Mackiewicz et al., 1984). Even the more massive beds of silty clay contain wispy laminae of silt. The deposits described above coarsen upwards and are overlain unconformably by pebbly fine-grained sand; this sand underlies a raised beach with its back-feature at 21.6 m above OD (Firth, 1989a).

Sediments at the Kirkton Section dip gently towards the south-west, but at Jamieson's Pit [7939 5616], 950 m to the south-east, there is evidence of severe postdepositional disturbance. The latter site, re-excavated in 1991, is the type section of the **Baddock Till**, which comprises 2 m of stiff, stratified stony clay diamicton. This till rests unconformably on over 6.6 m of silty, fine- to medium-grained micaceous sand with sparse seams of clay. The sands immediately beneath the till have been moved along subhorizontal thrust planes, while frozen, to form a stack of thrust slices (schuppen), whereas lower in the section the sands have been homogenised, folded and dislocated along steeply dipping clay-lined thrusts. This assemblage of glacitectonic structures indicates that the sandy deposits were pushed towards the south-east and then partly overridden by a glacier that deposited the Baddock Till; whether the till accumulated subglacially or as debris flows at the ice front is not clear.

No fossils were recovered during re-excavations of the Kirkton Section and Jamieson's Pit, but Jamieson's (1874) collection of sparse shell fragments included Arctic mollusc species from a deposit of clay at the latter site. Wallace (1883) noted that specimens examined by Jamieson included *Nuculana pernula*, *Macoma calcarea* and *Tridonta elliptica*. Robertson (in Wallace, 1883) identified *Astarte sulcata* and several species of ostracod and foraminifera. The full faunal list is given in Appendix 4. Until the shell-bearing clay discovered by Jamieson is relocated and shown indisputably to be in place, or other fossiliferous deposits are found, a glaciomarine origin for the Ardersier Silts cannot be established beyond doubt. As none of Jamieson's collection of shells has been traced, there are no specimens available for radiocarbon or amino-acid dating.

Syndepositional deformation in the form of flame structures, dish-structures, ball-and-pillow structures and convolute bedding commonly affect graded beds of silt and very fine-grained sand within the Ardersier Silts. These structures appear to become less common north-eastwards, so were probably formed by processes associated with the glacial advance rather than being the result of palaeoseismicity (cf. Davenport and Ringrose, 1987; Ringrose, 1989). Spectacular syndepositional convolute bedding and load structures, together with postdepositional folding and thrusting, are displayed in a cliff section [7803 5598] behind the village of Ardersier (Gordon and Merritt, 1993; Plate 15). These 'Contorted Silts of Ardersier' (Firth, 1990b) comprise beds of silt 0.2 to 0.4 m thick interbedded with thinly laminated fine-grained sand, laminae of clay and lenses of fine gravel. Well-developed asymmetrical ripple-drift cross-lamination indicates south-westward and south-eastward palaeocurrent flow (not north-eastward as stated by Firth, 1989b and 1990b). The beds exposed in the Ardersier cliff section are folded into two asymmetrical synclines separated by an asymmetrical anticline. The folding has an amplitude of about 3 m, with a wavelength of approximately 6 m. The axis of the anticline plunges 16° towards 150°. J S Smith (1977) concluded that the contorted silts formed part of the subaerial terminal push-moraine of the Ardersier Readvance, but Firth maintained that the folding resulted from subaqueous slumping close to a tidewater–glacier margin. Firth's conclusion was based mainly on the supposed absence of major thrust structures, and the supposed

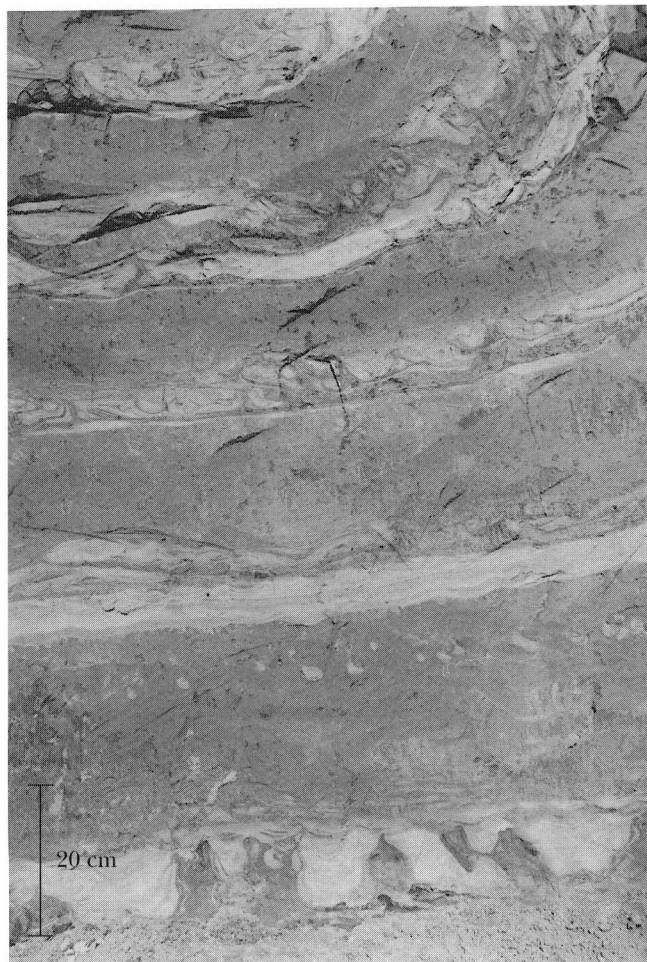

Plate 15   The contorted silts of Ardersier: glacitectonic folding of subaqueous soft-sediment deformation (ball and pillow) structures developed in glaciomarine silts. Cliff section behind Ardersier [7803 5598] (D4818).

absence of till overlying the contorted beds, features considered to be typical of large push-moraines (cf. Moran, 1971; Banham, 1975). However, thrusts do indeed occur within the contorted silts described above; also, in the core of the anticline, the hinge and lower limb of the fold are displaced north-eastwards along a thrust plane that dips steeply towards the south-west. Furthermore, at a nearby cliff section [7821 5565] at Hillhead, a thin bed of till, which may be a lateral equivalent of the Baddock Till, caps a sequence of very compact silty sands cut by numerous low-angle thrust planes. Early thrusts, which dip 10 to 20° towards the south-west, are cut by later dislocations dipping 3 to 5° towards the north-east; the former are the product of ice-push from the south-west, whereas the latter are probably due to subsequent gravitational sliding and collapse.

Other sections displaying convoluted graded beds of silt and very fine-grained sand within the Ardersier Silts occur along the abandoned cliffline backing the Carse of Delnies [8322 5590 and 8286 5570]. Shear planes within silts at the base of the former exposure indicate that these sediments also have been pushed towards the east. In an old sand pit [7939 5453], 550 m north-east of Viewhill, 2 m of deposits typical of the Ardersier Silts and displaying splendid ball-and-pillow structures grade upwards into 2.5 m of pebbly sand ascribed to the Alturlie Gravels Formation. The latter is capped unconformably by a thin, discontinuous bed of diamicton, possibly flow till. A broadly similar sequence 16 to 17 m thick, has been proved in site-investigation boreholes sunk for a new road bridge [8209 5350] crossing the Inverness–Nairn railway line near Gollanfield, where sand and gravel of the Alturlie Gravels grade downwards into silt resting on bedrock. The silt is ascribed tentatively to the Ardersier Silts.

The land backing the coastline between Ardersier and Castle Stuart [742 497] is unusual because, for the most part, it is highest close to the Main Flandrian Cliffline and shelves inland towards the Gollanfield Depression (Figure 28). Synge (1977a) and Synge and Smith (1980) interpreted this broad, elongate coastal ridge as a lateral moraine formed during the Ardersier Readvance. On evidence gathered during the resurvey, it is more likely to be a continuation of the push-moraine forming the Ardersier promontory, particularly as glacitectonic thrusts affect deposits exposed in a cliff section [7422 5120] near Balnaglack. This stretch of cliffline is formed mainly of pebbly mud and stiff, stony, stratified diamicton interdigitating with deposits more typical of the Ardersier Silts. The matrix of the diamictons is predominantly whitish silt and fine-grained sand, in which angular to well-rounded clasts up to 30 cm in diameter are generally well dispersed; in places they are concentrated in horizontal clusters and lenses. The clasts are composed mainly of psammite with schistose semipelite, quartzite, sandstone and gneissose semipelite. Diamictic mud with dropstones occurs near the top of one good section [7395 5067].

Sections revealing the Ardersier Silts occur in a disused sand pit [7105 4720] near Milton of Culloden, midway between Inverness and Alturlie Point. Diamictic deposits similar to those described above appear to pass upwards into plastic pebbly clay overlain by fine- to medium-grained sand. The latter includes beds of finely interlaminated silty sand, clay, and pebbly muddy diamicton. The sequence is capped here by cemented gravel of a raised beach. Towards Inverness, the Ardersier Silts seem to pass laterally into stiff, stony, stratified, matrix-supported diamicton; this overlies well-rounded, matrix-rich gravel in an old pit [6963 4600] near Seafield. Site-investigation boreholes around Kessock Bridge [665 475] proved unbottomed deposits of stiff pebbly silty clay, possibly of glaciomarine origin, beneath gravels ascribed to the Alturlie Gravels Formation (see cross section B–B' on the 1:50 000 geological map).

Alturlie Gravels Formation

The Alturlie Gravels Formation lies seaward of the Late Devensian marine limit, forming a broad spread from Alturlie Point north-eastwards to the eastern margin of the district. It is difficult to subdivide stratigraphically, owing to a lack of sections, but includes many thin patches of blown sand, beach gravel and silt in addition to deltaic sand and gravel. It also includes marine-planated glaciofluvial gravels, especially towards the Flemington esker system, where the Late Devensian marine limit is itself difficult to place locally.

Bothyhill Gravels Member

The gravels forming Alturlie Point and extending inland to Morayhill [752 489] contain clasts of a distinctive granitic augen gneiss derived from the Carn Chuinneag–Inchbae Complex in central Ross-shire, some 40 km north-west of Inverness (Wilson and Shepherd, 1979). The presence of these erratics, which have almost

certainly been carried by ice via the Beauly Firth (Horne and Hinxman, 1914; Sissons, 1967; Figure 23), allows these gravels, named here the Bothyhill Gravels Member, to be distinguished from the rest of the Alturlie Gravels Formation in this district, although they are not shown separately on the 1:50 000 geological map.

DETAILS

The type section of the Bothyhill Gravels in Bothyhill Pit [715 491] at Alturlie has been described in detail by Merritt (in Firth, 1990b). The uppermost 10 m or so of the pit section reveal well-sorted, subrounded to well-rounded, clast-supported cobble-gravel that is locally openwork, but generally has a matrix of silty sand. The gravels, composed mainly of psammite and quartzite, are capped by up to 1.5 m of gravelly diamicton filling north-east-oriented channels. The gravels are mainly massive, but cross-stratification in lenses of sand indicates a north-eastward palaeocurrent flow. In 1989, the lowermost 10 m of the pit revealed loose shingle interstratified with thickly bedded medium- to coarse-grained sand, disposed in very large-scale, tangentially based, deltaic foresets dipping to the east-north-east. There is evidence of both syndepositional and post-depositional slumping and faulting, the latter being related to the formation of kettleholes.

The sequence does not resemble that of a typical Gilbert-type delta, because no distinctive topset gravels overlie the foresets (cf. Fairbridge 1978; Clemmensen and Houmark-Neilsen, 1981). Indeed, the silty matrix of much of the gravel and the presence of channel-constrained, unstratified deposits of pebbly sand are more typical of subaqueous fans than deltas (Cheel and Rust, 1982; Nemec and Steel, 1984; Powell, 1990). Deposits similar to, but more sandy than those exposed in the lower part of the pit at Bothyhill, are found also at Morayhill Pit [7546 4961]. The palaeocurrent indicators there also point to east-north-east-directed flow, but there is no evidence of any postdepositional collapse.

Alturlie Gravels, undifferentiated

No erratics of Inchbae Rock granitic augen gneiss have been found in gravels of the Alturlie Formation lying north-east of Morayhill. In that region, it is probable that the Alturlie deposits formed a little earlier than the Bothyhill Gravels, in the period before ice from the Great Glen had retreated sufficiently to allow meltwaters derived from ice flowing out of the Beauly Firth to enter the area. A good exposure in these older gravels is at Mid Coul Pit [774 502], where over 8 m of dense, clast-supported gravel with subordinate lenses of sand have been worked. The gravels occur as massive horizontal beds, 30 cm to 2 m thick, comprising subrounded to well-rounded clasts up to cobble size, mainly of micaceous psammite and quartzite. Channels filled with cross-bedded sand or sandy gravel are most prevalent at the top of the sequence and indicate palaeocurrent directions towards the north to east-south-east sector.

Near Milton of Balnagowan [816 550], Firth (1989a; 1990b) recognised a flight of raised shorelines with levels at 9.8, 11.8, 12.8 and 22 m above OD, culminating at a shingle ridge [812 548] with a crest lying at 24.9 m above OD. These features, and any beach deposits associated with them, have not been distinguished from the Alturlie Gravels on the 1:50 000 geological map. A roadside section [8145 5505] in a flat-topped mound revealed 2.3 m of gravel in horizontal and cross-bedded units, 20 to 30 cm thick, intercalated with thin beds of sand. The clasts are moderately sorted, generally well rounded and imbricated in places, but the depositional environment (glaciofluvial or littoral) is unclear. The Alturlie Gravels that crop out north-east of Castle Stuart [742 497] and on Ardersier Peninsula are probably also of mixed genesis, being partly glaciofluvial and partly littoral in origin. However, unlike the deposits near Milton of Balnagowan, they are mostly thin.

The mounds of sand and gravel lying north-east of Mid Coul represent a major resource of aggregate, and intervening hollows and flats are mostly underlain by fine-grained sand that passes downwards into silt and clay; this silt and clay unit overlies gravels in a well [8136 5356] dug in low-lying ground at Pooltown. These lower gravels are known to be at least 11.5 m thick in boreholes sunk for the nearby Gollanfield road bridge. Trimming of the gravel mounds is evident [823 542] near Tomhommie, where flights of marine benches are cut into the sides of several mounds at elevations between 20 and 27 m above OD. Such benches are best developed on the north-eastern side of each mound, facing the direction of maximum fetch along the coast; particularly fine examples occur [820 540] near Tomhommie.

Kettleholes, up to 8 m wide and 1.5 m deep and infilled with silty clay, are exposed in cross-section in a small working in the Alturlie Gravels [8240 5496], 700 m north of Tomhommie. The silty clay is hard, pale olive-brown to yellowish brown, contorted and extensively faulted. In undisturbed faces, the gravel, which is interbedded with yellow fine-grained silty sand, forms tabular cross-bedded units with channelled bases. Individual beds of gravel are up to 1 m thick and moderately to very well sorted. Clasts, which reach cobble size, are generally well rounded and predominantly comprise psammite and brown sandstone. Close to kettlehole rims, the stratification of the gravels has been disturbed by slumping and faulting; these can be attributed to ice collapse, which has produced some vertical fabrics.

MARINE DEPOSITS

Apart from sediments occurring at depth offshore and coastal to sea-bed marine deposits, two distinct sets of marine deposits are preserved in the district. The older, Late Devensian, set formed during late-glacial times, that is, from the initial decay of the Main Late Devensian Ice Sheet in this region until the end of the Loch Lomond Stadial (Gray and Lowe, 1977). These marine deposits are intimately associated with glacial and glaciofluvial sediments and are backed by discontinuous degraded cliff-lines that relate to a series of shoreline remnants lying at up to 35 m or more above OD around Inverness (Figure 24a, ILG-1 to 10). These shorelines are commonly truncated by the Main Flandrian Cliffline, which is almost continuous along the southern coast of the Inverness and Moray firths, and backs the younger, Flandrian (post-glacial) set of raised marine deposits and associated shore-lines (Figure 24b, IF-1 to 5). The latter were largely formed during the postglacial transgression, which culminated between 5775 and 7100 years ago in the Inverness area (Cullingford et al., 1991) and reached about 8 m above OD.

Both series of raised shorelines have been tilted isostatically towards the north-east, the younger, lower-lying ones being tilted less than the older ones (Figure 24). Deposits relating to both series have been subdivided lithologically into shoreface and beach deposits (mainly shingle and sand) and a quiet-water facies (mainly fine-grained sand and silt). The latter are interpreted to be tidal flat deposits in the case of the Late Devensian set, and brackish lagoon deposits in the case of the Flandrian set. The deposits of a raised marine fan-delta of Late Devensian age have also been mapped.

Pleistocene sediments occurring at depth offshore

Little information is available regarding Quaternary deposits within the Inverness Firth. Limited seismic data from a small area towards the north-eastern end of the firth reveals some buried channels with complex infills. The thickness of the channel-fill deposits is locally over 50 m and several seismo-stratigraphical units can be recognised, ranging from acoustically transparent to well layered. Acoustic basement is either an opaque unit presumably of glacial material, or rockhead with an irregular surface. Acoustic blanking on some sections suggests that gas is present locally within the sediment.

In the Moray Firth, boreholes in the Nairn Basin, a Quaternary basin complex that lies just north-east of the district, cored diamictons overlain by soft muds (Chesher and Lawson, 1983; Andrews et al., 1990) ranging in age from Late Devensian to Flandrian. Sediment thickness within the Nairn Basin is variable, but in places is over 70 m. However, limited seismic and sample data suggest that Quaternary deposits beneath the part of the firth lying within this district consist mainly of till (Chesher, 1984) only locally exceeding 10 m in thickness. Fine- to medium-grained sand with shell fragments and thin interbeds of silt and gravel was recorded in boreholes sunk on intertidal flats prior to the development of the Ardersier platform construction yard. The sandy deposits pass downwards into grey sandy and clayey silt with interbeds of sand. The sequence, which was unbottomed at a maximum recorded depth of 30.5 m, probably includes Late Devensian marine deposits, but a more precise stratigraphical subdivision is not possible on the available evidence.

Late Devensian Raised Marine Deposits

RAISED MARINE FAN-DELTA DEPOSITS

A large remnant of a former marine fan-delta composed of sand and gravel occurs on the western side of Chanonry Peninsula on the Black Isle. Its upper surface slopes towards the south-west, indicating that its apex lay at the entrance to Rosemarkie Glen [736 588] (Figure 29). Only the south-western part of the original fan-delta is preserved, and this has been cut into benches and covered for the most part by Late Devensian shoreface and beach deposits. Its upper surface forms a gently sloping planar feature stretching from Rosemarkie [7355 5762] to the railway bridge at Bishop Road, Fortrose [7234 5650].

RAISED TIDAL FLAT DEPOSITS

Spreads of fine-grained sand, silty clay and clayey silt infill broad topographical depressions in the Alturlie Gravels on the southern side of the Inverness Firth, particularly between Inverness and the Alturlie Peninsula, and to the east of Ardersier. These fine-grained deposits, which are

Figure 29
Glacially
sculptured
benches and
raised
shorelines
on the
Black Isle.

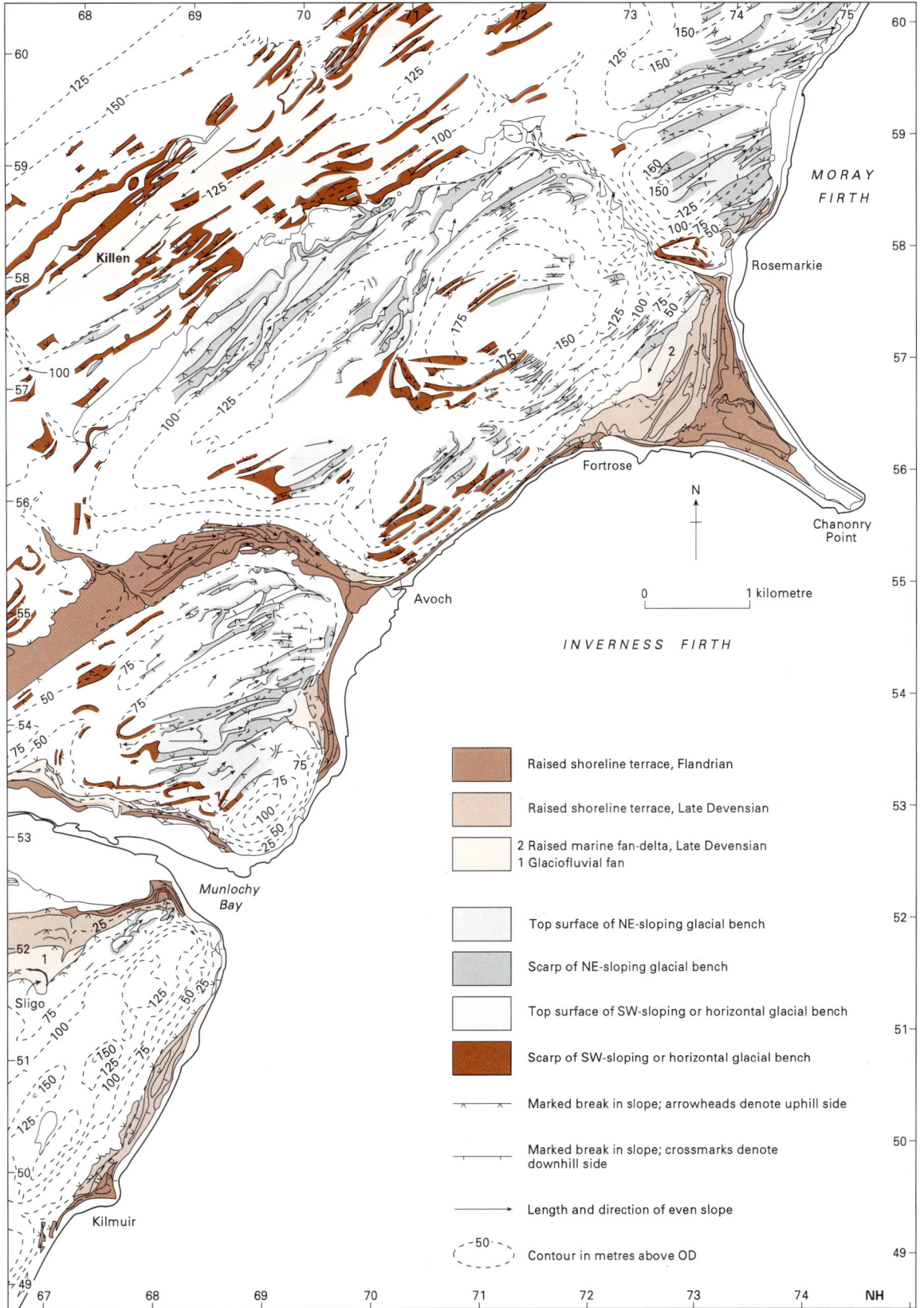

Raised shoreline terrace, Flandrian

Raised shoreline terrace, Late Devensian

2 Raised marine fan-delta, Late Devensian
1 Glaciofluvial fan

Top surface of NE-sloping glacial bench

Scarp of NE-sloping glacial bench

Top surface of SW-sloping or horizontal glacial bench

Scarp of SW-sloping or horizontal glacial bench

Marked break in slope; arrowheads denote uphill side

Marked break in slope; crossmarks denote
downhill side

Length and direction of even slope

Contour in metres above OD

MORAY
FIRTH

Rosemarkie

Fortrose

Chanonry
Point

Avoch

INVERNESS FIRTH

N

0   1 kilometre

Killen

Munlochy
Bay

Sligo

Kilmuir

93TF14AC

generally less than 5 m thick, mostly infill kettle-holes (shown in black in Figure 28) that developed within the gravels while relative sea level was falling rapidly. Although palaeontological evidence has yet to be found, the similarity of these fine-grained deposits to those underlying present-day tidal flats in the district, and their association with raised marine shorelines, point to a lagoonal, tidal or intertidal origin.

DETAILS

An extensive spread of fine-grained raised marine deposits stretches north-eastwards from Inverness towards Alturlie Peninsula (Figure 28), where it forms flattish ground, gently rising to about 35 m above OD within the Late Devensian marine limit. The deposits, mostly pale olive-grey, massive clayey silt or silty clay, appear to interdigitate with sands and gravels forming the glaciofluvial Raigmore Terrace. As the latter is partly deltaic in origin, it follows that the silts probably accumulated in shallow marine/estuarine conditions in front of the delta. Samples of the silt contain leaf cuticle, wood, pollen and fungal remnants, but as yet there is no conclusive evidence for a marine origin.

Exposures are scarce, but ditch sections south of Allanfearn [716 475] revealed whitish-weathering, blocky jointed, stiff, massive, very silty clay containing well-dispersed subrounded to rounded pebbles up to 40 mm in diameter. These deposits, typical of many in the area, are overlain by up to 1 m of ferruginous sand and gravel filling shallow channels. The silts and clays are probably generally less than 10 m thick, but the only available borehole information is of deposits close to their attenuating margin. A borehole [7214 4679] drilled 300 m north of Culloden House, for example, proved 1.25 m of silty sand on 2.25 m of blue-grey clayey silt on 0.5 m of dark grey, silty, clayey, fine-grained sand (unbottomed). Broadly similar deposits were dug for brick-making in a pit [7270 4790] near Balloch, where 6.7 m of clay with subordinate layers of sand are reported to have overlain at least 10.7 m of sand (Horne, 1923).

Most of the flat-lying ground between 20 and 25 m above OD that backs the Main Flandrian Cliffline between Ardersier and the eastern margin of the district is underlain by about 1 m of fine-grained sand. The sand, which is probably mainly aeolian, is generally quartzose and well sorted, but micaceous silty units occur locally. Its colour ranges from dark orange or pale yellow, to white or pale grey. Beneath poorly drained ground, for example between Viewhill [791 541] and Tomhommie [823 542], the sand passes down into pale grey to olive-grey, soft to firm, silt or silty clay. This downward transition is exposed in a 50 m-long drainage ditch [838 551], 600 m south of Wester Delnies, where the lower micaceous sand and silt were augered to a depth of 2.5 m. The colour, structureless appearance and softness of the silt, which has also been dug from many drainage ditches between Pooltown [814 537] and Ballinreich [812 544], distinguishes it from the stiff, brown, laminated silt and clay alluvium that infills some nearby kettleholes.

RAISED SHOREFACE AND BEACH DEPOSITS

Remnant raised-beach deposits backed by degraded clifflines of Late Devensian age are common on both sides of the firths, and particularly well displayed along the southern shores. In general, the shoreline correlations made by Firth (1984; 1989a) and summarised in Figure 24 have been corroborated, although not all shorelines identified by him are associated with mappable deposits.

Black Isle The Late Devensian raised-beach deposits in this part of the district are mainly preserved along the coastline, but they also extend inland at Avoch, forming a gravelly terrace along the northern side of the Suddie–Rosehaugh Gap, a valley that stretches south-westwards from Rosehaugh [680 550] towards Munlochy Bay and beyond the western margin of the district. When late-glacial relative sea level stood at about 34 m above OD, the land to the south-east of this gap (Figure 29) formed an island. Sections along Suddie Burn [6696 5435, 6707 5443, 6717 5451 and 6805 5497] reveal a discontinuous cover of silt and sand with gravel lenses generally resting on cobble gravel.

The coastal raised-beach deposits are generally thin and rest on benches cut in till or bedrock. They comprise mainly gravels of varying grade, with a cover of sand or silt. In the south, a prominent raised beach with its backfeature at about 30 m above OD lies below East Craigbreck. Remnants of the same shoreline are preserved [6728 4971 to 6793 5058] behind Kilmuir, where it is 2 m higher. Two lower raised beaches occur in the same area, one falling from about 24 m above OD [at 6806 5080] to 20.4 m above OD [at 6820 5100], the other occurring to the north-east [6823 5094] at 14.4 m above OD (Firth, 1989a, table 2).

Along the southern side of Munlochy Bay from Craigiehowe to near Sligo [671 516], the highest raised beach, formed mainly of boulders in a silty sand matrix, has its backfeature at 35 m above OD. Individual raised beaches are not well developed hereabouts and bouldery material extends down towards the coast, probably having been transported downslope by erosion and cultivation. Remnants of three raised shorelines are recognisable on the northern side of Munlochy Bay. The highest is at 41 m above OD near Chapelhead [6831 5322], the second [at 6804 5326] has its backfeature at 37 m above OD and the third, the most prominent and extensive of the three, has its backfeature at 30 m OD.

A narrow, yet prominent, shoreline lying at about 35 m above OD stretches along the coast between Munlochy Bay and Avoch. It is underlain by 1 m of silt on gravelly sand and densely packed gravel in a good section [697 539] near Castleton. Farther north, the Fortrose–Rosemarkie–Chanonry Point peninsula is formed of both Late Devensian and Flandrian shoreface and beach deposits resting upon benches cut in till and Late Devensian marine fan-delta deposits (Figure 29). Four Late Devensian shorelines are recognisable. The uppermost, with its backfeature at 26 m above OD, forms the ground on which much of Fortrose is built, and is underlain by up to 2 m of coarse cobbly gravelly sand; this rests on till at the top of the cliff [7288 5631]. The next lower shoreline, at about 21 m above OD, stretches north-eastwards from just seaward of the Academy [730 563]. A third shoreline has its backfeature at about 15 m above OD [between 6320 5635 and 6353 5724]; sections at the southern end of this raised beach [7317 5631] exhibit 2.5 m of well-sorted, imbricated shingle. The lowest raised beach has been considerably destroyed by erosion and only two prominent benches are preserved [7355 5670 and 7358 5715], with backfeatures between 9 and 12 m above OD.

Two Late Devensian shorelines are preserved on the northern side of Rosemarkie. The higher one near Kincurdy House [7385 5820] has its backfeature at about 34 m above OD; the lower one at about 20 m above OD is a short bench [7410 5833] cut in rock with only a thin cover of pebbly sand.

Inverness to Castle Stuart At Inverness, the glaciofluvial Raigmore Terrace, upon which much of the southern part of the city is built, grades north-eastwards towards an abandoned shoreline at 27.5 m above OD (Firth, 1989a). This bench is underlain by a veneer of gravelly beach deposits that crops out along the top of the Main Flandrian Cliffline. A more prominent, lower-lying, shelving raised beach lies between Seafield [6975 4605] and Milton of Culloden [7075 4690]. This much dissected feature rises to about 15.8 m above OD (Firth, 1989a) and is underlain by up to 1.5 m of shingle locally bound by ferruginous cement. The beach, backed by a former storm-beach ridge [at 7000 4590], is capped by massive silty clay just east of Milton of Culloden, where a sample from a ditch section [7091 4690] yielded four cysts of the marine dinoflagellate *Operculodinium centrocarpum* and a single cyst of *Protoperidinium conicum* (Harland, 1988).

Fragmentary raised beaches and beach ridges occur at altitudes of up to 29.3 m above OD around Alturlie Point (Firth, 1989a), but few are sufficiently extensive to be shown on the 1:50 000 geological map. A remnant of a raised beach (now destroyed) between 20 and 25 m above OD at the entrance to Bothyhill Pit [7141 4922], comprised at least 50 cm of horizontally bedded, imbricated openwork gravel, the pebbles being well rounded, with good size and shape sorting. A similar deposit, lying at about 20 m above OD, occurs just north of the pit [at 7164 4935]. In the absence of fauna, the pockets of imbricated, discoidal or bladed, well-rounded clasts in these deposits are taken to indicate beach gravel (cf. Bourgeois and Leithold, 1984; Nemec and Steel, 1984); more specifically, they are features typical of gravel deposited in the 'imbricated zone' of berm ridges (Bluck, 1967). Both raised-beach deposits described above rest on erosion surfaces; they clearly postdate, and are not related to, the Alturlie Gravels exposed in the pit.

Ardersier eastwards   Some of the highest raised beaches in the district occur on Ardersier Peninsula, which probably formed an island until relative sea level had fallen to about 11 m above OD (Firth, 1989 a, b). The 2 km-long shingle ridge that passes beneath the cemetery of Ardersier Church [7907 5560] and reaches 28 to 31 m above OD, is generally taken to be a beach ridge (J S Smith, 1977; Synge, 1977a; Firth, 1989a, b), but its origin might result from glacitectonic thrusting associated with the Ardersier Oscillation. Firth has also recognised a staircase of shorelines lying between 9.8 and 22 m above OD west of Milton of Balnagowan [815 550], but as these shoreline terraces appear to be mostly erosional features, they are not shown as raised beaches on the 1:50 000 geological map.

Raised beaches formed of fine-grained silty sand are truncated by the Main Flandrian Cliffline between the road to the Ardersier platform construction yard and the eastern boundary of the district. The beaches slope gently seawards and terminate at steep bluffs 2 to 3 m high at their seaward margins; those near Wester Delnies [839 557] are narrow and backed by cliffs up to 8 m high. In contrast, the landward edge of the 200 m-wide raised beach [860 561] east of Easter Delnies is marked by an almost imperceptible break of slope.

Flandrian Raised Marine Deposits

RAISED BRACKISH LAGOON DEPOSITS

DETAILS

During mid-Flandrian times, the sea invaded the low-lying ground between Castle Stuart and Viewhill [791 541], here named the **Gollanfield Depression** (Figure 28), when higher ground along the coast at Fisherton formed an island. Shingle ridges around Treeton [783 538] lie on each side of the northern outlet of the former brackish lagoon; they extend to heights of between 8.2 and 9.6 m above OD (Firth, 1984).

The Gollanfield Depression, now largely occupied by Inverness Airport, is variably underlain by fine-grained sand, micaceous sandy silt and sandy peat, but very little is known of the underlying succession. Boreholes drilled for a runway extension [783 527] bottomed in medium- to fine-grained organic sand at depths of 2 m. A trial pit [7799 5326] sited on a terrace lying adjacent to the main flat revealed 35 cm of soil overlying a 2 cm-thick layer of shell debris of unknown origin resting on fine-grained sand.

RAISED SHOREFACE AND BEACH DEPOSITS

DETAILS

Black Isle   A series of terraces below 12 m above OD is well developed along the Black Isle coastline. In the south, at Kilmuir, two raised beaches are preserved. The higher one is present on each side of the village [at 6710 4945 and 6757 4985]; the lower one shelves towards the sea south of St Mary's Church. At the southern side of the entrance to Munlochy Bay [6825 5240], three distinct raised beaches formed of sandy gravel are found below 10 m above OD, but on the northern shore only the higher two seem to have been preserved. The best exposure [6890 5278] in the highest beach shows ochreous stony sand resting on over 3 m of cobble gravel with sparse boulders. An exposure in the deposits of the lower beach [6855 5302] shows 2 m of sand and shingle resting on 80 cm of grey sulphurous organic silt with rootlets and iron pan.

Two prominent raised beaches underlain by fine shingle occur between the shore at Ormond [6975 5338] and the mouth of Rosehaugh Burn [7005 5485], where they lie between 11 and 9 m above OD respectively. The lower raised beach has been worked for shingle [at 6993 5363] and a shore section [at 6990 5351] exposes 80 cm of shingle, on 40 cm of cobbles, on red till. The main road from Avoch to Fortrose runs along a low Flandrian shoreline for about 1 km [to 7130 5567], then rises to a higher shoreline that continues to a point [7234 5630] near Fortrose Harbour. The sediments forming both terraces appear to be mainly reworked pale red stony clay on cobbly gravel [as at 7151 5576].

On Chanonry Peninsula, three gravelly Flandrian raised beaches are recognisable (Figure 29). The highest forms a narrow bench with its backfeature sloping from about 11 m above OD at Rosemarkie [7372 5779] to about 9 m above OD near west Scorrielea [7360 5666]. A lower shoreline backs an extensive area of flat-lying ground [7378 5745 to 7418 5635]. The lowest beach forms a prominent flat below 5 m above OD between Wester Greengate caravan site [7355 5631] and the Links of Fortrose [7445 5588]; it is masked by blown sand near Chanonry Point. North of Rosemarkie, a narrow gravelly raised beach remnant with a backfeature at 11 m above OD rests on rock [7386 5808 to 7427 5856].

Inverness   Two low-lying terraces identified tentatively on the 1:50 000 geological map as Flandrian raised beaches underlie the north-eastern part of the city. The higher and less extensive of the two (**Millburn Terrace**) abuts against the abandoned Main Flandrian Cliffline at about 10 m above OD and is underlain near the city centre by at least 15 m of gravel, capped locally by deposits of silty clay and peat. The more extensive **Longman Terrace**, which underlies Longman Industrial Estate and rises from 2 to 3 m above OD at the coast, to between 4 and 6 m above OD inland (Firth, 1984), is underlain by cobble gravel.

Peacock (1977) suggested that the deposits underlying Longman Terrace are the topset beds of a delta that formed during the Loch Lomond Stadial, when relative sea level was at about the same as it is at present. Evidence from boreholes drilled prior to the construction of Kessock Bridge [665 475] indicates that the deposits reach 30 m in thickness, extend at depth across the Kessock Narrows, and consist of coarse gravelly sediment with little silt and clay. Both Sissons (1981a) and Firth (1984) maintain that such material could not have been associated with a delta that built up gradually, and suggest instead that it was formed as a single event during the Loch Lomond Stadial, when the 260 m above OD-level ice-dammed lake in Glen Roy–Glen Spean drained catastrophically through Fort Augustus, Loch Ness and the Ness Valley before debouching into the sea at Inverness. Drilling results confirm that the gravel thins towards the south and generally becomes less coarse in that direction (see cross section B–B' on the 1:50 000 geological map).

It is difficult to adjudicate between these alternative views on the origin of the deposit, but the presence of both vertical and horizontal variations in grading and a general absence of a distinct coarsening-upward sequence tend to support a non-catastrophic origin.

The gravel unit forming Longman Terrace commonly overlies a few metres of grey to black, organic silty pebbly clay with shells including *Chlamys islandica* and *Mölleria costulata*, both of which are cold-water molluscs absent from Scottish waters today; a microfauna comprising foraminifera and ostracoda similarly indicates cool, but not Arctic conditions (Graham, 1974a, b; Peacock, 1977). The shelly deposit appears to be partially reworked and overlies a lag gravel developed on an underlying deposit of brown, silty, fine-grained sand (see cross section B–B' on the 1:50 000 geological map). The latter is at least 30 m thick locally, and beneath Kessock Narrows rests upon an unknown thickness of stiff, grey, fissile clayey silt tentatively correlated with the Ardersier Silts. From the slope of rockhead (derived from borehole evidence), Quaternary deposits in the Longman area may reach 200 m or more in thickness (Peacock, 1977).

Inverness to Ardersier   The Main Flandrian Cliffline is fronted by a single, shelving raised beach along most of the coast between Inverness and Ardersier (Frontispiece). An excavation towards the back of the raised shoreline [7138 4867] revealed 70 cm of well-sorted, well-rounded shingle resting unconformably on cross-stratified sand of the Alturlie Gravels Formation; it is likely that the Flandrian beach deposit is generally thin. Remnants of higher Flandrian raised beaches and associated shingle ridges reaching about 10 m above OD occur seaward of Blackhill [7183 4807], at Alturlie Point and near Lonnie [7370 4913].

East of Ardersier   Tabular deposits of sand, silt and shingle flank the coast from above the level of Mean High Water Springs to between 7 and 8 m above OD. The most extensive spreads, up to 3 km wide, occur beneath the Carse of Delnies and the Carse of Ardersier, where they are partially concealed by blown sand. Narrow, higher raised beaches composed of sand and shingle occur adjacent to the Main Flandrian Cliffline. Four raised beaches occur near Muir of Balnagowan [817 556], but only three are recognised farther east.

Exposures of raised-beach material are sparse, but well-rounded shingle is visible beneath blown sand that blankets Flandrian deposits in a drainage ditch [8481 5655] at the western end of Nairn Golf Course. Shingle also forms beach ridges trending north-west (shown as lines on the 1:50 000 geological map) to the north-west of Baddock [797 561], where they are partially concealed beneath blown sand. Ridges 2 to 4 m high, trending east–west and composed of cobbles and shingle, form the eastern end of the Whiteness Head spit. The ridge material exhibits well-developed bimodal sorting, being predominantly a mixture of well-rounded cobbles (30 to 70 mm in diameter), and finer pebbles (5 to 10 mm grain size). The ridges are asymmetrical in profile, with steeper seaward-facing slopes; their orientation, shape and gravel composition are all similar to that of present-day storm-beach ridges nearby.

Present Marine Deposits

SALTMARSH DEPOSITS

Two saltmarshes are developed in the district: one lies along the southern side of Munlochy Bay [west from 6813 5253], where silty mud underlies grassy marshland; the other lies at the head of a creek [738 497] near Castle Stuart. The latter passes landwards into a freshwater peat bog.

TIDAL FLAT, SHOREFACE AND BEACH DEPOSITS

DETAILS

Black Isle   The most extensive area of tidal-flat deposition is in Munlochy Bay, where dark grey muddy silt is exposed along both shores at low tide. In Avoch Bay, tidal flats are over 300 m wide at low tide. Farther north, a small area of intertidal mud and sand lies between Fortrose Harbour and Craig an Roan skerry [7291 5614]. Elsewhere, the narrow intertidal zones are underlain by beach sand and gravel, especially at the top of the tidal range. Sea walls have been erected in places to prevent coastal erosion, notably at Rosemarkie. At Chanonry Point gravelly beach sand is stabilised by marram grass.

Inverness to Ardersier   Intertidal flats are widespread in the relatively sheltered bays on both sides of Alturlie Point, and in Ardersier Bay. Shingle generally forms a narrow fringing upper beach adjacent to extensive mudflats strewn with large angular boulders derived from glacial deposits, or deposited in lateglacial times by floating ice (Ritchie et al., 1978).

East of Ardersier   A sandy foreshore backed by a storm beach of coarse shingle extends from Whiteness Head, eastwards to a wave-cut platform formed of Nairn Sandstone [850 569]. East of the platform, the foreshore sediments are a mixture of sand and mud; similar sediments are presently accreting on the seaward margins of Whiteness Head spit, which is prograding north-westwards into the sea.

Tidal-flat deposits underlie a brackish intertidal marsh to the east of the platform construction yard at Ardersier. Ephemeral sections (up to 1 m high) in the banks of the adjoining creek expose finely laminated, olive-grey and orange sandy silt with *Arenicola* burrows. The silt is thinly interbedded with white and pale orange sand; lenses of pebble gravel are present locally below 80 cm depth.

SEA-BED SEDIMENTS

The sea bed is generally covered by a patchy layer of unconsolidated sediment less than 1 m thick, interspersed with partially buried lag deposits that were probably formed during the postglacial rise in sea level.

The thickest deposits are marked by bedforms such as sand waves. Size distribution and sediment transport are strongly influenced by the strength and duration of tidal currents, and more rarely by storm-induced wave and current action.

Sea-bed sampling within the district indicates that sand is widespread at the north-eastern end of the Inverness Firth and also in the Moray Firth (Hydrographic Office

1981, 1989). Side-scan sonar data from the north-eastern part of the Inverness Firth show areas of small sandwaves up to 1 m high with wavelengths of between 10 and 20 mm. In some areas, the bedforms are asymmetrical in profile and appear to indicate that sediment transport occurs both to the south-west and to the north-east, at least within this northern part of the firth.

The sediments are a mixture of terrigenous lithic and biogenic carbonate material. The terrigenous component consists partly of reworked material derived from underlying glacial and postglacial deposits and partly from the surrounding land areas through coastal erosion and river discharge. The mean annual sediment discharge from the River Ness, for example, is estimated to be about 22 000 tons per year (Reid and McManus, 1987). The biogenic component probably consists mainly of Holocene shell debris. Data from the Moray Firth imply that the carbonate content of the sediments beyond the nearshore zone is likely to be less than 10 per cent (Chesher, 1984).

A detailed study of modern sediments within the Moray Firth by Reid and McManus (1987) suggests that there is a net influx of material into the Inverness Firth, both from onshore and from offshore, and that all the small firths along the western side of the Moray Firth act as sediment traps. For example, the net effect of hydraulic activity on the relatively soft coastal deposits along the southern side of the Moray Firth is probably a long-term westerly transport of material towards the entrance of the Inverness Firth. There is also an influx of carbonate material into the district from sources farther offshore, with an estimated 80 000 tons of mud-grade carbonate material settling in the inner part of the Moray Firth per year. The same study estimates that something of the order of 460 000 tons of terrigenous material are discharged from rivers into the Moray Firth annually. Almost 25 per cent of this total is discharged into the small inner firths, where most of it remains trapped.

POSTGLACIAL TERRESTRIAL DEPOSITS

River Terrace Deposits

Dissected remnants of former floodplains flank the alluvium of most rivers and burns in the district. These remnants form terraces that slope gently down valley and are composed mainly of stratified clast-supported cobble gravel. The gravel was deposited as bars and channel infills by braided and meandering streams migrating across floodplains. Terrace aggradation probably occurred mainly during the Loch Lomond Stadial and early Flandrian times. The gravel is locally overlain by thin spreads of sand and silt, laid down as overbank deposits during periodic flood events. These fine-grained deposits are widespread on the broad terraces of the major rivers, where they produce well drained, light sandy soils.

In narrow upland valleys, such as those of Moy Burn, Riereach Burn and Allt Dearg, only the larger terraces can be shown on the 1:50 000 geological map; much of the mapped alluvium in these valleys is dissected to form flights of terraces, some rising 5 to 6 m above the level of the river or burn.

DETAILS

Terraces of the rivers Nairn and Findhorn   The broadest and most continuous spreads of river terrace deposits occur in Strath Nairn, downstream of the road bridge [743 430] between Blacktown and Castletown. A terrace lying 2 to 3 m above the floodplain underlies the archaeological site at Clava [757 444], where the stones of the chambered cairns and other piles of boulders derive mainly from the boulder gravel in the terrace; a borehole [7658 4490] drilled immediately downstream proved 7.9 m of coarse gravel on bedrock. For some 10 km downstream, the terraces shelve towards the river and appear to be mostly erosional features cut into till with little gravel on them.

Strath Nairn widens downstream of Kilravock Castle [814 494] and flat-lying terraces up to 500 m wide are developed on both sides of the river near Newton of Budgate [828 499]. Sections in the southern bank, downstream of the farm, show up to 6 m of horizontally bedded, rounded, boulder and cobble gravel. Dark brown, silty clay overbank deposits up to 2 m thick, resting on cobble gravel, are exposed in the upper part of a terrace bluff [8471 5211] on the northern side of the river opposite Rosefield.

The largest terrace upstream of Daviot Gorge is crossed by the re-aligned A9 road, where boulder gravel, at least 1 m thick, is exposed and capped by a veneer of silt. Another broad low-lying terrace underlain by cobble gravel lies close to Old Cross Cottage [6898 3489], beneath the new houses. The continuation of this terrace between Dalvourn [6880 3440] and Inverarnie Church [6875 3403] is strewn with large blocks of gneissose rock forming a train of boulders crossing the valley.

Remnant terraces are present on both sides of the Findhorn Valley. Fine examples occur near Ruthven [818 331], Shenachie [827 349] and Ballachrochin [847 368] and a flight of four terraces, between 2 and 6 m above floodplain level, is preserved on the eastern side of the river at Quilichan (see Plate 14; Figure 27). An excellent exposure of sand and gravel underlying a river terrace occurs on the south-eastern side of the valley [8171 3279] opposite Ruthven, where up to 7.5 m of imbricated, cross-bedded, well-rounded cobble gravel is truncated by a 1.5 m-deep channel infilled with medium- to coarse-grained sand, the upper part of which is truncated by a second channel filled with trough cross-bedded gravel. The highest part of the sequence is composed of up to 1.5 m of rounded-boulder gravel (with clasts up to 1 m in diameter) resting unconformably on the underlying deposits. Only the boulder gravel, however, was deposited by the north-east-flowing River Findhorn, because cross-stratification and imbrication within the underlying deposits indicate opposing palaeocurrent directions and must be of glaciofluvial origin.

Terraces flanking narrow valleys   Terraces occur in three areas along burns that drain the northern side of the Nairn–Findhorn watershed; in the Allt Dearg valley, between Dalcharn and Achindown [838 479], between Inchyettle [838 483] and the Allt Dearg–Riereach Burn confluence, and in the Riereach Burn valley around Glengeoullie [855 471]. Small fragmentary terraces are also mappable along the lower course of Moy Burn and in the Allt na Beinne valley. Good exposures of imbricated terrace gravel, overlying stiff reddish brown till, occur downstream of Dalcharn [at 8140 4542 and 8165 4548]; gravel terraces, resting directly on conglomerates of the

Inverness Sandstone Group, are well developed downstream of Inchyettle.

Black Isle The most extensive deposits lie in the Rosehaugh Burn valley and in the Suddie–Rosehaugh Gap, particularly immediately downstream of Rosehaugh Mains [681 553], where four distinct terraces are developed; the highest is probably composed of reworked raised Late Devensian marine deposits.

A dissected river terrace, standing about 6 m above the flood-plain, occurs in Rosemarkie Glen [7175 5910 to 7275 5843]. It is underlain by silt resting on cobbly gravelly sand over 1.5 m thick.

Alluvial Fan Deposits

Fans composed of sand, gravel and gravelly diamicton have accumulated since deglaciation, where tributary streams with steep gradients debouch into the major river valleys.

DETAILS

Several fans are present on the north-western side of the Findhorn Valley; two good examples occur near Ballaggan (Plate 14; Figure 27) and a third at Ruthven, at the mouth of Allt nam Meannan. A very large fan has formed where Moy Burn enters Moy Gap. This merges with another formed by Dalriach Burn to the north; both deposits become finer grained downstream, where they are mostly composed of silty coarse-grained sand. Braeval Burn, Allt na Loinne Moire, Allt na Slanaich, Tullochclury Burn and Allt a' Chuil also have formed fans within Moy Gap [at 7440 3922, 7715 3435, 7485 3425, 7860 3340 and 7840 3230 respectively]. Fans have formed within Strath Nairn, where tributary streams join the main valley [at 7400 4325, 7490 4365 and 7665 4495] and a piedmont fan has developed where streams debouch on to the glaciofluvial Raigmore Terrace [at 6840 4340], just south of Inverness. This feature, and another fan to the south-west [at 6765 4270] are formed of clayey clast-supported gravel.

Several alluvial fans are preserved in the Black Isle area, but are generally small. Two of the most prominent occur north of Fortrose. The older one underlies Wards farmstead [7270 5715] and is composed of a mixture of sand and red stony clay. It rests on the hillside, well above the present-day burn, and is flanked on its north-eastern side by a 15 m cliff in rock. This fan may have formed in an ice-marginal lake during deglaciation. Farther downslope, a much larger fan with a Late Devensian cliff 16 m high on its north-eastern flank, is situated at the mouth of a deep ravine [7266 5700] near Platcock; it comprises gravelly clay that oversteps the Late Devensian marine fan-delta.

Alluvium

The alluvial deposits in the district are of two types; fluvial deposits underlying the floodplains of rivers and burns, and lacustrine deposits within enclosed basins. Floodplain sediments are predominantly cobble and boulder gravels, whilst lacustrine alluvium is mainly fine- to medium-grained, humic sand, silt and clay infilling kettleholes and fringing small lochs.

DETAILS

Floodplain alluvium Wide ribbons of alluvium flank the river courses of the Nairn and Findhorn along most of their lengths, forming low-lying ground liable to flooding. Generally, only the top few metres are exposed in the river banks. These sections typically reveal clast-supported river gravel capped by overbank deposits comprising one or two metres of thinly laminated silty sand, locally intercalated with peat, as in the western bank [8281 3452] of the Findhorn. In general, the alluvial gravel is well rounded to angular and in places exhibits a pronounced imbrication dipping upstream; it is generally coarse, horizontally bedded and well sorted and forms linear bars where river channels bifurcate. Vegetated braid-bars, associated with abandoned river channels, are well developed on the eastern side of the Nairn downstream of its confluence with Cawdor Burn.

Thicknesses of alluvium are difficult to assess and likely to be very variable. For example, it is probable that only a few metres of alluvial sand and gravel overlie bedrock where the Findhorn passes through Streens Gorge, but 15.8 m of sand and gravel resting on till were recorded in Borehole 83SW4 drilled through the floodplain about 1 km upstream of Ruthven. It is likely, however, that at least the lower portion of the sand and gravel proved in the borehole is of glaciofluvial origin.

The fluvial deposits of the upland burns are similar in many respects to those of the main river valleys, although their flood-plains are much narrower, and the deposits generally thinner and coarser. The alluvium of Moy Burn within the Moy Gap, however, is probably thick and mainly fine grained. Augering has shown that the floodplain is generally underlain by interca-lated silt, fine-grained sand and peat unbottomed at 1.1 m.

On the Black Isle, only the Rosehaugh and Rosemarkie burns have floodplains of any great extent. The alluvium generally comprises a fining-upward sequence no more than 2 m thick with cobble gravel at the base.

Lacustrine deposits Flat-lying spreads of intercalated humic sand, silt and clay lie within poorly drained, semi-enclosed basins. They are most common within the undulating spreads of Alturlie Gravels and glaciofluvial ice-contact deposits lying between Alturlie Point and the eastern margin of the district. Most of these boggy hollows are large kettleholes, probably containing sequences of glaciomarine sediment concealed beneath the lacustrine deposits. The most extensive recent deposits occur within the 'Kildrummie Depression' [856 546] and flank Loch Flemington and the Loch of the Clans.

In contrast, lacustrine deposits on the Black Isle mostly occur within elongate ice-scoured depressions in bedrock, aligned parallel to the former north-easterly direction of ice flow. Many of these hollows would have been ponded until drained artifi-cially. The most widespread deposits occur at Killen [676 585], upstream of the narrow Killen Burn gorge, in the Bog of Shannon (now drained) upstream of the narrow defile [677 564] of Goose Burn, in the flat [705 576] near Wester Craiglands upstream of Carse of Raddery gorge [7085 5796], and in the flat [715 590] at the head of Rosemarkie Glen.

Shell Marl and Calcareous Tufa

Although nowhere exposed as an outcrop, a prominent development of white calcareous marl occurs in the Bog-giewell Burn valley, where over 1 m of marly material rests on peat [at 7013 5850]. Rare shell fragments have been noted.

Peat

Most of the high ground forming the Nairn–Findhorn watershed is blanketed by hill peat, but only the thickest and most widespread deposits are shown on the 1:50 000 geological map. Hill peat locally exceeds 3.5 m in thickness on the flanks of Carn nan Tri-tighearnan and has been eroded into peat hags covering much of the upland. Roots, stumps and broken trunks of Scots Pine and scattered roots of birch are found commonly within the deposit. In general, pine stumps occur at only one level, close to the base of the peat, but in places a second, higher layer of stumps is present. A single layer of pine stumps, close to the base of hill-peat overlying till, is well exposed in a cliff section [7978 3678] of Caochan Odhar. A former dense cover of pine trees is suggested by the abundant pine stumps exposed in deflation hollows in peat just beyond the eastern margin of the district, about 1.5 km south-east of Daless [861 385]. It is unlikely, however, that all of the trees were coeval.

Waterlogged peat overlies glacial and glaciofluvial sediments infilling ice-marginal and subglacial drainage channels on the interfluves. Pine stumps are sparse in these deposits as marshy ground hindered tree growth. Exposures are rare, but waterlogged peat has been augered to a depth of up to 1.3 m in the floor of many of the channels that trend east–west across the Allt na h-Athais–Allt Dearg–Riereach Burn interfluves. Peat also infills small basins on the lower ground. More-extensive spreads occur at Blàr nam Fiadh [835 535], Carr Bàn [670 365] and around Loch Bunachton [665 350]. Peat over 4 m thick was recognised in boreholes drilled for the construction of the new bridge [821 535] over the railway at Gollanfield west of Blàr nam Fiadh, and peat infills many of the kettleholes within the Flemington esker system and also around Cnoc-an-t Sidhean [7664 5108].

On the Black Isle, peat is mostly found as a capping to alluvium that fills poorly drained ice-scoured depressions, such as those on the Blairfoid flat [between 6910 5788 and 7045 5850, and at 7160 5713, 7180 5864 and 674 510].

Blown Sand

Thick deposits of blown sand, derived mainly from exposed shoreface and beach deposits, form extensive sand dunes mantling many of the Flandrian raised beaches east of Fort George. The dunes are generally 2 to 3 m high, but dune ridges 5 m high occur locally. In many instances, the dunes take the form of crescentic barkhans, with gentle slopes facing the direction of the prevailing wind and with steep faces to leeward (Horne, 1923). Blown sand also mantles raised-beach ridges composed of gravel, giving them a dune-like appearance. Linear ridges of both kinds are well developed in Carse Wood [815 565], and the Main Flandrian Cliffline is concealed by ridges of blown sand between Baddock [797 561] and Muir of Balnagowan [817 556], and at Easter Delnies [855 562]. The aeolian sand is predominantly quartzose, fine grained and well rounded: judging by the results of a detailed petrographical study of the nearby Culbin Sands (Mackie, 1899) it may also include a significant proportion of feldspar.

Apart from a previously unrecorded spread that lies about 1.5 km west of Milton of Balnagowan, the mapped extent of blown sand has changed little from that recorded during the primary survey, suggesting that there has been relatively little migration of the dunes in the last 70 years and that they have been stabilised by vegetation.

Dunes occur at two places along the coast of the Black Isle, on the promontory of Chanonry Point [745 560] and on the shore [7435 5885] below the cliffs at Flowerburn.

Head, including Downwash and Solifluxion Deposits

In this account, the term Head is applied to a thick weakly coherent mass of weathered bedrock or drift deposit prone to movement on waterladen slopes. On many of the steeper slopes in the district, weathered drift and bedrock have been mobilised by flash floods and seepages to form debris flows and ramp-like spreads of clayey lithology. Such material is common, but in this district only certain particularly thick spreads on the Black Isle have been delineated. Along the base of the cliffs below Flowerburn [at 7429 5860 and 744 592], screes intercalate with mudflows derived from clayey till and have been mapped as head. More commonly, however, head comprises the downflowings of unstable wet till alone, as at Rosehaugh Glen [6720 5560], or of gravelly sands, as at Rosemarkie Dens [7345 5775] and Bog of Shannon Wood [680 565]. In all cases, the resulting spreads have very different engineering properties from those of the parent deposits.

On the 1:50 000 geological map, deposits mapped as scree in the Findhorn Valley also include some head. They comprise frost-shattered regolith and till that have been reworked by debris-flow activity and by ephemeral streams to form steeply sloping debris cones at the bases of precipitous gullies. Many of the debris cones are still active. Fairbairn (1967) has discussed the general widening of gullies and the deposition of reworked debris during periods of high rainfall during the last 100 years or so. He outlined the initiation and growth of the 'Tirfogrean Gully' and its debris cone [865 387] on the eastern side of the Findhorn Valley opposite Daless, just beyond the eastern margin of the district. No trace of erosion is recorded here on the 1869 Ordnance Survey map, but a gully, somewhat smaller than that seen at present, is recorded on the 1903 edition. Similar, but less well documented examples are commonly developed on the steep drift-covered lower slopes of the Findhorn Valley and many less spectacular examples, too small to be shown on the 1:50 000 geological map, are present in tributary valleys across the upland parts of the district.

Scree

Scree deposits on the lower, steeply sloping, sides of the Findhorn Valley have formed by the progressive accumulation of rockfall debris beneath almost sheer rock cliffs, which are locally between 200 and 250 m high. Some of the talus deposits, notably those on the south-eastern side of Streens Gorge, may be relict features, formed immediately after deglaciation under periglacial conditions and again during Loch Lomond Stadial times (Ballantyne, 1991). The talus blocks are commonly over 2 m long and covered by patches of vegetated soil. At

Creag a' Chròcain [829 359], a zone of shattering along the Findhorn Fault has left the bedrock susceptible to freeze–thaw action and angular talus is still actively accumulating. This scree is composed of smaller blocks and is bare of soil and vegetation. Similar talus overlies diamicton exposed in gullies west of Kincraig [839 365].

On the Black Isle, conspicuous rocky screes are developed under the sandstone cliff adjacent to Avoch Harbour [7015 5513 to 7090 5542].

Landslips

Numerous landslips of various scales are developed on the Black Isle, most being associated with springs in solifluated drift deposits on steep slopes, as on the right bank of Craigbreck Burn [668 496 and 6690 4995] and on the southern side of Munlochy Bay [6810 5231]. In the Killen Burn valley, five major slips occur [between 6695 5744 and 6723 5612]; farther downstream, in the Rosehaugh Burn valley, slips in till are present [at 6730 5543, 6835 5505 and between 6858 5587 and 6893 5595]. Farther east, similar slips occur in the Shaltie Burn valley [7000 5590] and in the north in the Den Burn of Raddery [7158 5938]; major slips in till are also developed in Rosemarkie Glen [7248 5872, 7260 5840, and 7280 5825] as well as in nearby Little Red Den [7266 5812] and Muckle Red Den [7277 5797].

South of the Inverness Firth, numerous small landslips occur along the Main Flandrian Cliffline, particularly in the vicinity of Easter and Wester Fisherton and at the back of Ardersier village, but none is extensive enough to be shown on the 1:50 000 geological map. No major landslips have been identified farther inland.

Made Ground

Domestic and other waste is being dumped on the foreshore to the north-east of Inverness, where a substantial area of ground has been reclaimed. Areas of tidal-flat deposits were also reclaimed by tipping prior to the establishment of the platform construction yard at Ardersier.

Many of the sand and gravel workings are partially filled with waste; none of those shown on the 1:50 000 geological map has been reinstated fully. Many small areas of made ground are shown on the 1:10 000 geological maps of the district, including backfilled quarries, railway and road embankments, harbour installations and spoil tips.

FEATURES OF GLACIAL EROSION

Glacial striae and *roches moutonnées*

Evidence of glacial erosion, in the form of scratched and striated bedrock, indicates that ice has moved across the district in several directions, but that the predominant direction, presumed to be related to the Main Late Devensian glaciation, was towards the north-east (Figure 22). This direction of movement cuts across the tectonic fabric of the basement rocks, but lies parallel to the

strikes of the Old Red Sandstone and the major faults. Ice-plucked and ice-moulded bedrock features are best developed where ice moved along the grain of the crystalline basement rocks, as in Strath Nairn and at Creag an Daimh [834 443]; in the former area whole mountainsides take the form of *roches moutonnées*. The alignment of ridges of glacially abraded bedrock in the floor of major valleys, such as those of the Findhorn and Nairn, indicates that ice movement was also topographically controlled locally, particularly during the final phases of glaciation.

Few of the striae recorded by Horne (1923) and his co-workers during the primary survey are now visible, but those that are, together with several discovered recently, generally confirm the pattern of ice movement originally recognised. Sparse striae aligned north and north-north-west, on high ground in the southern part of the district, were recorded by Horne and taken to indicate ice movement from the south; the resurvey provided no confirmation of these observations.

Good examples of *roches moutonnées* occur in Strath Nairn [at 7130 3758, 7215 3826, 6803 3322, 6830 3447 and 7746 4445]; excellent striae are exposed at the top of Daviot Quarry [7168 3906 and 7172 3924] and crag-and-tail features can be seen in the same general area [720 352 and 724 367]. Ice-plucked and abraded ridges of bedrock at Tom na Slaite, Creag a' Chròcain and at a locality [850 372] north-east of Ballachrochin indicate former ice movement north-eastwards along the Findhorn Valley.

On the Black Isle, north-east-oriented striae on a *roche moutonnée* are well exposed [6718 5100] just north-west of Pitlundie.

Glacial meltwater channels

Channels cut by glacial meltwaters, many of them not associated with any significant modern drainage, are very common throughout the district. Three genetic types can be recognised: subglacial channels, ice-marginal channels and proglacial spillways. Although the best examples of each type are distinctive, many have a long and complicated history, possibly spanning more than one glaciation, and cannot easily be categorised.

SUBGLACIAL CHANNELS

This type of channel typically formed after the glacial maximum, but while even the highest ground in the district remained buried beneath ice. Subglacial meltwaters were constrained to flow parallel to the direction of regional ice movement, i.e. towards the north-east, irrespective of the subglacial topography. In general, the only evidence of this early suite is where meltwater channels are preserved cutting across interfluves that lie at an oblique angle to the direction of ice movement.

Subglacial drainage channels cut into bedrock, such as those at Cnocan Mór [834 348] on the eastern side of the Findhorn Valley and those on the flanks of Carn a' Mhàis Lethain [843 410], are generally steep-sided, winding features. They have steep, undulating long profiles, may branch and reunite repeatedly, and commonly cut across the strike of geological structures. Channels of this type are also well developed between Dallaschyle Wood [823 486] and Newlands of Budgate [820 475]. More rarely, subglacial channels link with englacially formed eskers that cross valleys between the interfluves; a good example is the Allt a' Chuil Esker [7680 3160], which links with a subglacial meltwater channel [at 7596 3050] just south of the district (Young, 1977, 1978, 1980).

Subglacial channels cut into drift are characteristically broad, open features, with a more gently sloping, undulating long profile; many are partially infilled with glaciofluvial deposits and till, in places capped by peat. This category is typified by a drainage channel south of Creag an Daimh [833 442], which is 2.2 km long and trends east–west. Excellent examples also occur at Dalcross [774 482] and Croy [796 498].

As the ice sheet thinned, meltwater became increasingly directed towards preglacial valleys which remained buried beneath the ice (such as those now occupied by the rivers Nairn and Findhorn), and which acted as major conduits. Examples of such channels occur near Daviot Wood [720 415] and between Beinn nan Cailleach [725 325] and Meall Mór [738 355].

ICE-MARGINAL CHANNELS

These channels formed once the highest ground in the area was free of ice, and meltwaters flowed between the ice-free uplands and the ice that still occupied low ground. As the ice sheet shrank, channels formed at progressively lower altitudes. Such staircases of channels occur near and north-east of Lower Muckovie [707 436]. A much more extensive system of deep ice-marginal channels occurs along the southern side of Strath Nairn downstream of Daviot Gorge. These channels are associated locally with hummocky glacial deposits of the Cantray Suite. Unlike subglacially formed channels, in which meltwaters flowed uphill in places (as in a syphon) to produce undulating profiles, the ice-marginal channels generally have regular long profiles, sloping gently towards the north-east. They cut across minor interfluves and typically are aligned at right angles to the modern drainage; they may be occupied locally by misfit streams.

PROGLACIAL SPILLWAYS

The two narrow linear channels up to 30 m deep and 300 m long at Creag a' Chròcain [829 359], on the western side of Streens Gorge, are spillways. They were cut by proglacial drainage of meltwater, derived at least in part from a temporary ice-dammed lake ponded at Shenachie within the gorge (Auton, 1990a). In contrast to the subglacial and ice-marginal drainage channels described above, the Creag a' Chròcain spillways have a pronounced V-shaped cross-section, and interlocking spurs of bedrock protrude from their sides. Towards their downstream ends, the floors of both spillways slope steeply north-eastwards, but at their upstream ends, the floors of both have gentle gradients. The Kincraig Terrace deposits (Auton, 1990a) were laid down as a glaciofluvial fan by meltwater that debouched from the lower of the two spillways.

On the Black Isle, Rosemarkie Glen and the Suddie–Rosehaugh Gap also acted as major spillways during deglaciation, although the former was almost certainly in existence before the onset of the Late Devensian glaciation.

Ice-sculptured benches

Subglacially sculptured benches and terraces occur on the Black Isle and on the south-eastern side of Strath Nairn between Meall Mór [7445 4088] and Drummournie [8245 4650]. Those on the Black Isle (Figure 29) are divisible into two sets, one set sloping towards the south-west, the other towards the north-east. The origin of these features is not entirely clear, but their north-easterly alignment, parallel to the direction of ice movement deduced from glacial striae and ice-moulded landforms, indicates a subglacial origin. The features are developed on till as well as bedrock, so appear to have been formed by combined processes of glacial erosion and deposition. Benches sloping south-westwards coincide with an uphill direction of ice flow, whereas those sloping north-eastwards were developed by ice flowing downhill.

Benches in Strath Nairn are generally much larger features than those on the Black Isle, but are similarly aligned. They shelve towards the main valley as well as having a gentle down-valley profile. The lower benches are underlain in part by hummocky glacial deposits of the Cantray Suite and, as they are intimately associated with ice-marginal drainage channels, they probably formed at the ice margin during deglaciation. The higher ones depicted on the 1:50 000 geological map are underlain by up to 40 m of glacigenic deposits, mostly till, and were mainly formed by subglacial accretion.

Glacial erratics

The general north-easterly direction of ice movement deduced from striae and streamlined landforms is corroborated by the distribution of erratic blocks, of which there are many fine examples in the district. The detailed descriptions of erratics made by local observers (Fraser, 1879; Wallace, 1883; Jolly, 1880; Mackie, 1905) were summarised in the earlier memoir (Horne, 1923, pp. 101–103) and placed in a wider context by Sissons (1967). The north-easterly movement of ice out of the Great Glen is recorded by the north-eastward carry of erratics from the outcrop of grey Foyers Granite (Figure 23) across the district, but the ice-stream evidently turned eastwards towards Elgin. This movement is also recorded by the similar distribution of boulders of the pink, fine-grained Abriachan Granite and conglomerate derived from Stratherrick, the latter containing distinctive pebbles of liver-coloured quartzite. The distribution of erratics of the Inchbae Rock granitic augen-gneiss across the Black Isle and in glaciomarine fan-delta gravels at Alturlie Point indicates that ice from central Ross-shire deflected the ice that streamed out of the Great Glen, causing it to flow along the southern coastlands of the Inverness and Moray firths. The widespread occurrence of erratics of Old Red sandstone, siltstone and limestone in tills across high ground south of the firths (Figure 23) shows that this southward deflection of ice emanating from the Great Glen was more pronounced during an early stage in the Late Devensian glaciation and in at least one previous glaciation.

On the Black Isle, blocks of Inchbae granitic augen-gneiss are scattered between Hillockhead [7456 6003] and Blairfoid [6814 5805], west of Rosemarkie, and between Avoch and Munlochy Bay. A train of large erratic blocks of gneissose rock derived from south-west of Loch a'Chlachain [665 323] crosses Strath Nairn between Ballone [6715 3275] and Dalvourn [6880 3440], forming an impressive feature. Farther down the valley, a large boulder of gneiss called Clach na h-Ulaidh [7371 4234] rests on a crag overlooking the river near Nairnside. Notable erratic blocks of the conglomerate described above form the Brownie Stone east of Loch Bunachton; Tom Riach [7789 4550], south-west of Cantraydoune; Cumberland's Stone [7498 4526] near Culloden; Clach na Sanais [7947 4921] near Croy; and the Grey Stone [8415 4890], near Cawdor.

FEATURES OF MARINE EROSION

Abandoned clifflines and the Late Devensian marine limit

Distinct sets of abandoned clifflines are associated with Late Devensian and Flandrian raised shorelines (Figure 24; Firth, 1989a). The most prominent feature, the Main Flandrian (Postglacial) Cliffline extends almost continuously from behind Inverness city centre eastwards, around the Alturlie and Ardersier peninsulas, to beyond the eastern boundary of the district (Figure 28). This major cliffline backs Flandrian raised beaches and truncates fragmentary Late Devensian raised beaches resting on deposits of the Ardersier Silts and Alturlie Gravels. The Main Flandrian Cliffline is also present, but less extensive, on the Black Isle coast, where it backs the Flandrian raised beaches at Kilmuir, around Munlochy Bay, at Avoch and on Chanonry Peninsula. Although the Main Flandrian Cliffline was certainly trimmed at the climax of the postglacial transgression (Haggart, 1986), stretches of it are definitely older features (Sissons 1981; Firth 1984; Firth and Haggart, 1989), as is shown by cross-section B–B' on the 1:50 000 geological map, which shows that deposits of Late Devensian age are banked up against it.

Degraded, fragmentary Late Devensian (Late-glacial) clifflines occur on both sides of the Inverness Firth, but these, unlike the Flandrian ones, are most prominent on the Black Isle coast, where the best examples occur between Rosemarkie and Fortrose [at 7280 5727], Castleton [6958 5408], on the northern side of Munlochy Bay [6700 5360] and at Kilmuir [6778 5033].

South of the firths, Late Devensian clifflines back raised beaches between Inverness and Alturlie Point, and on Ardersier Peninsula. Other less well-developed clifflines, not shown on the 1:50 000 geological map occur as far inland as the Late Devensian marine limit (Figure 28). The latter is the highest level that the sea is thought to have reached locally during deglaciation. It reached this level diachronously, as the ground was progressively uncovered by the receding ice. Only rarely can the marine limit be precisely identified; usually no more may be inferred than that the limit lies no lower than the highest local marine deposit or shoreline feature.

NINE

Economic geology

METALLIFEROUS MINERAL DEPOSITS

Little metalliferous mineralisation is known in the district. At surface, notable occurrences of manganese, iron and copper have been recorded, but the limited investigations carried out so far indicate little economic potential.

Although no major mineral surveys have been undertaken, the geochemical atlas of the Great Glen (British Geological Survey, 1987) highlights the element anomalies of this district. These anomalies may provide the foci of future exploration, especially those coinciding with the base of the Old Red Sandstone, where concentrations of manganese and baryte are apparent.

Manganese and iron

The only well-documented occurrence of metalliferous mineralisation in the Fortrose district is a manganese deposit at Dalroy [766 448] discovered in 1919 (Wallace, 1919, 1922) in Dalroy Burn close to the basal Old Red Sandstone unconformity [7665 4481]. Wallace reported the disposition of manganese ore extending 1.5 miles both to east and west of the discovery site. A shaft sunk 'twenty five to thirty yards east' of the original site revealed 'five feet' of manganese ore resting on 'seven feet' of haematite (Horne, 1923, p.111), and another trial shaft 'fifty yards east of Dalroy Burn' [7670 4481] encountered 'seven feet' of manganese ore above 'eight feet' of haematite ore. The haematite ore was reported to 'die out against the schist' across the shaft (Dewey and Dines, 1923). Boreholes drilled in 1922–23 [766 449 and 766 447] encountered manganese ore within the basal Old Red conglomerates, although the deposit there also varied in thickness and grade over short distances.

More recent detailed investigations in the Dalroy area were undertaken by Nicholson (1983), Parnell (1983) and Mulgrew (1985). Parnell (1983, p.205) considered the occurrence as an enriched zone within a ferricrete developed within a topographic depression. Nicholson (1983, 1987) described the geochemistry and 'complex mineralogy' of specimens from the spoil of the old trial shaft. He considered that the mineralisation was of hydrothermal origin and resulted from precipitation of Fe-Mn ore, either on to a land surface, or into an aqueous environment. Mulgrew, however, interpreted the mineralisation as the result of hydrothermal alteration of lithified Old Red conglomerates. He related the veining in Saddle Hill Granite, and the wide-spread occurrence of baryte near the unconformity, to this hydrothermal activity. Minor traces of alluvial gold in Easter Town Burn were also tentatively related to the Fe-Mn mineralisation. Both Mulgrew and Nicholson found

little evidence for the extensive spread of blocks of manganese ore in the vicinity that was reported by Wallace.

Further exploration has been undertaken by BGS as part of its Mineral Reconnaissance Programme and shallow boreholes have been drilled [7684 4506, 7677 4512 and 7722 4533]. This work is still in progress (Coats, in preparation). Disseminated manganese oxide grains are a common feature of ultracataclasite veins in the Rosemarkie Metamorphic Complex (p.00).

Bog-iron ore occurs within Quaternary sediments in the low-lying ground east of Gollanfield. Iron ore has been reported [at 814 537 and 823 535]; the recent resurvey, however, has not confirmed the latter occurrence. The presence of iron within waterlogged alluvial and marine silts and clays between Easter Glackton [825 535] and Pooltown [814 537] is indicated by seepages of chalybeate groundwater. The iron-bearing strata probably extend beneath the peaty moss at Blàr nam Fiadh [835 535], but their overall extent is unknown.

Mulgrew (1985) reported the occurrence of minor pyrite veins in the Old Red Sandstone in Cawdor Burn and on the foreshore near Nairn, although the localities were not specified.

Haematite and pyrite are common as disseminated grains in both fenitised breccia veins and ultracataclastite veins in the Rosemarkie Metamorphic Complex.

Copper

Green malachite patches within red arkosic Old Red sandstones are prominently displayed on the track [7274 5739] to Platcock on the Black Isle. Their proximity to the faulting in the Rosemarkie Glen area may indicate copper mineralisation in that region. Small amounts of native copper have been recorded from dump material at the Dalroy manganese trial site (Nicholson, 1987).

Baryte

Mykura (1980; in Gallagher, 1984) has described a baryte deposit at Balfreish [796 467]. It occurs in the form of veins and lenticular patches within limestone and breccio-conglomerate at the base of the Old Red Sandstone succession. 'A large lenticular mass' of baryte lies at the base of a 3 m-thick limestone, whereas the upper part of the baryte contains small angular felsite clasts and coarse baryte crystals. Breccio-conglomerate overlying the limestone is cut by a horizontal lenticular 'vein' of coarse pink baryte 30 cm thick that passes laterally into breccio-conglomerate with a baryte matrix; this is overlain by finer-grained breccia with no apparent baryte. The quarry described by Mykura contains many loose blocks generally composed of 'breccia with a limestone matrix forming up to 60% of the rock volume', although there are also limestone layers devoid of clasts. Coarsely crystalline baryte, comprising less than one per cent of the rock volume, forms patches up to 5 cm by 15 cm within the matrix. Limestone adjacent to the baryte is described as having a 'porcellanous texture'.

A more recent quarry lies some 30 m north-east of that just described. About 6 m above the base of the south-western face, there is a 2 m-high exposure of breccio-conglomerate with limestone matrix containing an irregular, near-horizontal zone of baryte marked by coarse aggregates of radiating crystals up to 20 cm long. The thickness of the baryte varies over short distances from 0 to more than 1.8 m, whilst the overlying breccio-conglomerate contains up to 15 per cent baryte in coarse patches within the matrix.

More recent trial diggings in breccio-conglomerate to the north-east of the old quarry and BGS boreholes at Dalroy have revealed a baryte matrix forming up to 50 per cent of the rock together with irregular patches exclusively composed of baryte. Total baryte content may be of the order of 20 to 30 per cent with subordinate limestone.

The Old Red Sandstone at Balfreish Quarry was deposited close to a steep north-west-facing pre-Devonian hillside composed of felsite, which now has thin baryte films along its joints. Mykura did not speculate on the origin of the baryte, although he noted that the matrix baryte is not a later infill of spaces between clasts, since the breccio-conglomerate is matrix supported. Mulgrew (1985), however, in a brief description of the Balfreish baryte, considered that mineralisation postdated lithification of the conglomerates.

Minor baryte veining and baryte in the matrix of breccio-conglomerates are widespread close to the basal Old Red Sandstone unconformity, for example in Cawdor Burn [843 491]. Baryte also infills fractures in psammite below the unconformity in Rierach Burn north of Glengeoullie Bridge [856 477].

On the Black Isle, baryte blades are conspicuous in an Old Red sandstone near Mount Pleasant Farm [at 7143 5676] without any sign of local faulting.

BUILDING STONE

As the costs of quarrying, dressing and transporting natural stone have increased, so the importance of stone as a building material has diminished. In the past, there were over 30 functioning stone quarries in this district, but none is now working and many have been at least partially backfilled or are flooded. All such quarries may be regarded as holding reserves, especially important should renovation of major local buildings be undertaken. By far the most used stone, and the most easily worked, is sandstone within the Old Red Sandstone succession, although various older rocks have also been used. The following is a list of all known stone quarries:

Rosemarkie Metamorphic Complex — Rosemarkie [7290 5835], Brown Hill [7474 6033], Flowerburn [7366 5852 and 7384 5874].

Central Highland Migmatite Complex — Gneissose psammite is quarried at Daviot [719 392] as a source of aggregate, but the rock was used in the past as a local building stone.

Moy Granite — Fresh grey porphyritic granodiorite in local buildings suggests the source is the Moy Quarry [7545 3683].

Old Red Sandstone — East Craigbreck [6700 4952], Millbuie [6730 6004 and 6729 5959], Pitlundie [6742 5080], Bay [6780 5325], Corrachie [6794 5432], Killen [6813 5804], Rosehaugh [6840 5595], Arkendeith [6930 5592], Avoch [6945 5578], Weston [7045 5905], Knockmuir [7075 5625], Newton [7075 5558], Croft House [7075 5623], Mid Craiglands [7128 5769, 7156 5789 and 7188 5785], Raddery [7144 5911], Broomhill [7184 5685], Platcock [7254 5724], Rosemarkie [7290 5835], Janefield Road [7293 5992], Wester Delnies [832 559], Cawdor [846 498], Leanach [748 446], Leys–Culduthell [6838 4277–6833 4263], Muckovie [7105 4382], Nairnside-Newton [7355 4240], Chapelton [7328 4647], Morayston (7527 4882], Balfreish [7960 4678], Hillhead [7776 4981] and Easter Delnies [8322 5595]. The most important of these are Bay Quarry on the northern shore of Munlochy Bay, where the stones of Fort George were hewn, and the Millbuie quarries that supplied stone for buildings in the Fortrose–Rosemarkie area.

CRUSHED ROCK AGGREGATE

The overriding property determining the suitability of a rock for making aggregate is its crushing strength, which is mostly governed by the nature of the rock, but can be affected quite significantly by the state of weathering and alteration. The mechanical properties, other than crushing strength, that are critical in more specific applications are resistance to polishing and wear. These govern the suitability of rock chippings for road surfacing, whilst drying shrinkage governs the versatility of aggregate for concreting purposes.

The south-eastern half of the district is well endowed with resources of rock suitable for producing aggregate. The granites have potential for most road-making and concreting applications, but none is being exploited at present. The grey porphyritic granodiorite of the Moy Granite was also formerly worked for roadstone [7545 3683] near Moy. Gneissose psammites of the Central Highland Migmatite Complex are also potentially suitable for most applications, although they tend to be too hard and fresh for skid-resistant road surface dressings. Such rock was probably wrought for railway ballast in a quarry [7374 3983] beside the former station at Daviot, and is presently exploited in a large quarry [719 392] to the north of the village for granular sub-base, coated road aggregate and armourstone.

SAND AND GRAVEL

The aggregates industry is very important to the local economy, but sand and gravel extraction, in particular, is placing increasing pressure on planning authorities to balance necessary mineral exploitation with competing uses of land and with conservation of the environment. Although extensive spreads of sand and gravel occur within the district (see Chapter Eight), no detailed quantitative assessment of these resources has yet been undertaken; it should be noted that the resurvey of Quaternary deposits indicates that sand and gravel resources in the district are considerably less than those previously suggested by the primary mapping published on the first one-inch geological map. Qualitative appraisals of the deposits were made as part of a study of the sand and gravel resources of the Inner Moray Firth area by Harris and Peacock (1969) and this information was included within an overall evaluation of the sand and gravel resources of the Highland Region (Mykura et al., 1978). Both reports provide brief descriptions of the composition and distribution of geographically distinct deposits thought to have the greatest resource potential, but a more comprehensive appraisal is required.

Until recently, sand and gravel has been exploited at widely scattered localities across the district, but with costs of haulage making up a substantial proportion of the delivered price of the commodity, there has been some reduction of the industry. Potential resources occur within glacial, glaciofluvial, glaciomarine, alluvial, raised-beach and aeolian deposits. Mid Lairgs Pit [710 365] is the largest active quarry in the district, producing a full range of aggregate products, including concrete and coated asphalt. Here, gravels of the Littlemill esker system are excavated together with adjacent glaciofluvial deposits, which form part of the moundy glaciofluvial outwash within Strath Nairn. Moundy sand and gravel deposits also constitute a major resource in Moy Gap.

The largest unsterilised sand and gravel resource lies within the low ground backing the coast between Inverness and Nairn. It principally comprises sand and gravel of littoral, glaciomarine and glaciofluvial origin, together with spreads of blown sand. The sands and gravels have been worked in numerous small 'dry' pits, notably at Bothyhill [715 491], Morayhill [753 495], Drumdivan [840 547], Mid Coul [776 503] and Moss-side [863 552]. Extensive water-saturated alluvial gravels occur beneath the floodplain and alluvial terraces of the River Nairn; less extensive resources of this type are present in the Findhorn Valley. The discontinuous spreads of terraced and moundy glaciofluvial deposits that overlie till on the northern margins of the Nairn/Findhorn interfluve are generally thinner and of relatively minor interest in resource terms, although they have been worked on a piecemeal basis in many small pits.

In general, the gravels of the district are clean and of good quality. They contain few friable clasts or material that could have deleterious effects on concrete made from the aggregate. However, the deposits tend to contain large proportions of cobbles and boulders.

PEAT

Hill peat blankets much of the upland in the southern and south-eastern parts of the district. The thickest deposits are probably those on the flanks of Carn nan Tri-tighearnan, but most fall within a large Site of Special Scientific Interest (SSSI), and thus are not likely to be available for exploitation. The widespread presence of roots, stumps and trunks of ancient pine trees within these peat deposits, as well as their remoteness and altitude, limit their potential for mechanical extraction. Other resources occur within basins on the lower ground, notably at Blàr nam Fiadh [835 535] and Carr Bàn [670 364], where waterlogging reduces their potential. Piecemeal hand digging of peat to be used as fuel has not been important recently, but persists in some areas, and has a specialised application in the local whisky industry. Commercial production for domestic fuel and horticultural usage has been undertaken recently [at 757 382] near Moy, where hill peat was stripped mechanically.

On the Black Isle, only two areas of potential peat working occur, one in the source region [694 581] of the Goose and Rosemarkie burns and one [675 511] near East Lundie; both coincide with waterlogged depressions.

BRICK CLAY

Late Devensian raised tidal-flat deposits, comprising silty clay with subordinate layers of sand, were formerly worked in a pit [7270 4790] near Balloch for brick and tile making. Although resources undoubtedly remain in the vicinity, particularly to the south-east, a combination of relatively high firing costs and remoteness from the major market in central Scotland, has led to rationalisation of the industry and the closure of this and all other brickworks in the Highlands.

LIMESTONE

Limestones are limited to the Old Red Sandstone outcrop, where they occur mainly as small concretions in dark bituminous silty mudstones in the Inverness Sandstone Group. In some places, for example north-east of Altlugie [at 735 423] and in the Leys–Culduthel quarries [6832 4278], there are indications of lime burning. Whether such operations were successful is not known, but they were very small undertakings. Thin limestones in the Daviot Conglomerate, e.g. at [757 422, 797 468 and 8020 4735], are the only bedded limestones; they are probably unsuitable as building stones.

WATER SUPPLY

The drainage of the district is into the Inverness and Moray firths. On the Black Isle, drainage is largely via the Killen and Rosemarkie burns, and on the south-eastern side of the firths is via the Nairn and Findhorn rivers and minor streams in a narrow coastal belt.

The district is generally served by the Water and Sewerage Department of the Highland Regional Council in Inverness, with separate public supplies in each of the

District Council areas. The north-west Ross and Cromarty District area is supplied from Loch Glass, well beyond the district boundary. Inverness-shire is supplied from Loch Ashie [630 350], 3 km west of the district boundary, with the provision for top-up from nearby Loch Duntelchaig. In the east, the Cawdor region, lying within the Nairn District Council area, is supplied from the dammed Clunas Reservoir [860 460], just east of the district boundary.

Standby facilities are available for Inverness from springs on the northern side of Bogbain Wood [705 416], but no borehole water is utilised in this district.

Few hydrogeological studies have been undertaken in the district and the available data are meagre. The general lack of wells is probably an indication of the limited resources of readily available groundwater.

The district lies within the Caithness and Moray Firth Old Red Sandstone Province (Robins, 1990), where the prevailing winds are westerly and rainfall is largely orographic. Lying on the eastern side of the Scottish landmass, it is somewhat drier than the west, receiving between 600 and 800 mm per year of rainfall, with the Black Isle considerably the driest area. Annual evaporation is between 380 and 420 mm and the estimated recharge is 100 to 300 mm per year.

South of the firths there are numerous upland catchments suitable for the collection and storage of surface water and there are considerable surface-water resources in the form of lochs, lochans, rivers and streams.

None of the bedrock outcrops is considered a good aquifer, but groundwater is nevertheless a local resource of some value to the rural community. The Rosemarkie and Central Highland basement rocks are replete with complex structures and the contorted distribution of the various lithologies greatly restricts the areal extent of any permeable strata. However, the rock matrices are for the most part poorly permeable, with permeability derived almost entirely from cracks, joints and brittle-fracture zones in the near-surface weathered zone. Having intergranular porosities generally less than 0.01, both the transmissivity of these systems and their storage capacity are low. Consequently, only small-scale shallow groundwater flow occurs within these rocks, a phenomenon illustrated by rare, often ephemeral spring discharges from selective fissures that intersect the ground surface in suitable topographical lows.

Although less subject to contortions and discontinuity of strata, the igneous bodies are comparable to the basement rocks in having a low potential for groundwater storage and thus are regarded as poor aquifers.

In many other parts of this Old Red Sandstone province, the sandstones and conglomerates have been highlighted as prime aquifers. However, the limited data available suggest that Old Red Sandstone rocks in this district are anomalous; no successful bedrock water boreholes have been recorded, and recoverable groundwater is considered negligible. No major reasons have been determined, but it is thought that the rocks are more indurated, fine grained and weakly jointed compared with rocks to the east, beyond Nairn, where slightly younger strata give worthwhile artesian yields. The hydraulic conductivity of the strata may be less than 0.1 m per day in contrast to that farther east, where conductivity values of 1 m per day have been obtained (Robins et al., 1989, table 1).

Sources from the Quaternary superficial deposits are largely controlled by the disposition of sand bodies, and springs usually coincide with sand to clay boundaries. Such sands are commonly lenticular in form and of variable thickness. Consequently, their storage capacity is highly variable and difficult to establish. Generally, once located, their yield-drawdown characteristics tend to be only favourable at moderate pumping rates (Robins, 1987). The relatively coarse granular sands and gravels offer good potential for recharge and transport of groundwater, and the hydraulic conductivity may attain 100 m per day. However, large parts of these strata are unsaturated as they naturally drain to lower elevations. There are a number of springs in the district as well as several shafts, but yields from these sources are generally modest and rarely exceed 1 litre per second. The blown sand of Ardersier is unsaturated, but trial pits have shown it to be moist throughout most of its depth. There is little groundwater movement or storage in most of the fluvial deposits.

The quality of bedrock groundwater is generally poor, and limited circulation in confined fissure systems produces reducing conditions, so that iron and manganese may be taken up in solution. Groundwater in the Old Red Sandstone is hard and moderately mineralised. Most of the groundwaters are dominated by calcium, magnesium and bicarbonate ions, especially in the Old Red Sandstone, where bicarbonate concentrations range between 150 and 250 mg per litre. Waters in superficial deposits are weakly mineralised and poorly buffered, as indicated by an analysis from a spring at Carse of Delnies [829 557] (Table 3).

Table 3 Chemical analysis of spring water at Carse of Delnies.

	mg/l		mg/l
Ca	61	HCO_3	90
Mg	7	SO_4	44
Na	17	Cl	74
K	3	NO_3-N	9

The pH is 6.9. This is a weakly mineralised, calcium bicarbonate-type water, having an elevated nitrate concentration, which reflects the vulnerability of such a shallow aquifer to local agricultural activity.

A notable feature of the Black Isle is the naming of certain spring wells. In addition to the common practice of assigning saintly names, as at St Boniface Well [7235 5658], Fortrose, two wells at Ormond Castle [6962 5357] and near Bay Farm Quarry [6793 5317] are Clootie (Devil or Witch) wells that annually are bedecked with a variety of torn coloured rags; both waters emanate from Old Red Sandstone strata.

APPENDIX 1

Modal analyses of granitic rocks from the Fortrose district south of the Moray Firth

Table 4   Location details of analysed rocks of Table 5.

Registered number in BGS (Edinburgh) collection	National Grid reference [NH]	Locality
S 79364	8428 3789	Interfluve between Allt Breac and Ballaggan Burn
S 79400	7687 4038	300 m W of Carn Dubh Beag
S 76411	8331 3817	Cnoc a' Chinn Leith
S 81922	7354 3622	Beinn a' Bheurlaich
S 81923	7371 3646	550 m NE of Beinn a' Bheurlaich
S 83390	7541 3684	500 m S of Loch a' Chaorainn
S 83470	7745 3688	Moy Burn
S 83471	8067 3741	Allt Odhar
S 83473	7970 3266	1 km ENE of Dalmagarry
S 83475	7574 3746	150 m E of Loch a' Chaorainn
S 83477	7663 3823	Dalriach Burn
S 83478	7845 3782	Beinn Bhreac
S 79398	7728 4042	Carn Dubh Beag
S 79363	8581 3816	Creag Ruadh
S 83509	8533 3743	Quilichan
S 83524	8173 3300	Ruthven
S 83525	8173 3300	Ruthven
S 83526	8173 3300	Ruthven
S 83527	8173 3300	Ruthven
S 83527A	8127 3238	850 m SW of Ruthven
S 79383	7889 4336	Saddle Hill
S 79384	7863 4330	SW Ridge of Saddle Hill
S 79385	7871 4328	SW of Saddle Hill
S 79395	7872 4051	Beinn Bhuidhe Mhór
S 83474	7883 3358	Tullochclury
S 95139	7396 3775	900 m WSW of Meallmore Lodge

Table 5   Modal analyses of granite rocks south of the Moray Firth.

	MOY GRANITE													QUILICHAN VEIN COMPLEX	
S numbers	79364	79400	76411	81922	81923	83390	83470	83471	83473	83475	83477	83478	79398*	79363	83509
Quartz	32.4	33.0	28.1	28.1	32.2	34.5	36.2	34.6	34.6	33.8	22.0	38.8	28.1	38.2	34.2
K-feldspar	15.8	17.9	20.4	27.1	8.7	15.4	17.3	14.1	15.0	7.4	13.7	8.1	34.6	6.2	11.4
Plagioclase	42.0	40.8	44.2	35.7	43.5	40.6	37.2	40.5	37.0	47.8	51.4	39.4	33.5	44.0	4.35
Biotite	7.0	4.1	6.8	7.3	13.5	7.6	8.9	9.2	11.3	7.7	10.0	12.6	2.9	8.7	7.4
Recrystallised biotite	—	4.0	—	—	—	—	—	—	0.8	—	—	—	0.6	—	—
Muscovite	—	—	—	0.9	1.1	0.8	—	0.7	—	3.0	2.0	0.7	0.3	2.6	3.3
Hornblende	0.4	—	—	—	—	—	0.1	—	—	—	0.7	0.1	—	—	—
Myrmekite	1.4	—	—	—	0.1	0.8	—	—	0.2	—	—	—	—	0.3	—
Opaque ore	0.5	—	0.4	—	—	—	—	0.5	0.7	—	—	—	—	—	—
Sphene	0.2	—	—	—	—	—	—	—	—	—	—	—	—	—	—
Apatite	—	—	0.1	—	0.1	—	—	—	—	—	—	—	—	—	—
Chlorite	—	—	—	0.3	0.5	—	—	—	—	—	—	—	—	—	—

Table 5   Modal analyses (continued)

	FINDHORN GRANITE COMPLEX					SADDLE HILL GRANITE					AUCHNAHILLIN MONZODIORITE
S numbers	83524	83535	83526	83527	83527a	79383	79384	79385	79395	83474	95139
Quartz	9.1	6.1	4.8	17.8	22.7	39.9	37.0	29.9	36.2	29.9	4.9
K-feldspar	13.4	2.7	4.6	5.3	16.8	29.7	31.5	36.1	32.7	31.6	22.7
Plagioclase	45.8	58.5	59.1	57.1	44.2	28.7	29.8	26.4	29.40	35.6	53.3
Biotite	11.8	16.9	7.2	14.2	12.8	1.6	1.7	1.2	1.2	2.0	13.9
Recrystallised biotite	—	—	—	—	—	—	—	—	—	—	—
Muscovite	—	—	—	—	—	—	—	—	—	—	—
Hornblende	16.1	13.9	20.9	4.1	1.3	—	—	—	—	—	2.7
Myrmekite	—	—	—	—	—	—	—	—	—	—	—
Opaque ore	2.2	0.4	1.2	0.1	1.6	0.1	—	—	0.2	0.6	0.8
Sphene	1.0	0.4	1.5	0.4	0.3	—	—	—	—	—	1.0
Apatite	—	0.3	0.3	0.3	—	—	—	—	—	—	—
Chlorite	0.1	0.3	—	0.1	—	—	—	—	—	—	0.2
Pyroxene	—	—	—	0.1	—	—	—	—	—	—	—

* S 79398 is considered to be part of the Moy Granite on the basis of field aspect, although its geochemistry and modal analysis show some similarities with the adjacent Saddle Hill Granite.

APPENDIX 2

XRF geochemical analyses of metamorphic and igneous rocks south of the Moray Firth

Major oxide and trace element analysis was performed by X-ray fluorescence spectrometry (XRF), using sequential, fully microprocessor-controlled wavelength dispersive XRF spectrometers (Philips PW1404/10 and PW 1480/10 with PW1500 sample changers). Computer hardware is a DEC Micro VAX 2000 using DEC VMS operating system, Philips X40 version 3.0b XRF application package and in-house reformatting software.

Major elements were determined from 5g samples dried for 24 hours at 1050°C. Fused beads were made from 0.9g sample + 9g Spectroflux-100 ($Li_2B_4O_7$) at 1200°C and cast in a platinum dish. Lithium iodide is added to the sample prior to fusion to act as a releasing agent. Loss-on-ignitition was determined from weight loss during sample drying.

Analyte angles are calibrated from international and in-house standards prepared as fused beads. Drift correction is applied using an external monitor and background correction applied where necessary. The 10 major elements are calibrated as oxides using the de Jongh algorithm and alpha coefficients generated empirically with one regression-based line overlap (Ca on Mg). No lower detection limited are quoted, but oxides are reported to 0.01wt% (0.001wt% for MnO).

Trace elements were determined on 40 mm pressed-powder pellets prepared from 12g of sample +3g Elvacite 2013 ground for 45 mins in agate.

Analyte angles are calibrated using 1000 mg/kg single-element standards in a silica matrix. Background factors are calculated from angular difference on or 'specpure' oxide blanks. Line overlap factors are calculated from high-concentration single-element standards and internal ratio is used for matrix correlation. Multi-element synthetic and natural, international and in-house standards are measured and curves fitted using the de Jongh algorithm producing saved calibration constants. Drift correction is applied using an external monitor.

Theoretical detection limited (3 standard deviations above background) are given below:

V, Cr, Co, Ni, Cu, Zn, Rb, Sr, Y, Zr, Nb, Mo, Ag, Sn, W, Bi, Th, U ....1 ppm
S, La ....2 ppm
Pb ....3 ppm
Ba ....4 ppm
Ce ....6 ppm

Table 6 Location details of XRF analysed rocks, Tables 7, 8, 9 and 10.

Registered number in BGS (Edinburgh) collection	National Grid reference [NH]	Locality
S 79367	8354 3718	Ballaggan Burn
S 81940	7068 3254	SW of Beinn Dubh
S 79379	8212 3343	c.500 m NE of Ruthven
S 79380	8204 3342	c.500 m NE of Ruthven
S 80362	8301 3423	c.1.8 km NE of Ruthven
S 79398	7728 4042	Carn Dubh Beag
S 79498	7673 4034	Ridge between Carn Dubh Beag and Beinn a' Bhuchanaich
S 83470	7673 3688	Moy Burn
S 83473	7745 3266	1 km ENE of Dalmagarry
S 83475	7970 3746	150 m E of Loch a' Chaorainn
S 83478	7574 3782	Beinn Bhreac
S 79384	7845 4330	SW ridge of Saddle Hill
S 79385	7863 4328	SW of Saddle Hill
S 79357	7871 4590	500 m S of Rehiran
S 79360	8318 4655	Allt Dearg
S 81960	6623 3163	Creag a' Chait, 300 m NNW of Creag Bhuide
S 81954	7318 3416	Ciste Creag an Eoin
S 79361	8231 4621	Allt Dearg
S 79378	8146 4169	750 m WNW of Carn Odhar

Table 7   XRF geochemical analyses of metamorphic rocks south of the Moray Firth.

	Migmatitic semipelite	Garnet amphibolite	Metagabbro		
	S 79367	S 81940	S 79379	S 79380	S 80362
SiO$_2$	59.74 %	45.46	48.63	48.78	48.52
TiO$_2$	1.03	5.01	1.62	1.25	1.73
Al$_2$O$_3$	17.36	12.38	16.15	17.14	15.86
Fe$_2$O$_3$	8.08	17.09	10.13	9.41	10.25
MnO	0.12	0.27	0.16	0.15	0.17
MgO	2.55	6.31	7.84	8.23	7.84
CaO	2.29	8.72	9.20	9.49	10.10
Na$_2$O	3.16	0.61	3.31	3.09	3.17
K$_2$O	3.87	2.25	1.43	1.11	0.76
P$_2$O$_5$	0.27	0.71	0.19	0.13	0.21
LOI	1.18	0.81	1.08	0.83	1.06
Total	99.65	99.62	99.75	99.60	99.66
V	127 ppm	518	209	173	220
Cr	68	148	143	111	193
Co	18	40	39	41	39
Ni	35	65	85	108	86
Cu	16	39	28	34	45
Zn	111	159	68	61	69
Rb	205	89	43	39	24
Sr	248	69	295	319	345
Y	44	102	33	28	35
Zr	242	445	161	117	163
Nb	20	12	6	3	4
Ba	1013	369	108	101	78
La	64	43	<7	4	15
Ce	95	131	<7	49	47
Pb	18	10	8	4	3
Th	16	5	<1	<1	<1
U	1	<1	<1	<1	<1
Ga	22	24	17	17	18
Mo	—	—	1	<1	—
Sn	7	<2	<2	4	<2
W	—	—	2	4	—

Table 8   XRF geochemical analyses of the Moy Granite.

	S 79398*	S 79498	S 83470	S 83473	S 83475	S 83478
SiO$_2$	73.66 %	71.55	71.86	71.87	71.99	72.18
TiO$_2$	0.18	0.39	0.40	0.52	0.37	0.43
Al$_2$O$_3$	14.22	14.91	14.73	13.96	14.96	14.69
Fe$_2$O$_3$	1.66	3.03	2.62	3.22	2.52	2.76
MnO	0.03	0.09	0.09	0.09	0.08	0.09
MgO	0.43	0.64	0.70	0.90	0.70	0.79
CaO	0.67	1.22	1.56	1.74	1.54	1.40
Na$_2$O	3.81	3.06	3.37	2.97	3.28	3.33
K$_2$O	4.70	4.36	3.53	3.65	3.40	3.64
P$_2$O$_5$	0.01	0.02	0.03	0.07	0.04	0.03
LOI	0.53	0.63	0.68	0.61	0.91	0.76
Total	99.91	99.91	99.57	99.61	99.80	100.10
V	20 ppm	26	34	42	28	33
Cr	189	40	29	51	32	46
Co	5	5	6	7	5	7
Ni	6	5	8	10	7	8
Cu	<2	2	2	4	1	3
Zn	16	48	38	46	43	43
Rb	130	108	93	92	92	100
Sr	310	347	355	361	371	315
Y	25	27	19	10	18	21
Zr	30	83	148	205	127	153
Nb	11	14	8	7	9	8
Ba	810	1099	886	1298	858	926
La	20	10	14	21	16	17
Ce	48	30	45	42	47	47
Pb	31	25	25	21	22	22
Th	4	6	5	4	6	6
U	2	3	1	2	3	3
Ga	14	—	—	—	—	—
Mo	73	—	—	—	—	—
Sn	<2	—	—	—	—	—
W	2	—	—	—	—	—

* S 79398 is considered to be part of the Moy Granite on the basis of field aspect, although its geochemistry and modal analysis show some similarities with the adjacent Saddle Hill Granite.

Table 9   XRF geochemical analyses of the Saddle Hill Granite.

	S 79384	S 79385		S 79384	S 79385
SiO$_2$	75.12 %	75.52	V	15 ppm	13
TiO$_2$	0.22	0.21	Cr	191	125
Al$_2$O$_3$	13.09	13.12	Co	<2	<2
Fe$_2$O$_3$	1.67	1.28	Ni	6	6
MnO	0.04	0.04	Cu	1	<2
MgO	0.38	0.34	Zn	30	20
CaO	0.50	0.60	Rb	228	219
Na$_2$O	3.48	3.64	Sr	114	102
K$_2$O	5.06	5.09	Y	12	14
P$_2$O$_5$	0.04	0.04	Zr	109	110
LOI	0.45	0.33	Nb	15	14
Total	100.06	100.21	Ba	331	339
			La	64	14
			Ce	99	81
			Pb	26	30
			Th	32	26
			U	3	4
			Ga	18	17
			Mo	<1	<1
			Sn	<2	5
			W	<2	3

Table 10 XRF geochemical analyses of minor intrusions south of the Moray Firth.

	Appinite S 79357	Lamprophyre		Microdiorite ──►		Microgranite
		S 79360	S 81960	S 81954	S 79378	S 79361
SiO$_2$	51.53 %	52.70	57.32	55.98	63.99	70.77
TiO$_2$	1.23	0.80	1.01	1.53	0.79	0.39
Al$_2$O$_2$	14.95	10.19	15.09	15.84	15.05	14.57
Fe$_2$O$_3$	9.75	9.02	6.26	7.12	3.93	2.51
MnO	0.17	0.18	0.15	0.15	0.08	0.05
MgO	6.02	12.36	5.75	2.48	2.52	0.99
CaO	7.93	7.56	5.51	3.52	3.13	1.15
Na$_2$O	2.88	1.55	3.88	4.09	4.18	4.30
K$_2$O	3.74	3.58	2.23	4.15	3.73	4.11
P$_2$O$_5$	0.35	0.23	0.27	0.66	0.23	0.13
LOI	1.16	1.43	1.77	3.68	2.44	1.03
Total	99.72	99.59	99.27	99.20	100.07	100.00
V	214 ppm	161	106	94		
Cr	159	937	234	25		150
Co	27	38	21	16		4
Ni	27	317	91	17		17
Cu	45	19	17	6		2
Zn	79	76	83	104		
Rb	87	91	60	110		
Sr	420	469	793	444		
Y	29	28	11	25		
Zr	149	122	160	374		
Nb	11	8	9	23		13
Ba	789	1669	1047	1006		651
La	33	23	22	80		29
Ce	79	85	57	135		126
Pb	9	11	18	54		
Th	3	5	4	10		
U	<1	2	3	4		
Ga	16	13	—	—		20
Mo	1	<1	—	—		<1
Sn	3	<2	—	—		
W	2	3	—	—		

APPENDIX 3

Quaternary sites of international importance

a Dalcharn interglacial site Sediments containing compressed and disseminated biogenic matter are exposed beneath a thick sequence of tills in river cliffs at Dalcharn [815 452], 6 km south-west of Cawdor. The organic deposits, which lie at an altitude of about 200 m above OD, have been cryogenically and glacitectonically disturbed, but contain pollen of full interglacial affinity reflecting the middle and later stages of an interglacial cycle (Walker 1990a). The overlying till sequence provides evidence of at least two separate glacial episodes.

The lithostratigraphy at Dalcharn was established by Merritt and Auton (1990), based on three closely spaced sections: Dalcharn East [8157 4537], Dalcharn West-A [8146 4521] and Dalcharn West-B [8143 4516] (Figure 31):

		Thickness m
7	Humic soil	c. 0.3
6	Glaciofluvial gravel	up to 2.5
5	Dalcharn Upper Till	8.5–10.0
4	Dalcharn Lower Till	8.5–9.5
3	Dalcharn Biogenic Formation	1.3–1.6
2	Dalcharn Gravel	up to 3.0
1	Dearg Till	at least 1.0

The **Dearg Till** is a very stiff, massive, sandstone-rich diamicton beneath the Dalcharn Gravel, seen only at Dalcharn East. The **Dalcharn Gravel** is a poorly sorted, matrix-rich gravel, bleached in its upper part and containing a high proportion of decomposed and unsound clasts. The bleached part of the deposit has a matrix that includes vermiculite, interpreted as being the product of subaerial weathering (Bloodworth, 1990).

The **Dalcharn Biogenic Formation** is subdivided into an upper *Biogenic Member* (0.5 to 0.6 m), and a lower *Cryoturbate Member* (0.8 to 1.0 m). The Cryoturbate Member is a massive, matrix-supported, clayey, gravel diamicton with a matrix of white, silty, fine-grained sand and contains small, sparsely disseminated fragments of organic material. Clasts within the diamicton are mainly of yellowish grey or red sandstone, most with white-weathering rinds. The uppermost 10 to 20 cm of the Biogenic Member comprises very compact, laminated, sheared, olive-grey sandy and clayey silt, with discontinuous wisps of pebbly sand and disseminated peaty matter. This overlies very compact carbonaceous sandy silt and diamicton containing fibres and lumps of dark peaty material as well as thin lenses of compressed sandy peat. The deposits have been disturbed glacitectonically, as indicated by tension cracks, and thrusts lined with silt and orange sand, that penetrate from above the base of the overlying till, through the biogenic deposits and into the underlying gravel.

The **Dalcharn Lower Till** comprises reddish brown, sandy diamictons with abundant clasts of Devonian sandstone. The **Dalcharn Upper Till** is divided into upper and lower members. Both comprise brown lodgement till, with clasts of psammite, semipelite and granite derived from basement rocks that form high ground to the south-west of the site. The overlying **Glaciofluvial gravel** comprises poorly sorted, clast-supported, cobble gravel with an imbrication indicating north-eastward palaeocurrent flow.

The results of pollen analyses on the Biogenic Formation have been reported by Walker (1990a), Whittington (1990) and Walker et al. (1992); an infinite radiocarbon age of >41 300 ^{14}C BP (GU-2340) has been obtained from compressed peat close to its base.

Interpretation The occurrence of brown till with few sandstone erratics, overlying reddish brown till with abundant sandstone clasts is a commonly found relationship across the high ground to the south of the coastal lowlands between Inverness and Nairn. At Dalcharn (East), the discontinuity separating the two tills is associated with pull-apart structures and tension cracks within which the Upper Till has been squeezed down into the Lower Till, suggesting that the two formations are the products of separate glacial episodes (Figure 30). The Upper Till is presumed to be of Late Devensian age and the Lower Till is probably early Devensian, if not older. The highly decomposed nature of the gravel underlying the biogenic deposits indicates that the gravel has been subjected to prolonged weathering under warm humid conditions and that the gravel and the associated organic material is of considerable antiquity.

The origins of the biogenic materials are uncertain, but they probably represent the remains of a destroyed palaeosol. Despite disturbance of the polleniferous sediment, the pollen record shows that while the organic sediments accumulated, closed pine forest with birch, alder and holly was succeeded by pine-heathland and that this was replaced initially by birch and later by heath and open grassland. As pine woodland has been the climax forest of the north-central Grampians during Flandrian times (Bennett, 1984), the Dalcharn sequence is thought to reflect a warm episode of interglacial rather than interstadial status. Furthermore, as the Dalcharn site lies near to the present northern limit of holly in Britain, the relative abundance of *Ilex* pollen in the profile almost certainly reflects a climate somewhat warmer than that of today.

Although an Ipswichian age is not ruled out, the pollen assemblage from Dalcharn is similar in some respects to that described from Fugla Ness on Shetland by Birks and Ransom (1969). The Fugla Ness site has been equated with the Gortian of Ireland, and hence with the Hoxnian of southern England. However, a Cromerian age for the Fugla Ness record was not excluded. On present evidence, it is not possible to firmly attribute either the Dalcharn or the Fugla Ness record to a particular interglacial period within the mid or late Quaternary and precise correlations with other interglacial or interglacial/interstadial sites in Britain have not been attempted. It is certain that the Dearg Till formed in a pre-Devensian glaciation, but its precise age is unknown.

b Allt Odhar interstadial site The site [798 368] is a river cliff of Allt Odhar located 16 km south-east of Inverness and lying at about 370 m above OD. A deposit of compressed peat containing pollen, insect remains and plant macrofossils of interstadial affinity occurs above a weathered till. There is at least one till formation higher in the sequence, although this is exposed in an adjacent section (73 NE 17, Figure 30). The lithostratigraphy of the Allt Odhar site given below is that estab-

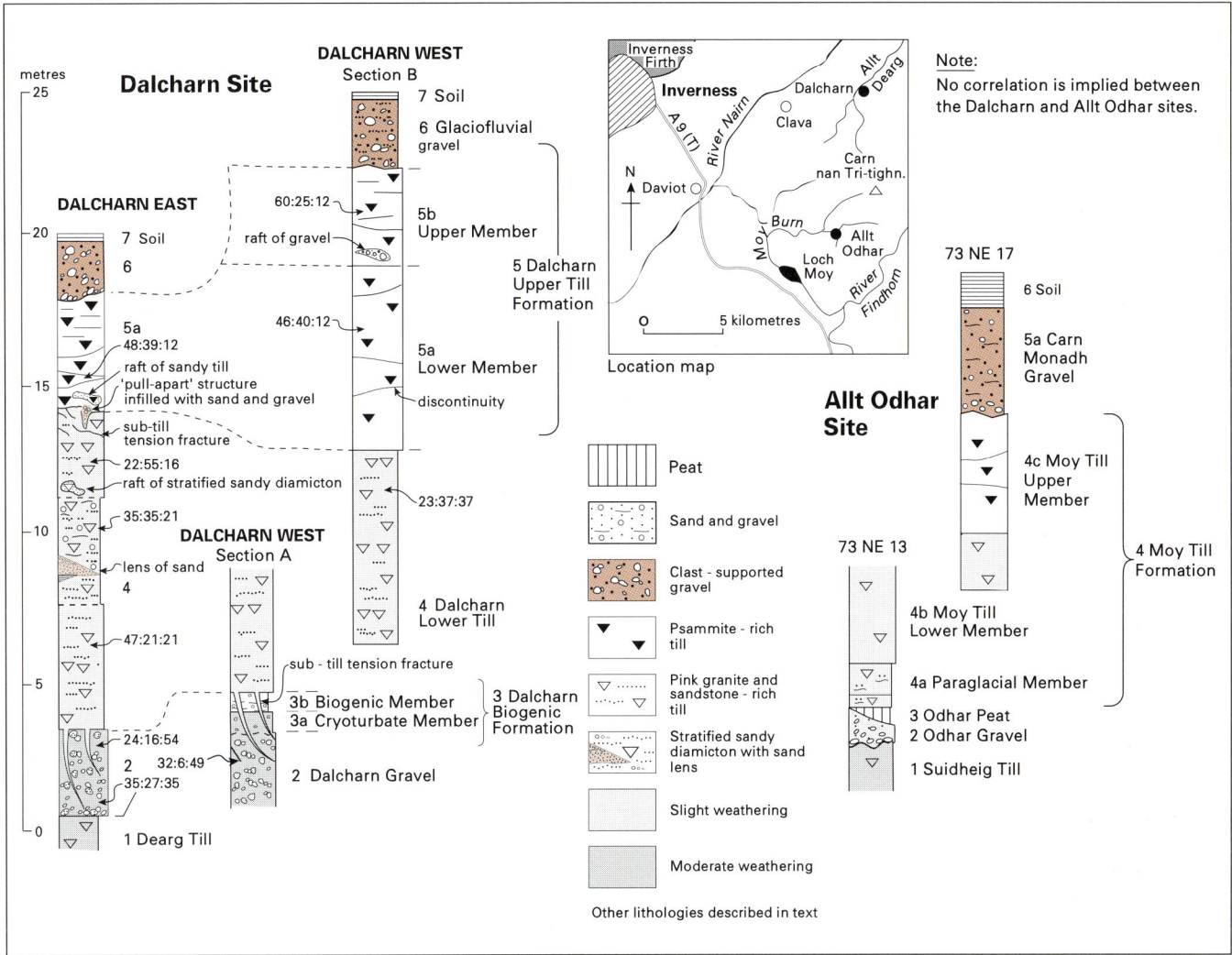

Figure 30 Sediment logs and stratigraphy of the Dalcharn and Allt Odhar sites. Stone-count data (e.g. 48: 39: 12) are percentages of metamorphic, igneous and Old Red Sandstone clasts respectively.

lished by Merritt (1990b), with minor modifications after Walker et al. (1992):

		Thickness m
6	Blanket peat	up to 2
5b	Sheet-wash gravel	up to 1.5
5a	Carn Monadh Gravel	up to 10
4	Moy Formation:	
c	Upper Till Member	up to 10
b	Lower Till Member	up to 6
a	Paraglacial Member	up to 2.2
3	Odhar Peat	up to 0.6
2	Odhar Gravel	up to 1.5
1	Suidheig Till	at least 1.5

The **Suidheig Till** is a very stiff lodgement till. Many of the clasts are decomposed and have orange-weathering rinds. The **Odhar Gravel** is a dense, poorly sorted, cobble gravel of fluvial or glaciofluvial origin with many decomposed clasts. The **Odhar Peat** lies within a shallow depression at the top of the underlying gravel. Four distinct beds are apparent:

		Thickness m
iv	pebbly, peaty sand	0.2–0.3
iii	black amorphous peat with sand wisps	0.15–0.3
ii	compressed, felted, fibrous peat	0.35
i	interlaminated sand and peat	0.2

The **Moy Formation** comprises *Upper Till* and *Lower Till* members and a *Paraglacial Member*. The latter is an extremely compact unit of pebbly, clayey, silt diamicton and silty sand with lenses of sand and gravel. Its upper and lower contacts are gradational, and the sediments have been sheared subglacially. The Upper and Lower tills can be distinguished throughout the area; the Upper Till is psammite-rich and the Lower Till is rich in pink granite and sandstone. Both are lodgement tills. The Lower Till is more sandy and contains clasts with orange-weathering rinds. The generally greater degree of weathering of the Lower Till may indicate that it is the product of an earlier glaciation than that which laid down the psammite-rich Upper Till.

The **Carn Monadh Gravel** comprises silty, sandy gravel with distinct planar subhorizontal stratification. It was probably deposited as an ice-marginal fan. The sheet-wash gravel is a coarse, poorly sorted deposit that caps most of the river cliffs in the area. The overlying blanket peat contains pine stumps near its base.

The results of pollen, plant macrofossil and fossil coleopteran analyses on the Odhar Peat are reported by Walker et al. (1992). Radiocarbon dating of samples from near the base of bed (ii) and from near to top of bed (iii) both gave age estimates >62 300 ^{14}C years BP (SRR-3677 and SRR-3678), indicating that the materials are older than the upper limit of radiocarbon dating (Harkness, 1990). A uranium-series disequilibrium age estimate of 124 ka ± 13 ka was initially obtained on a sample of peat from bed (ii) (Heijnis, 1990). Subsequently, based on additional measurements, a revised estimate of 106 ka +11/-10 ka was obtained (Heijnis and van der Plicht, 1992; Walker et al., 1992).

Interpretation Pollen analysis of the Odhar Peat (Walker, 1990b; Walker et al., 1992) reveals that as the peat accumulated, a landscape of birch woodland with juniper and willow scrub interspersed with open grassland was replaced first by grassland and heathland, and then by species-poor grass and sedge communities. The pollen record reflects an episode of climatic amelioration followed by a decline in temperature, perhaps accompanied by a change to wetter conditions, and finally to a colder climatic regime. The scarcity of pine pollen, relatively low arboreal pollen counts, the absence of thermophilous taxa and the presence of herbaceous taxa with northern or montane affinities, all indicate a climatic regime markedly cooler than that of a full interglacial. This interpretation of the palaeoenvironment is strongly supported by the fossil coleopteran assemblage which suggests a cool to cold climate, with mean July temperatures a little above 10°C and colder winters than at present (Coope, *in* Walker et al., 1992). The pollen and insect data taken together strongly indicate that the Odhar Peat is more likely to have formed during an interstadial than an interglacial period.

Radiocarbon dates from the Odhar Peat indicate that the deposit is older than mid Devensian (Harkness, 1990), whereas the uranium series date of about 106 ka places it firmly in the early Devensian. The interstadial episode is considered to be the terrestrial equivalent of Oxygen Isotope Substage 5c of the ocean record, which has been dated using the technique of 'orbital tuning' to 103 290 ± 3410 BP (Martinson et al., 1987). This early Devensian age is corroborated by the pollen and insect evidence. The nearest correlative of the Odhar Peat is the interstadial deposit at Chelford in Cheshire (Simpson and West, 1959; Coope, 1959), which has been dated on the basis of amino-acid geochronology as Substage 5c (Bowen, 1989) and by thermoluminescence dating of quartz and feldspar grains from the Chelford Sands to between 90 and 100 ka (Rendell et al., 1991).

c Clava The site at Clava [766 442], located at an altitude of about 150 m above OD, 9 km east of Inverness, comprises a series of sections along the lower reaches of the Cassie and Finglack burns, both tributaries of the River Nairn. The deposits, proved in sections and in boreholes, form a complex glacigenic succession and include the famous Arctic marine shelly clay. The latter, first described by Fraser (1882a, b) in an old claypit [7658 4411], excited considerable controversy in the last century when some regarded it to be an in-situ deposit indicating a great submergence of the country during the Pleistocene, while others considered that it was transported by ice from offshore. The site and its significance have been widely discussed in the literature (see Gordon, 1990; 1993b).

In view of the controversy over the origin of the shelly clay and its role in the whole concept of glacially related submergence in northern Britain, a Committee of the British Association for the Advancement of Science was convened to carry out investigations on the deposits in 1892 (Horne et al., 1894). They made two excavations near the claypit (the 'Main Pit') and sank seven boreholes (Figure 31). The clay was confirmed to be a marine deposit extending laterally for at least 170 m and reaching a maximum thickness of 4.9 m. It has a gentle dip and sharp contacts with adjacent beds above and below. There was little sign of disturbance, although cracks and fissures were noted. The shells are shallow-water marine species representing an Arctic or Subarctic faunal assemblage. They were unscratched and generally well preserved with periostraca intact.

The Clava site was subsequently re-investigated by Peacock (1975a). Although the original sections were no longer exposed, fresh ones had appeared along the Cassie Burn a few hundred metres to the south-west (Figure 31, I to VI). Here, Peacock described a till *(Clava Shelly Till)* with fragments of marine shells including *Portlandia arctica*, which had not been recorded previously at Clava.

The lithostratigraphy in the vicinity of the Clava site (Table 11) is that established by Merritt (1990a) with minor modifications after Merritt (1992).

The Clava site lies mainly on pebbly sandstones and breccio-conglomerates of the Daviot Formation, a part of the Old Red Sandstone.

The ***Cassie Till***, known only from the British Association boreholes, is probably a lodgement till. The ***Drummore Gravel***, exposed at the Finglack Section (Figure 31), comprises a weathered, stratified, matrix-supported, gravelly silty sand diamicton with clasts mainly of sandstone and flagstone. It was probably formed as subaerial sediment gravity flows, in an ice-marginal or supraglacial environment.

The ***Clava Shelly Formation*** includes three members, the unfossiliferous *Clava Sand*, the underlying *Clava Shelly Clay* and the *Clava Shelly Till*. The first two members form a conformable coarsening-upward sequence at the Main Pit, dipping gently to the west. The Shelly Till, which is overlain unconformably by Clava Sand at Section III, is essentially a glacially resedimented glaciomarine pebbly clay, but its stratigraphical relationship and age are not absolutely clear.

The *Clava Shelly Till Member* typically comprises a stiff, matrix-supported silty sandy clay diamicton containing clasts mainly of micaceous gneiss, with some of granite and sandstone. The sparse fauna of the till (see Appendix 4) is interpreted as having inhabited a shallow-marine environment in fully Arctic conditions during a stadial episode (Graham, 1990a, b; Graham et al., 1993). The Shelly Till has been folded at Section III, where a plane of glacitectonic *décollement* separates it from the overlying Clava Sand, which has been folded also.

The *Clava Sand Member* is a very compact silty fine- to medium-grained sand with sparse dropstones. At the Main Pit, it shows poorly defined subhorizontal lamination and a well-developed system of tension cracks, which are probably the product of brittle fracture, when the material was frozen and either overridden by, or transported within glacial ice. At Section III, the Clava Sand is cut by microfaults and shear planes as well as being folded.

The uppermost metre of the *Clava Shelly Clay Member*, exposed during the reopening of the Main Pit in 1990, comprises poorly defined, graded beds of dark grey silty clay, clayey silt and silty very fine-grained sand. This unit passes down into 60 cm of a much stiffer silty clay containing a network of black carbonaceous streaks, closely spaced shears, and micro-faults that ramify through the deposit. The lowermost 40 cm consist of pale grey plastic clay, but the base of the deposit was not seen. The lithologies of the clasts are as reported by Horne et al. (1894), mainly gneissose psammite of

Figure 31 The Clava site showing locations of sections and boreholes.

E–stream-bed exposure of Clava Shelly Till
FS–Finglack Section
SE–small excavation 146 m south-west of the Main Pit
I to VI–sections first described by Peacock (1975a).

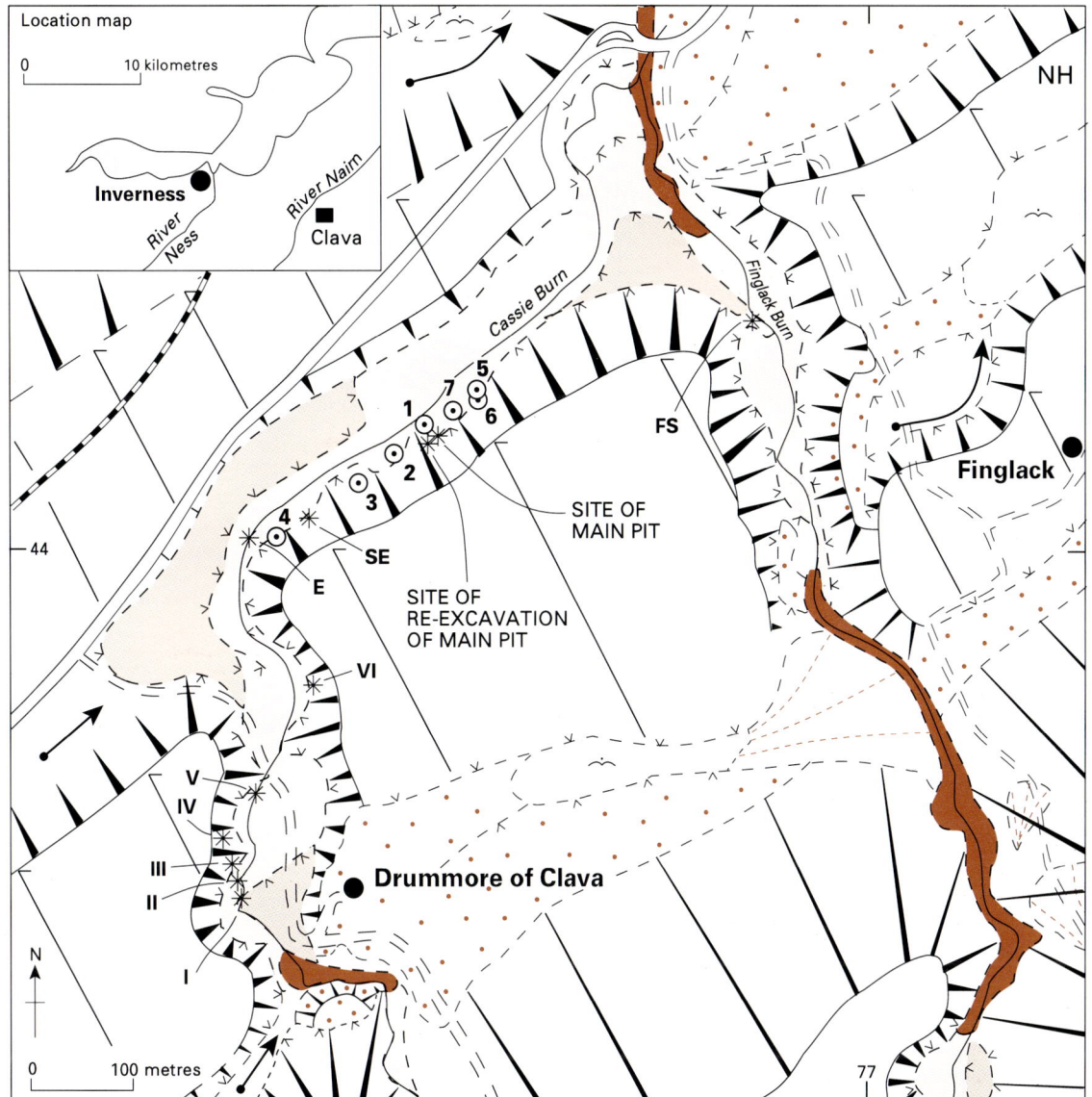

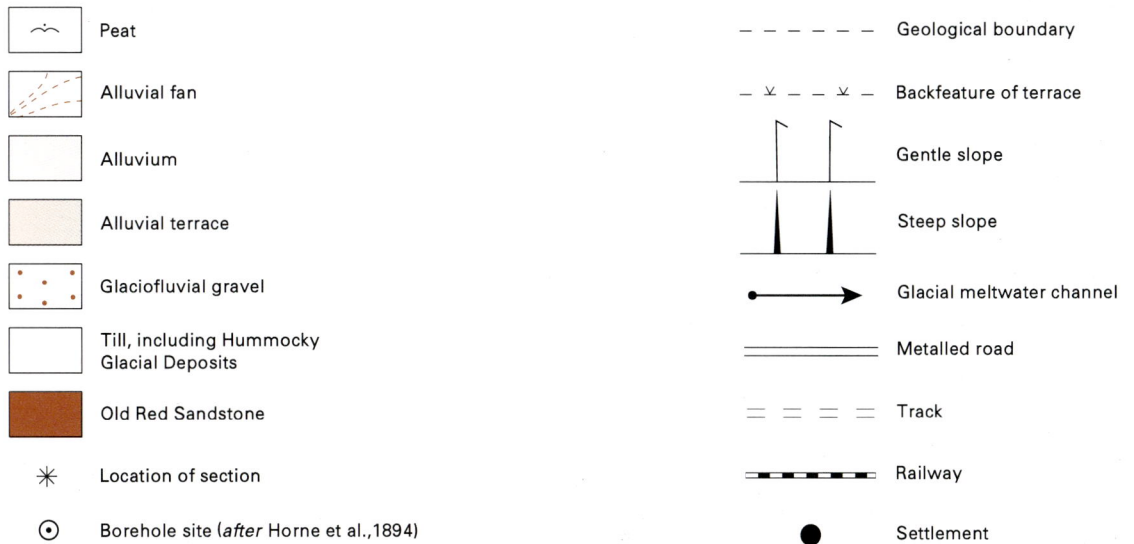

⌒ Peat	– – – – – Geological boundary
Alluvial fan	– ⌄ – – ⌄ – Backfeature of terrace
Alluvium	Gentle slope
Alluvial terrace	Steep slope
Glaciofluvial gravel	Glacial meltwater channel
Till, including Hummocky Glacial Deposits	Metalled road
Old Red Sandstone	Track
✳ Location of section	Railway
⊙ Borehole site (*after* Horne et al.,1894)	● Settlement

93TF14AE

Table 11   Lithostratigraphy at Clava.

Full succession	Main Pit	Finglack Section	Section III	Section VI	Maximum proven thickness (m)
6 Finglack Till Formation					
flow-till facies				(+)	4
melt-out facies				(+)	4
lodgement-till facies	+	(+)	+		10
5 Glaciofluvial Ice-contact Deposits			+	?	8
4 Clava Shelly Formation					
c Clava Sand Member	(+)	?	(+)		6
b Clava Shelly Clay Member	(+)	?			4.9
a Clava Shelly Till Member		?	(+)		3
3 Drummore Gravel	+	(+)			5
2 Cassie Till	(+)		?		7
1 Bedrock	+		near		–

Type sections in brackets ()

a type not common south of the Great Glen and 17 per cent only of the local Old Red Sandstone.

Shells, chiefly molluscs, are most common in the uppermost metre of the deposit, being mainly as described by Horne et al. (1894). There are, however, many laterally discontinuous beds (1 to 5 cm thick) with scoured bases containing angular to sub-rounded pebbles up to about 30 mm in diameter and broken shells only. These diamictic beds, not previously noted, are an integral part of the originally deposited sequence, possibly the result of stirring during storms. Intact shells are generally located within the layers between the diamictic beds, where there are also scattered subrounded to well-rounded pebbles, some with an epifauna including barnacles. The pebbles may have dropped from floating ice, but transportation by floating seaweed is also a possibility (cf. Gilbert, 1990).

Recent palaeontological analyses (see Appendix 4) confirm that the Clava Shelly Clay formed in a shallow-marine environment in a high-Boreal to low-Arctic interstadial climate with conditions apparently becoming increasingly estuarine during its deposition (Graham, 1990a, b; Graham et al., 1993). The dinoflagellate cyst record suggests that there may have been two short, warmer interludes and, on the basis of calcareous nannofossils, the deposit can be dated tentatively as being no older than about 77 000 years.

The *Glaciofluvial Ice-contact Deposits* are partly waterlaid and partly formed by mass-flow processes at an ice front. The deposit comprises poorly sorted, clast-supported gravel that becomes more clayey upwards: it is cut by a series of shear planes probably caused by ice-push. The *Finglack Till Formation* contains clasts mainly of sandstone and flagstone, and it comprises lodgement, melt-out and flow-till facies.

Interpretation   The majority of the British Association Committee concluded that the Clava Shelly Clay in the Main Pit was in situ, indicating former submergence of the land up to about 150 m above OD (Horne et al., 1894). As evidence, they cited the assemblage of organic remains, their mode of occurrence, the extent of the deposits and their apparently undisturbed

character. A minority of the Committee (Bell and Kendall), however, argued that there was insufficient evidence to reach a firm conclusion and doubted that there was any substantial evidence at all in Scotland for a great submergence. They questioned the widespread absence of shell beds and other traces of submergence at such elevations elsewhere, and the lack of marine organisms in the overlying till. Although acknowledging certain difficulties, notably the extent of the deposit and the good preservation of the shells, they favoured an ice-rafted origin for the shelly clay, with a source area in Loch Ness, as indicated by ice-movement patterns inferred from striae and erratics.

Peacock (1975a) concluded from his investigations at Clava that the shelly clay had been transported by ice and was part of autochthonous melt-out till (cf. Boulton, 1968) comprising reworked sea-floor material. Sutherland (1981), however, accepted an in situ origin for the shelly clay and presented a model relating these deposits to glacio-isostatically induced submergence in front of an expanding ice sheet. However, the deposits at Clava fit only partly into the overall distribution pattern of the high-level shell beds in Scotland and are apparently at too great an altitude to be fully explained by this model.

Following the re-opening of the Main Pit in 1990 and a critical re-examination of the sections described by Peacock, there is now unambiguous evidence to indicate that the shelly deposits at Clava have been deformed by glacial processes, and there is a strong possibility that they have been transported glacially as rafts derived from the Great Glen (Merritt, 1992). The raft exposed at the localities investigated in the 19th Century has to a large extent maintained its integrity during transport. It comprises an internally conformable, coarsening-upward sequence (Clava Sand overlying Clava Shelly Clay) typical of deposition in a delta. This possibly developed during a period of glacial retreat and associated falling sea level within an interstadial episode (cf. Boulton, 1990), as indicated by the fauna and flora recovered from the shelly clay, and the presence of dropstones in it.

The smaller masses of shelly diamicton (Clava Shelly Till) occurring upstream of the Main Pit are also interpreted to be

Table 12   Radiocarbon dates on shells from the Clava Shelly Clay.

Radiocarbon Accelerator Unit laboratory reference	Species	^{14}C age (years BP)	$\delta^{13}C$ (‰ rel PDB)
Clava Shelly Clay			
OxA—2482	*Tridonta elliptica**	>39 400	- 0.7
OxA—2483	*Littorina littorea**	43 800 ± 3 300	- 0.1
OxA—2876	*Littorina littorea*	>43 000	+ 2.7
OxA—2877[†]	*Astarte sulcata*	>41 200	+ 1.7
Clava Shelly Till			
AA9410/AMS90	*Mya ?truncata*	>46 400	1.4 ± 0.1

*   museum specimens collected from the Main Pit in 1892; other specimens collected during re-excavations in 1990.
†   incorrectly quoted in Merritt, 1993.

rafts, but their Arctic fauna demonstrates that they are not simply lateral equivalents of the shelly clay in the main raft. The composition of the clast assemblages in the diamicton bodies and in the Glaciofluvial Ice-contact Deposits at Section III suggests that the rafts of Shelly Till originated as glaciomarine sediment in the Loch Ness Basin, but whether this was associated with the retreat of 'early' Devensian glaciers or with a subsequent glacial advance, cannot be ascertained on present evidence.

The rafts were all probably detached as a result of high pore-water pressure building up in laterally restricted aquifers beneath a confined glacier that flowed north-eastwards across the Loch Ness Basin (Merritt, 1992). This glacier was deflected eastwards and upwards towards Clava by ice flowing from the northern Highlands along the Beauly Firth during the build-up of the last Scottish ice sheet. The rafts were stacked at the ice margin when the glacier entered Strath Nairn before being overridden by the expanding ice sheet.

Amino-acid dating on specimens of *Littorina littorea* indicates that the Clava Shelly Clay is mid Devensian in age and the results of radiocarbon accelerator dating (Table 12) confirm that the deposit is not younger than the mid-Devensian. The rafting thus occurred during the build-up of the Late Devensian Ice Sheet in Scotland and it follows that the Finglack Till is also of Late Devensian age.

APPENDIX 4

List of Quaternary fossils recorded from the Fortrose district

BGS Technical Report numbers are in parentheses where appropriate.

CLAVA SHELLY CLAY

Macrofossils
D K Graham (WH/89/292R; WH/90/388R; WH/91/52R)

GASTROPODA
Boreotrophon clathratus (Linné) *gunneri* (Lovén)	I
Buccinum undatum Linné	IB
Littorina littorea Linné	IBN
L. saxatilis (Olivi) *rudis* (Maton)	B
Littorina sp.	B
Lunatia pallida (Broderip & Sowerby)	B
Lunatia sp.	B
Margarites helicinus (Fabricius) [*Trochus helicinus*]	*
M. groenlandicus (Gmelin) [*Trochus groenlandicus*]	*
Neptunea antiqua (Linné) [*Fuses antiquus*]	*
Oenopota scalaris (Möller)	B
O. trevelliana (Turton)	B
O. turricula nobilis (Möller) [*Pleurotoma nobilis*]	*
O. turricula ss. (Montagu)	IBN
Omalogyra atomus (Philippi)	I
Rissoa parva? (da Costa)	B

BIVALVIA
Astarte sulcata (da Costa)	I N
Cerastoderma edule (Linné)	I
Lepton nitidum Turton	I
Macoma balthica (Linné)	IB
M. calcarea (Gmelin)	BN
Macoma sp.	B
Mytilus edulis Linné	I N
Nicania montagui (Dillwyn)	B
Nucula sp.	B
Nuculana pernula (Müller)	IBN
Nuculana sp.	B
Nuculoma tenuis (Montagu)	IBN
Thyasira flexuosa (Montagu) [*Axinus flexuosus*]	*
probably misidentification of a cold-water species of *Thyasira*	
Tridonta elliptica (Brown)	B
Tridonta sp.	B
Yoldiella lenticula (Möller)	IB
Yoldiella sp.	B
unidentified shell fragments	B

CIRRIPEDIA
Balanus balanoides Linné	I
B. crenatus? Bruguière	IB
Balanus sp. plates	B

DECAPODA
crustacean claw	B

Key: **I** specimens from the Fraser Collection (Inverness Museum).

B specimens from the Bennie Collection (British Geological Survey Edinburgh).

Although no repository was cited, Bennie's material is probably that referred to in the British Association's report (Horne et al., 1894) as 'Mr Bennie's list 1893' which was originally identified by G Sharman and later revised by D Robertson.

N material collected from the 1990 reopening of the Main Pit (British Geological Survey, Edinburgh).

* Taxa recorded under the names shown in parentheses in Horne et al. (1894), but not seen in recent fieldwork or in repositories visited, and hence unverified.

Microfossils

OSTRACODA I P Wilkinson (WH90/399R)
The following ostracods from the Clava Shelly Clay of No.2 Pit (Horne et al., 1894), collected by Bennie in 1892, have been re-identified by I P Wilkinson:

Acanthocythereis dunelmensis (Norman)
Cytheropteron nodosum Brady
Finmarchinella (Barentsovia) angulata (Sars)
Heterocyprideis sorbyana (Jones)
Leptocythere pellucida (Baird)
Roundstonia globulifera (Brady)
Sarsicytheridea bradii (Norman)
Sarsicytheridea punctillata (Brady)
Semicytherura nigrescens (Baird)

In addition, Dr J E Robinson of University College London has identified the following taxa (*personal communication*, Nov. 1990):

C. pyramidale Brady
Elofsonella concinna (Jones)
Finmarchinella (F.) finmarchica (Sars)

FORAMINIFERA I P Wilkinson (WH90/399R)
Ammonia batavus (Hofker)
Bolivina pseudopunctata Högland
B. pseudoplicata Heron-Allen & Earland
Buccella frigida (Cushman)
Buliminella borealis Haynes
Cassidulina reniforme Nörvang
C. teretis Tappan
Cibicides lobatulus (Walker & Jacob)
Elphidium asklundi Brotzen *
E. bartletti Cushman
E. ex. gr. clavatum Cushman
E. incertum (Williamson)
E. subarcticum Cushman
E. williamsoni Haynes
Fissurina lucida (Williamson)
Globigerina bulloides d'Orbigny
Guttulina sp.
Haynesina orbiculare (Brady)
Lagena substriata Williamson
Massilina secans (d'Orbigny) *
Pateoris hauerinoides (Rhumbler)
Pseudopolymorphina cf. *novangliae* (Cushman) *
Quinqueloculina seminulum (Linné)
Trifarina angulosa (Williamson)

* found only in original excavation of Main Pit

DINOFLAGELLATE CYSTS   R Harland (WH/91/391R)
Algidasphaeridium? minutum (Harland and Reid) Matsuoka and Bujak
Bitectatodinium tepikiense Wilson
Operculodinium centrocarpum (Deflandre and Cookson) Wall
Protoperidinium conicoides (Paulsen) Balech
P. conicum (Gran) Balech
Spiniferites lazus Reid
S. membranaceus (Rossignol) Sarjeant

Calcareous nannofossils
N Hine (WH/92/47R)

Emiliania huxleyi (Lohmann)
Geophyrocapsa sp. (Kamptner)

CLAVA SHELLY TILL

Materials collected from the type section (Figure 31, Section III; Plate 16)

Macrofossils

D K Graham (WH/89/318R)
GASTROPODA
Lunatia sp.

BIVALVIA
Portlandia arctica (Gray)

CIRRIPEDIA
Balanus sp. plates

Microfossils
I P Wilkinson (WH/89/335R)

OSTRACODA
Heterocyprideis sorbyana (Jones)

FORAMINIFERA
Elphidium bartletti Cushman
E. groenlandicum Cushman
Haynesina orbiculare (Brady)

ARDERSIER SILTS FORMATION

Fossils recorded from Jamieson's Pit [777 567] by Wallace (1883). The molluscan taxonomy has been revised by D K Graham on the assumption that the original identifications were correct. Modern names of the macrofossils are shown in parentheses. The foraminiferal and ostracod identifications have not been revised, because the taxa listed could apply to more than one species in present terminology and specimens were not available for examination.

Macrofossils

BIVALVIA
Astarte elliptica [Tridonta elliptica (Brown)]
A. sulcata (da Costa)
Leda pernula [Nuculana pernula (Müller)]
Tellina calcarea [Macoma calcarea (Gmelin)]

CIRRIPEDIA
Balanus crenatus (Bruguière)

Microfossils

OSTRACODA
Cytheridea papillosa (Bosquet)
Cytheridea punctillata (Brady)
Cytheridea sorbyana (Jones)
Cytherura sarsii (Brady)

FORAMINIFERA

Polymorphina acuminata (Williamson) [wrongly listed as *P. assumata*]
P. lactea (Walker & Jacob)
Polystomella arctica (Parker & Jones)
P. striatopunctata (Fitchel & Moll)

Plate 16   Examples of some marine species in the Clava Shelly Formation fauna:

a) *Littorina saxatilis rudis* (GSE 14846 × 3) (MNS 5213)
b) *Lunatia pallida* (GSE 14859 × 3) (MNS 5218)
c) *Nuculana pernula* (GSE 14845 × 3) (MNS 5212)
d) *Portlandia arctica* (GSE 14842 × 4) (MNS 5210)
e) *Tridontia elliptica* (GSE 14844 × 2.5) (MNS 5211)
f) *Acanthocythereis dunelmensis* (GSE 14851 × 50) (MNS 5734)
g) *Heterocyrideis sorbyana* (GSE 14853 × 50) (MNS 5733)
h) *Elphidium asklundi* (GSE 14855 × 50) (MNS 5732)
i) *Quinqueloculina seminulum* (GSE 14857 × 35) (MNS 5735)

APPENDIX 5

List of Geological Survey photographs

Copies of these photographs are deposited for public reference in the Library of the British Geological Survey, Murchison House, West Mains Road, Edinburgh EH9 3LA. Prints are available on application. The photographs belong to Series C, D, MNS, PMS and TS as indicated.

C 1408–9	Kame-terrace in the valley of Allt na Seanalaich, west of Shenachie.
C 1410–11	River and glaciofluvial terraces and the Creag a' Chròcain overflow channel viewed from the vicinity of Shenachie, Findhorn Valley.
C 1412	Southern end of the main Creag a' Chròcain overflow channel with talus and debris cones, Findhorn Valley.
C 1413–4	Panoramic view of the Findhorn Valley from the southern end of the Creag a' Chròcain overflow channel.
C 1415–6	Kame-terraces and glaciolacustrine delta near Ballachrochin, Findhorn Valley.
C 1417	The Findhorn Valley looking upstream towards the Streens Gorge from the vicinity of Ballachrochin, showing kame-terraces, river terraces and the northern end of the Creag a' Chròchain overflow channel.
C 1418 and D4269–71	Glaciofluvial and fluvial terraces, glaciofluvial fan and floodplain alluvium in the Findhorn Valley near Quilichan.
C 1419	Kame near Ballaggan, Findhorn Valley.
C 1420	River Findhorn undercutting cliffs in shattered granite, near Quilichan.
C 1421	River and glaciofluvial terraces, recessional moraine and deeply eroded gully in glacial drift near Quilichan.
C 1423 and 1426	Cliff sections in the Allt Breac Valley show glaciofluvial fan-delta deposits overlying till.
C 1424 and 1425	Gently sloping fan-delta flanking Allt Breac; exposures in the Cnoc a' Chinn Leith Fan-delta on the skyline.
C 1427–8	Degraded front of the Cnoc a' Chinn Leith Fan-delta at the head of the Allt Breac Valley.
C 1429–31	Deposits of the Cnoc a' Chinn Leith Fan-delta resting on till and decomposed granite.
C 1866–7 and D 4799–800	The delta of the River Ness at Inverness.
C 1868–9	Raised beaches, Munlochy Bay.
C 1870	Coast of the Black Isle, south of Fortrose.
C 1871–4	Chanonry Point and Fortrose.
C 1875–80 1878–80	Panoramas of twin forelands of Chanonry Point and Fort George at the mouth of the Inverness Firth.
C 1881–5 and D 1618	Earth pillars in till, Rosemarkie Glen.
C 1886	Raised beach, NE of Rosemarkie.
C 1887	Raised beach at base of cliff cut in gneiss, NE of Rosemarkie.
C 1888–9	View of raised beach and degraded cliffline cut in acid gneisses, NE of Rosemarkie.
C 1890–1	Banded psammite–semipelite intercalated with acid gneisses, cut by foliated granitic sheets and veins; all cut by carbonate veins. Shore of Moray Firth N of Rosemarkie.
C 1892–3	Cliffs in acid gneiss cut by low-angled reverse faults, Scatraig.
D 1848	Neolithic chambered cairns, Clava.
D 4269–71	The Findhorn Valley looking SW from Daless
D 4272	Psammite-rich till of the Dalcharn Upper Till Formation overlying sandstone-rich till of the Dalcharn Lower Till Formation. 'Dalcharn West' section, near Cawdor.
D 4273 4814	Dalcharn Gravel overlying Dearg Till, and 'Dalcharn East' section, near Cawdor.
D 4274	Dalcharn Gravel, 'Dalcharn East' section.
D 4275–7	Type section of the Dalcharn Biogenic Formation and Dalcharn Gravel, showing glacially and periglacially disturbed palaeosol including charcoal and compressed peat resting on whitened gravel; both deposits cut by fissures infilled with orange clayey sand. 'Dalcharn West' Section.
D 4278	Stratified matrix-supported diamicton, probably basal melt-out till, 'Dalcharn East' section.
D 4279	Glacially scratched boulders from sandstone-rich till, 'Dalcharn West' Section.
D 4280–1 4821–2	Sections through the Littlemill Eskers, and Mid Lairgs Gravel Pit, near Daviot.
D 4282	Gravel infilling former frost wedge in glaciofluvial sheet deposits lying between the Littlemill Eskers. Mid Lairgs Gravel Pit.
D 4283	Glaciofluvial gravel forming terraced spread between eskers. Mid Lairgs Gravel Pit.
D 4284	Mass-flow deposits interbedded with glaciolacustrine sediment and overlain by gravel. Mid Lairgs Gravel Pit.
D 4285–6	Esker gravels. Mid Lairgs Gravel Pit.
D 4287	Glaciolacustrine deposits with dropstone boulder. Ballachrochin, Findhorn Valley.
D 4288	Faulted and disrupted sands infilling glacial drainage channel. Riereach Burn, near Cawdor.
D 4289–90	Interbedded silt, sand and gravel infilling glacial drainage channel. Riereach Burn.
D 4291	Mass-flow deposit at margin of glacial drainage channel. Riereach Burn.
D 4292–3	Flat-topped kamiform ridge. Meikle Kildrummie.
D 4294–5	Kame-terraces and associated ice-marginal glacial meltwater channels on the south-eastern side of the valley of the River Nairn, near Cantraydoune.
D 4296	River terrace and flight of glaciofluvial terraces on SE side of the valley of the River Nairn, near Newton of Budgate.
D 4297–8	Glacial drainage channel at Dalcross, near Croy.
D 4299–301	Flemington Eskers, near Easter Lochend.
D 4302	Flandrian raised beaches and Fort George promontory viewed from the Main Flandrian Cliffline at Ardersier.
D 4303	Kettlehole at Redhill, near Balloch.
D 4304	Ice-moulded bedrock at Tordarroch, near Farr.
D 4305	Boulder train at Dalvourn, near Farr.

D 4306	*Roche moutonnée* at Scatraig, near Daviot.
D 4307 and 4817	Soft-sediment deformation structures in the Ardersier Silts, Ardersier
D 4308 and 4817	Large-scale folding in Ardersier Silts, Ardersier.
D 4309	Glacial striae on psammite at Dun Davie, near Daviot.
D 4345	Unconformity between Devonian breccio-conglomerate and Central Highland Migmatite Complex, in Glengeoullie Gorge.
D 4346	Unconformity between Devonian breccio-conglomerate and psammites of the Central Highland Migmatite Complex, Riereach Burn, near Glengeoullie.
D 4347	Migmatitic psammite with rare isoclinal folds, Central Highland Migmatite Complex, Riereach Burn near Glengeoulie.
D 4348	Porphyritic granodiorite of the Moy Granite, Moy Burn.
D 4349	Coarsely migmatitic gneissose semipelite of the Central Highland Migmatite Complex, Creag Bhuide.
D 4775	Red Craig Gravels; an early Devensian fan deposit, Rosemarkie Glen.
D 4776	Ardersier Silts Formation, showing homogenisation, folding and dislocation along clay-lined thrust planes, due to glacier push during the Ardersier Readvance, Jamieson's Pit, Ardersier.
D 4777–8	Glacitectonic sheared and thrusted sands and silts of the Ardersier Silts Formation, overlain by Baddock Till, Jamieson's Pit, Ardersier.
D 4801–4A	Type section of the Clava Shelly Clay, Clava.
D 4805	Clava Sand cut by clastic veins at the type section of the Clava Shelly Clay, Clava.
D 4806	Excavating sections in the Ardersier Silts at Kirkton, near Ardersier.
D 4807–9	Type section of the Ardersier Silts Formation, Kirkton.
D 4810–11	Main Flandrian Cliffline, Kirkton.
D 4811–2	Clava Sand at its type section, Clava.
D 4813	Dalcharn Lower Till, 'Dalcharn West' section.
D 4815	The Odhar Peat Formation at the Allt Odhar site, near Moy.
D 4816	Pine stumps beneath blanket peat, cliff section in the valley of Allt Odhar, near Moy.
D 4818	The Contorted Silts of Ardersier.
D 4819	Flandrian raised-beach gravel overlying glaciomarine deltaic deposits, Alturlie Point.
D 4820	Shape sorting and imbrication of pebbles in Late Devensian raised beach, Alturlie Point.
D 4823	Head gravel beneath Late Devensian till, Allt Breac Valley, west of Daless.
D 4824	Topset gravel and cross-bedded foreset sand of the Cnoc a' Chinn Leith Fan-delta at the head of Allt Breac Valley.
D 4825	Cross-bedded delta foresets, Cnoc a' Chinn Leith.
D 4826	Hill peat overlying topset gravel on sand, Cnoc a' Chinn Leith.
D 4827	Flemington Eskers, Meikle Kildrummie.
D 4828	Late Devensian raised shorelines, Munlochy.
D 4829	Gravel pit at Milton of Balnagowan, near Ardersier.
D4886A	The Black Isle coastline looking over the Chanonry and Ardersier peninsulas towards Inverness and the Great Glen.
D4887	Millbuie Sandstone showing channel fills and trough cross-bedding. Killen Burn, Black Isle.
D4888	Killen Mudstone showing interbedded mudstones and sandstones with sole structures. Killen Burn, Black Isle.
D4889	Goose Sandstone showing channelling and trough cross-bedding. Goose Burn, Black Isle.
D4890	Feldspathic gneissose psammite and amphibolite–hornblende schist. Rosemarkie Metamorphic Complex. Black Isle.
D4892	Layered gneissose psammite cut by foliated leucogranite sheets. Rosemarkie Metamorphic Complex. Black Isle.
D4893	Fenite breccia. Rosemarkie coast.
MNS 5210	*Portlandia arctica* (Gray)
MNS 5211	*Tridonta elliptica* (Brown)
MNS 5212	*Nuculana pernula* (Müller)
MNS 5213	*Littorina saxatilis* (Olivi) *rudis* (Maton)
MNS 5218	*Lunatia pallida* (Broderip & Sowerby)
MNS 5732	*Elphidium asklundi* Brotzen
MNS 5733	*Heterocyprideis sorbyana* (Jones)
MNS 5734	*Acanthocythereis dunelmensis* (Norman)
MNS 5735	*Quinqueloculina seminulum* (Linné)
MNS 5696/1	Leanach Sandstone showing cross-bedding. Clava Viaduct.
MNS 5696/2	Leanach Sandstone showing soft-sediment deformation. Clava Viaduct.
MNS 5696/3	Leanach Sandstone showing aeolian bedding. Clava Viaduct.
MNS 5696/4	Kilmuir Conglomerate in cliff below Wood Hill, Black Isle.
MNS 5696/5	Daviot Conglomerate showing fining-upward pebbly sandstone units. River Nairn section.
MNS 5696/6	Daviot Conglomerate. Close-up view of junction between two sandstone units. River Nairn section.
MNS 5696/7	Daviot Conglomerate. Carbonate nodules in thin shaly mudstone. River Nairn section.
MNS 5696/8	Avoch Sandstone showing soft-sediment deformation. Fort George building stone.
MNS 5696/9	Avoch Sandstone showing trough cross-bedding. Fort George building stone.
MNS 5750	*Coccosteus cuspidatus* Miller *ex* Agassiz (GSE 15084, T3389a). Killen Mudstone Member, Killen Burn.
MNS 5751	*Homosteus milleri* Traquair (GSE 1209). Hillhead Sandstone Formation, Hillhead Quarry, Dalcross.
MNS 5752	*Millerosteus minor* (Miller) (GSE 15085, T3426d). Hillhead Sandstone Formation, Hillhead Quarry, Dalcross.
PMS 611	Photomicrograph of hornfelsed migmatitic semipelite with andalusite and fibrolitic silimanite. Central Highland Migmatite Complex.
PMS 612	Photomicrograph of hornfelsed migmatic semipelite with symplectite. Central Highland Migmatite Complex.
PMS 613	Photomicrograph of hornfelsed Moy Granite with symplectites pseudomorphing biotite.
PMS 614	Photomicrograph of symplectite in hornfelsed Moy Granite.
PMS 615	Photomicrograph of granoblastic polygonal quartz in hornfelsed Moy Granite, Carn Beag.
PMS 616	Photomicrograph of hornfelsed semipelite with sillimanite and biotite intergrowths in Central Highland Migmatite Complex. Tom a'Ghealagaidh.

PMS 617   Photomicrograph of hornfelsed semipelite with fibrolitic sillimanite in Central Highland Migmatite Complex, Tom a'Ghealagaidh.

PMS 618   Photomicrograph of granitic gneiss with irregular textures of antiperthite and myrmekite. Central Highland Migmatite Complex, Allt Seileach.

PMS 622   Photomicrograph of relict subophitic texture in metabasite. Rosemarkie Metamorphic Complex, Black Isle.

PMS 625   Photomicrograph of feldspathic gneissose psammite in Rosemarkie Metamorphic Complex with blue riebeckitic amphibole. Black Isle.

PMS 626   Photomicrograph of feldspathic gneissose psammite in Rosemarkie Metamorphic Complex with felted masses of riebeckitic amphibole. Black Isle.

PMS 629   Photomicrograph of breccia cutting gneissose psammite in Rosemarkie Metamorphic Complex. Black Isle.

PMS 630   Photomicrograph of carbonate vein cutting feldspathic gneissose psammite in Rosemarkie Metamorphic Complex. Black Isle.

TS 2399   Slide of gneissose psammite showing truncation of D2-fold limbs in Central Highland Migmatite Complex. Beinn nan Cailleach.

REFERENCES

Most of the references listed below are held in the Library of the British Geological Survey at Keyworth, Nottingham. Copies of the references can be purchased from the Library subject to the current copyright legislation.

ABER, J S, CROOT, D G, and FENTON, M M. 1989. *Glaciotectonic landforms and structures.* (Dordrecht: Kluwer Academic Publishers.)

AGASSIZ, J L R. 1844–45. *Monographie des poissons fossiles du Vieux Gres Rouge ou Systeme Devonian (Old Red Sandstone) des Iles Britanniques et de Russie.* (Neuchâtel: Soleure, chez Jent & Gassmann.)

ANDREWS, I J, LONG, D, RICHARDS, P C, THOMSON, A R, BROWN, S, CHESHER, J A, and McCORMACK, M. 1990. *United Kingdom offshore regional report: the geology of the Moray Firth.* (London: HMSO for the British Geological Survey.)

ARMSTRONG, M. 1964. The geology of the region between the Alness River and the Dornoch Firth. Unpublished PhD thesis, University of Newcastle upon Tyne.

— 1977. The Old Red Sandstone of Easter Ross and the Black Isle. 25–33 in *The Moray Firth area geological studies.* GILL, G (editor). (Inverness: Inverness Field Club.)

AUTON, C A. 1990a. The Middle Findhorn Valley. 74–96 in *Beauly to Nairn. Field guide.* AUTON, C A, FIRTH, C R. and MERRITT, J W (editors). (Cambridge: Quaternary Research Association.)

— 1990b. Early models for the late-Devensian glaciation and deglaciation of the upland areas. 3–5 in *Beauly to Nairn. Field guide.* AUTON, C A, FIRTH, C R, and MERRITT, J W (editors). (Cambridge: Quaternary Research Association.)

— 1992. Scottish landform examples — 6: The Flemington Eskers. *Scottish Geographical Magazine,* Vol. 108, No. 3, 190–196.

— 1993. Dalcharn. 154–159 in *Quaternary of Scotland. Geological conservation review series,* 6. GORDON, J E, and SUTHERLAND, D G (editors). (London: Chapman and Hall.)

— FIRTH, C R, and MERRITT, J W. 1990. *Beauly to Nairn. Field guide.* (Cambridge: Quaternary Research Association.)

BALLANTYNE, C K. 1991. Holocene geomorphic activity in the Scottish Highlands. *Scottish Geographical Magazine,* Vol. 107, 84–98.

BAMFORD, D, NUNN, K, PRODEHL, C, and JACOB, B. 1978. LISPB-IV. Crustal structure of Northern Britain. *Geophysical Journal of the Royal Astronomical Society,* Vol. 54, 43–60.

BANHAM, P H. 1975. Glacitectonic structures: a general discussion with particular reference to the contorted drift of Norfolk. 69–94 in *Ice ages: ancient and modern.* WRIGHT, A E, and MOSELEY, F (editors). (Liverpool: Seel House Press.)

BARTON, P J. 1992. LISPB revisited: a new look under the Caledonides of northern Britain. *Geophysical Journal International,* Vol. 110, 371–391.

BATES, R L, and JACKSON, J A. 1987. *Glossary of geology* (3rd edition). (Alexandria, Virginia: American Geological Institute.)

BEAUDRY, L M, and PRICHONNET, G. 1991. Late Glacial de Geer moraines with glaciofluvial sediment in the Chapais area, Québec (Canada). *Boreas,* Vol. 20, 377–394.

BELL, D. 1891. Phenomena of the Glacial Epoch: II. The "great submergence". *Transactions of the Geological Society of Glasgow,* Vol. 9, 100–138.

— 1893a. On the alleged proofs of submergence in Scotland during the Glacial Epoch. *Report of the British Association for 1892,* 713–714.

— 1893b. On the alleged proofs of submergence in Scotland during the Glacial Epoch. I. Chapelhall, near Airdrie. *Transactions of the Geological Society of Glasgow,* Vol. 9, Pt. 2, 321–344.

— 1895a. On the alleged proofs of submergence in Scotland during the Glacial Epoch. II. Clava and other northern localities. *Transactions of the Geological Society of Glasgow,* Vol. 19, 105–120.

— 1895b. Notes on "The Great Ice Age" in relation to the question of submergence. *Geological Magazine,* Vol. 2, 321–326, 348–355 and 402–405.

— 1896. The Ayrshire 'shell-beds'. *Geological Magazine,* Vol. 3, 335–336.

— 1897a. The "great submergence" again: Clava. *Geological Magazine,* Vol. 4, 27–30 and 63–68.

— 1897b. The high-level shelly clays and Mr Mellard Reade. *Geological Magazine,* Vol. 4, 189–190.

BENNETT, K D. 1984. The post-glacial history of *Pinus sylvestris* in the British Isles. *Quaternary Science Reviews,* Vol. 3, No. 2/3, 133–155.

BIRKS, H J B, and RANSOM, M E. 1969. An interglacial peat at Fugla Ness, Scotland. *New Phytologist,* Vol. 68, 777–796.

BLOODWORTH, A J. 1990. Clay mineralogy of Quaternary sediments from the Dalcharn Interglacial site. 60–61 in *Beauly to Nairn. Field guide.* AUTON, C A, FIRTH, C R, and MERRITT, J W (editors). (Cambridge: Quaternary Research Association.)

BLUCK, B J. 1967. Sedimentation of beach gravels: examples from South Wales. *Journal of Sedimentary Petrology,* Vol. 37, 128–156.

BOULTON, G S. 1968. Flow tills and related deposits on some Vestspitsbergen glaciers. *Journal of Glaciology,* Vol. 7, No. 51, 391–412.

— 1970. On the deposition of subglacial and melt-out tills at the margins of certain Svalbard glaciers. *Journal of Glaciology,* Vol. 9, 231–245.

— 1971. Till genesis and fabric in Svalbard. 41–72 in *Till: a symposium.* GOLDTHWAIT, R P (editor). (Columbus: Ohio State University Press.)

— 1972. The role of thermal regime in glacial sedimentation. *Institute of British Geographers Special Publication,* No. 4, 1–19.

— 1976. The origin of glacially fluted surfaces — observations and theory. *Journal of Glaciology,* Vol. 17, No. 76, 287–309.

— 1979. Processes of glacier erosion on different substrata. *Journal of Glaciology*, Vol. 23, 15–38.

— 1986. Push-moraines and glacier-contact fans in marine and terrestrial environments. *Sedimentology*, Vol. 33, 677–698.

— 1990. Sedimentary and sea-level changes during glacial cycles and their control on glacimarine-facies architecture. 15–52 in Glacimarine environments: processes and sediments. DOWDESWELL, J A, and SCOURSE, J D (editors). *Special Publication of the Geological Society of London*, No. 53.

— DENT, D L, and MORRIS, E M. 1974. Subglacial shearing and crushing, and the role of water pressure in tills from SE Iceland. *Geografiska Annaler, Series A Physical Geography*, Vol. 56A, No. 314, 135–145.

— and DEYNOUX, M. 1981. Sedimentation in glacial environments and the identification of tills and tillites in ancient sedimentary sequences. *Precambrian Research*, Vol. 15, 397–422.

— and PAUL, M A. 1976. The influence of genetic processes on some geotechnical properties of glacial tills. *Quarterly Journal of Engineering Geology*, Vol. 9, 159–194.

BOURGEOIS, J, and LEITHOLD, E L. 1984. Wave-worked conglomerates—depositional processes and criteria for recognition. 331–343 in Sedimentology of gravels and conglomerates. KOSTER, E H, and STEEL, R J (editors). *Memoir of the Canadian Society of Petroleum Geologists*, No. 10.

BOWEN, D Q. 1989. The last interglacial–glacial cycle in the British Isles. *Quaternary International*, Vol. 3/4, 41–47.

BRAND, P J. 1991. Report on fossil fish held in Edinburgh from 1 50 000 Sheet 84W. *British Geological Survey Technical Report*, WH/91/340R.

BREEMAN, O VAN, and PIASECKI, M A J. 1983. The Glen Kyllachy Granite and its bearing on the nature of the Caledonian Orogeny in Scotland. *Journal of the Geological Society of London*, Vol. 140, Pt. 1, 47–62.

BREMNER, A. 1939a. The glaciation of Moray and ice movements in the north of Scotland. *Transactions of the Edinburgh Geological Society*, Vol. 13, 17–56.

— 1939b. The River Findhorn. *Scottish Geographical Magazine*, Vol. 55, 65–85.

— 1942. The origin of the Scottish river system. *Scottish Geographical Magazine*, Vol. 58, 15–20, 54–59 and 99–103.

BRITISH GEOLOGICAL SURVEY. 1987. *Regional geochemical atlas: Great Glen*. (Keyworth, Nottinghamshire: British Geological Survey.)

BROWITT, C W A, BURTON, P W, and LIDSTER, R. 1976. Seismicity of the Inverness Region. *Institute of Geological Sciences, Global Seismology Unit, Report* No. 76.

CAMERON, T D J, STOKER, M S, and LONG, D. 1987. The history of Quaternary sedimentation in the UK sector of the North Sea Basin. *Journal of the Geological Society of London*, Vol. 144, Pt. 1, 43–58.

CHARLESWORTH, J K. 1956. Lateglacial history of the Highlands and Islands of Scotland. *Transactions of the Royal Society of Edinburgh*, Vol. 63, 769–928.

CHEEL, R J, and RUST, B R. 1982. Coarse-grained facies of glacio-marine deposits near Ottawa, Canada. 279–295 in *Research in glacial, glacio-fluvial and glacio-lacustrine systems: Proceedings, 6th Guelph Symposium on Geomorphology, 1980*. DAVIDSON-ARNOTT, R, NICKLING, W, and FAHEY, B D (editors). (Norwich: Geo Books.)

CHESHER, J A. 1984. Moray–Buchan. Sheet 57N-04W. Sea-bed sediments and Quaternary geology. British Geological Survey 1:250 000 Series. (Southampton: Ordnance Survey for British Geological Survey.)

— and LAWSON, D. 1983. The geology of the Moray Firth. *Report of the Institute of Geological Sciences*, No. 83/5.

CHURCH, M, and RYDER, J M. 1972. Paraglacial sedimentation: a consideration of fluvial processes conditioned by glaciation. *Geological Society of America Bulletin*, Vol. 83, 3059–3072.

CLEMMENSEN, L B, and HOUMARK-NIELSON, M. 1981. Sedimentary features of a Weichselian glaciolacustrine delta. *Boreas*, Vol. 10, 229–245.

COOPE, G R. 1959. A Late Pleistocene insect fauna from Chelford, Cheshire. *Proceedings of the Royal Society*, Vol. B151, 70–86.

COWAN, E A, and POWELL, R D. 1990. Suspended sediment transport and deposition of cyclically interlaminated sediment in a temperate glacial fjord, Alaska, USA. 75–89 in *Glacimarine environments: processes and sediments*. DOWDESWELL, J A, and SCOURSE, J D (editors). *Special Publication of the Geological Society of London*, No. 53

CULLINGFORD, R A, SMITH, D E, and FIRTH, C R. 1991. The altitude and age of the Main Postglacial Shoreline in eastern Scotland. *Quaternary International*, Vol. 9, 39–52.

DAVENPORT, C A, and RINGROSE, P S. 1987. Deformation of Scottish Quaternary sediment sequences by strong earthquake motions. 299–314 in Deformation of sediments and sedimentary rocks. JONES, M E, and PRESTON, R M F (editors). *Special Publication of the Geological Society of London*, No. 29.

DEANS, T, GARSON, M S, and COATS, J S. 1971. Fenite-type soda metasomatism in the Great Glen, Scotland. 145–147 in *Nature, London, Physical Science*, Vol. 234, No. 44.

DEWEY, H, and DINES, H G. 1923. Tungsten and manganese ores. *Special Report on the Mineral Resources of Great Britain, Memoir of the Geological Survey of Great Britain*, Vol. 1.

DONOVAN, R N. 1980. Lacustrine cycles, fish ecology and stratigraphic zonation in the Middle Devonian of Caithness. *Scottish Journal of Geology*, Vol. 16, 35–50.

— FOSTER, R J, and WESTOLL, T S. 1974. A stratigraphical revision of the Old Red Sandstone of north-eastern Caithness. *Transactions of the Royal Society of Edinburgh*, Vol. 69, 167–201.

DREIMANIS, A. 1989. Tills: their genetic terminology and classification. 17–83 in *Genetic classification of glacigenic deposits*. GOLDWAITE, R P, and MATSCH, C L (editors). (Rotterdam: Balkema.)

FAIRBAIRN, W A. 1967. Erosion in the Findhorn Valley. *Scottish Geographical Magazine*, Vol. 83, No. 1, 46–52.

FAIRBRIDGE, R W. 1978. Delta sedimentation. 240–244 in *The encyclopedia of sedimentology*. FAIRBRIDGE, R W, and BOURGEOIS, J (editors). (Stroudsburg, Pa.: Dowden, Hutchinson and Ross, Inc.)

FIRTH, C R. 1984. Raised shorelines and ice limits in the inner Moray Firth and Loch Ness areas, Scotland. Unpublished PhD thesis, Coventry Polytechnic.

— 1986. Isostatic depression during the Loch Lomond Stadial: preliminary evidence from the Great Glen, northern Scotland. *Quaternary Newsletter*, No. 48, 1–9.

— 1989a. Late Devensian raised shorelines and ice limits in the inner Moray Firth, northern Scotland. *Boreas*, Vol. 18, 5–21.

— 1989b. A reappraisal of the supposed Ardersier Readvance, inner Moray Firth. *Scottish Journal of Geology*, Vol. 25, 249–261.

— 1990a. Late-Devensian relative sea-level changes associated with the deglaciation of the Inverness Firth and Beauly Firths. 5–9 in *Beauly to Nairn. Field guide*. AUTON, C A, FIRTH, C R, and MERRITT, J W (editors). (Cambridge: Quaternary Research Association.)

— 1990b. An excursion to view the evidence of changes in relative sea level during, and following the deglaciation of the coastal lowlands bordering the northern shore of the Beauly Firth and the southern shore of the Inverness Firth. 96–116 in *Beauly to Nairn. Field guide*. AUTON, C A, FIRTH, C R, and MERRITT, J W (editors). (Cambridge: Quaternary Research Association.)

— and HAGGART, B A. 1989. Loch Lomond Stadial and Flandrian shorelines in the inner Moray Firth area, Scotland. *Journal of Quaternary Science*, Vol. 4, 37–50.

FLETT, J S. 1905. On the petrographical characters of the inliers of Lewisian rocks among the Moine gneisses of the north of Scotland. *Summary of progress for 1905, Geological Survey of Great Britain*. (London: Her Majesty's Stationery Office.)

FRASER, J. 1877. Report of the field excursion to the Nairn Valley. *Transactions of the Inverness Scientific Society and Field Club*, Vol. 1, 63–64.

— 1879. The recent formations and glacial phenomena of Strathnairn. *Transactions of the Inverness Scientific Society and Field Club*, Vol. 1, 211–223; also in *Transactions of the Edinburgh Geological Society*, Vol. 4, 1883, 55–66.

— 1882a (1883). The shell-bed at Clava. *Transactions of the Inverness Scientific Society and Field Club*, Vol. 2, 168–176.

— 1882b (1883). First notice of a post-Tertiary shell-bed at Clava in Nairnshire, indicating an Arctic climate and a sea-bed at a height of 500 feet. *Transactions of the Edinburgh Geological Society*. Vol. 4, 136–142.

GALLAGHER, M J. 1984. Barite deposits and potential of Scotland. *Transactions of the Institution of Mining and Metallurgy*, Sect. A: Mining Industry, Vol. 93, A130–A132.

GARSON, M S, and PLANT, J. 1973. Alpine type ultramafic rocks and episodic mountain building in the Scottish Highlands. 34–38 in *Nature, London, Physical Science*, Vol. 242, No. 116.

— COATS, J S, ROCK, N M S, and DEAN, S, T. 1984. Fenites, breccia dykes, albitites and carbonatitic veins near the Great Glen Fault, Inverness, Scotland. *Journal of the Geological Society of London*, Vol. 141, Pt. 4, 711–732.

GILBERT, R. 1990. Rafting in glacimarine environments. 105–120 in Glacimarine environments: processes and sediments. DOWDESWELL, J A, AND SCOURSE J D (editors). *Special Publication of the Geological Society of London*, No. 53.

GLOVER, B W, and WINCHESTER, J A. 1989. The Grampian Group: a major late Proterozoic clastic sequence in the Central Highlands of Scotland. *Journal of the Geological Society of London*, Vol. 146, Pt. 1, 85–96.

GORDON , G. 1859. On the geology of the lower or northern part of the province of Moray: its history, present state of enquiry and points for future examination. *Edinburgh New Philosophical Journal*, New Series, Vol. 9, 14–58.

GORDON, J E. 1990. The high-level 'Arctic' shelly clay of Clava, Inverness-shire: a century of research. 17–20 in *Beauly to Nairn. Field guide*. AUTON, C A, FIRTH, C R, and MERRITT, J W (editors). (Cambridge: Quaternary Research Association.)

— 1993a. Littlemill. 181–184 in *Quaternary of Scotland. Geological conservation review series;* 6. GORDON, J E, and SUTHERLAND, D G (editors). (London: Chapman and Hall.)

— 1993b. Clava. 165–170 in *Quaternary of Scotland. Geological conservation review series;* 6. GORDON, J E, and SUTHERLAND, D G (editors). (London: Chapman and Hall.)

— and AUTON, C A, 1993. The Kildrummie Kames. 176–181 in *Quaternary of Scotland. Geological conservation review series;* 6. GORDON, J E, and SUTHERLAND, D G (editors). (London: Chapman and Hall.)

— and MERRITT, J W. 1993. Ardersier. 170–174 in *Quaternary of Scotland. Geological conservation review series;* 6. GORDON, J E, and SUTHERLAND, D G (editors). (London: Chapman and Hall.)

GRAHAM, D K. 1974a. Palaeontological report on the macrofauna from Kessock Bridge Borehole No. 2. *Palaeontological Report of the British Geological Survey*, PDS 74/20. (Unpublished).

— 1974b. Palaeontological report on the macrofauna from Kessock Bridge boreholes Nos. 1–8. *Palaeontological Report of the British Geological Survey*, PDS 74/30. (Unpublished).

— 1989a. Palaeontological report on Quaternary material from the Clava area held in the collections of the Edinburgh Office of the British Geological Survey. *British Geological Survey Technical Report*, WH/89/292R.

— 1989b. Palaeontological report on Quaternary material from Drummore of Clava. *British Geological Survey Technical Report*, WH/89/318R.

— 1990a. Palaeontological report on sediment samples from the Clava area. *British Geological Survey Technical Report*, WH/90/77R.

— 1990b. Palaeontological report on recently collected fossil material from a temporary exposure at Clava. *British Geological Survey Technical Report*, WH/90/388R.

— 1990c. The fauna of the Clava shelly clay and shelly till. 20–23 in *Beauly to Nairn. Field guide*. AUTON, C A, FIRTH, C R, and MERRITT, J W (editors). (Cambridge: Quaternary Research Association.)

— 1991a. Report on the Fraser Collection of macrofossils from Clava held by the Inverness Museum and Art Gallery. *British Geological Survey Technical Report*, WH/91/52R.

— HARLAND, R, HINE, N, ROBINSON, J E, and WILKINSON, I P. 1993. The biostratigraphy of the Clava shelly clay and till. Appendix 1. 776–779 *in* The high-level, shell-bearing deposits of Clava, Inverness-shire, and their origin as glacial rafts. MERRITT, J W. *Quaternary Science Reviews*, Vol. 11, 759–779.

GRAY, J M, and LOWE, J J (editors). 1977. *Studies in the Scottish Lateglacial environment*. (Oxford: Pergamon Press.)

GREGORY, J W. 1926. The moraines, boulder clay, and glacial sequence of south-western Scotland. *Transactions of the Geological Society of Glasgow*, Vol. 17, 354–376.

HAGGART, B A. 1982. Flandrian sea-level changes in the Moray Firth area. Unpublished PhD thesis, University of Durham.

— 1986. Relative sea-level changes in the Beauly Firth, Scotland. *Boreas*, Vol. 15, 191–207.

— 1987. Relative sea-level changes in the Moray Firth area, Scotland. 67–108 *in* Sea level changes. SHENNAN, I, and TOOLEY, M J (editors). *Institute of British Geographers Special Publication*, Vol. 20.

— 1988.   The stratigraphy, depositional environment and dating of a possible tidal surge deposit in the Beauly Firth area, northeast Scotland.   *Palaeogeography, Palaeoclimatology, Palaeoecology*, Vol. 66, 215–230.

HALL, A M.   1986.   Deep weathering patterns in north-east Scotland and their geomorphological significance.   *Zeitschrift für Geomorphologie, Neue Folge*, Vol. 30, 407–422.

— 1991.   Pre-Quaternary landscape evolution in the Scottish Highlands.   *Transactions of the Royal Society of Edinburgh: Earth Sciences*, Vol. 82, 1–26.

HARKNESS, D D.   1990.   Radiocarbon dating of the Odhar Peat. 69–70 in *Beauly to Nairn. Field guide*. AUTON, C A, FIRTH, C R, and MERRITT, J W (editors). (Cambridge: Quaternary Research Association.)

HARLAND, R.   1988.   Dinoflagellate cyst analysis of Late-Glacial marine clay from Scotland 1:50 000 Sheet 84W.   *British Geological Survey Technical Report*, WH/88/102R.

— 1991.   Dinoflagellate cyst analysis of the Clava Shelly Clay. *British Geological Survey Technical Report*, WH/91/391R.

HARRIS, A L.   1978.   Metamorphic rocks of the Moray Firth District.   9–24 in *The Moray Firth area geological studies*. GILL, G (editor). (Inverness: Inverness Field Club.)

— and PEACOCK, J D.   1969.   Sand and gravel resources of the Inner Moray Firth.   *Report of the Institute of Geological Sciences*, No. 69/9.

— and PITCHER, W S.   1975.   The Dalradian Supergroup. 52–75 *in* A correlation of Precambrian rocks in the British Isles. HARRIS, A L, HOLLAND, C E, and LEAKE, B E (editors). *Special Report of the Geological Society of London*, No. 6.

— BALDWIN, C T, BRADBURY, H J, JOHNSON, H D, and SMITH, R A. 1978.   Ensialic basin sedimentation: the Dalradian Supergroup.   115–138 *in* Crustal evolution in north-western Britain and adjacent regions. BOWES, D R, and LEAKE, B E (editors). *Geological Journal Special Issue*, No. 10.

— PARSONS, L M, HIGHTON, A J, and SMITH, D I.   1981. New/Old Moine relationships between Fort Augustus and Inverness.   *Journal of Structural Geology*, Abst. 3, 187–188.

HASELOCK, P J.   1984.   The systematic geochemical variation between two tectonically separate successions in the southern Monadhliaths, Inverness-shire.   *Scottish Journal of Geology*, Vol. 20, Pt. 2, 191–205.

— and LESLIE, A G.   1992.   Polyphase deformation in Grampian Group rocks of the Monadhliath, defined by a ground-magnetic survey.   *Scottish Journal of Geology*, Vol. 28, Pt 2, 81–87.

— WINCHESTER, J A, and WHITTLES, K H.   1982.   The stratigraphy and structure of the southern Monadhliath Mountains between Loch Killin and upper Glen Roy.   *Scottish Journal of Geology*, Vol. 18, Pt. 4, 275–290.

HEIJNIS, H.   1990.   Dating of the Odhar Peat at the Allt Odhar site by the Uranium Series Disequilibrium Dating Method. 72–74 in *Beauly to Nairn. Field guide*. AUTON, C A, FIRTH, C R, and MERRITT, J W (editors). (Cambridge: Quaternary Research Association.)

— and VAN DER PLICHT, J.   1992.   Uranium/thorium dating of Late Pleistocene peat deposits in NW Europe, uranium/thorium isotope systematics and open-system behaviour of peat layers.   *Chemical Geology (Isotope Geoscience Section)*, Vol. 94, 161–171.

HIGHTON, A J.   1986.   Pre-and post-Caledonian events in the Moine south of the Great Glen.   Unpublished PhD thesis, University of Liverpool.

— 1992.   The tectonostratigraphical significance of pre-750 Ma metagabbros within the northern Central Highlands, Inverness-shire.   *Scottish Journal of Geology*, Vol. 28, Pt. 1, 71–76.
— In preparation.   The geology of the Aviemore and mid-Strathspey district.   *Memoir of the British Geological Survey*, Sheet 74E (Scotland).

HILLIER, S, and MARSHALL, J E A.   1992.   Organic maturation, thermal history and hydrocarbon generation in the Orcadian Basin, Scotland.   *Journal of the Geological Society of London*, Vol. 149, Pt. 4, 491–502.

HINE, N.   1992.   Calcareous nannofossil analysis of samples of Clava Clay, Inverness.   *British Geological Survey Technical Report*, WH/92/47R.

HODGSON, D M.   1986.   A study of fluted moraines in the Torridon area, NW Scotland.   *Journal of Quaternary Science*, Vol. 1, 109–118.

HOPPE, G.   1959.   Glacial morphology and inland ice recession in northern Sweden.   *Geografiska Annaler, Series A Physical Geography*, Vol. 41, No. 4, 193–212.

HORNE, J.   1880.   Geology of the Nairn and Findhorn. *Transactions of the Inverness Scientific Society and Field Club*, Vol. 7, 283–286.

— 1923.   The geology of the Lower Findhorn and Lower Strath Nairn.   *Memoir of the Geological Survey of Great Britain*, Sheet 84 and part of 94 (Scotland).

— and HINXMAN, L W.   1914.   The geology of the country around Beauly and Inverness.   *Memoir of the Geological Survey of Great Britain*, Sheet 83 (Scotland).

— ROBERTSON, D, JAMIESON, T F, FRASER, J, KENDALL, P F. and BELL, D.   1894.   The character of the high-level shell-bearing deposits at Clava, Chapelhall and other localities.   *Report of the British Association for the Advancement of Science for 1893*, 483–514; also in *Transactions of the Inverness Scientific Society and Field Club*, Vol. 4 (1888–1895), 300–339.

HYDROGRAPHIC OFFICE.   1981.   *Approaches to Cromarty Firth and Inverness Firth*.   Admiralty Chart 1107. (Taunton: Hydrographer to the Navy.)

— 1989.   *Inverness Firth*.   Admiralty Chart 1078. (Taunton: Hydrographer to the Navy.)

IMBRIE, J, HAYS, J D, MARTINSON, D G, MCINTYRE, A, MIX, A C, MORLEY, J J, PISIAS, N G, PRELL, W L, and SHACKLETON, N J.   1984. The orbital theory of Pleistocene climate: support from revised chronology of the marine ^{18}O record.   269–306 in *Milankovic and climate*. BERGER, A L, IMBRIE, J, HAYS, J, KUKLA, G J, and SALTZMAN, B (editors). (Dordrecht: Reidel.)

JAMIESON, T F.   1874.   On the last stage of the Glacial Period in North Britain.   *Quarterly Journal of the Geological Society of London*, Vol. 30, 317–338.

— 1882.   On the red clay of the Aberdeenshire coast and the direction of ice-movement in that quarter.   *Quarterly Journal of the Geological Society of London*, Vol. 38, 160–177.

— 1906.   The glacial period in Aberdeenshire and the southern border of the Moray Firth.   *Quarterly Journal of the Geological Society of London*, Vol. 62, 13–39.

JOHNSTONE, G S.   1975.   The Moine succession.   39–42 *in* A correlation of Precambrian rocks in the British Isles. HILLAND, C E, and LEAKE, B E (editors). *Special Report of the Geological Society of London*, No. 6.

— and MYKURA, W.   1989.   *British regional geology: The Northern Highlands* (4th edition, revised).   (London: HMSO for British Geological Survey.)

JOLLY, W. 1876. Excursions to Strathnairn. *Transactions of the Inverness Scientific Society and Field Club,* Vol. 1, 211–223.

— 1880. Notes on the transportation of rocks found on the south shores of the Moray Firth. *Proceedings of the Royal Society of Edinburgh,* Vol. 10, 178–185.

KIRK, W, RICE, R J, and SYNGE, F M. 1966. Deglaciation and vertical displacement of shorelines in Wester and Easter Ross. *Transactions of the Institute of British Geographers,* Vol. 39, 65–78.

LAMBERT, R ST J, and POOLE, A B. 1964. The relationship of Moine schists and Lewisian gneisses near Mallaigmore, Inverness-shire. *Proceedings of the Geologists' Association,* Vol. 75, 1–14.

LE MAITRE, R W (editor). 1989. *A classification of igneous rocks and glossary of terms.* (Oxford: Blackwell Scientific Publications.)

LINDSAY, N G, HASELOCK, P J, and HARRIS, A L. 1989. The extent of Grampian orogenic activity in the Scottish Highlands. *Journal of the Geological Society of London,* Vol. 146, Pt. 5, 733–735.

LINTON, D L. 1951. Problems of Scottish scenery. *Scottish Geographical Magazine,* Vol. 67, 69–79.

MACDONALD, K. 1881. Glacial drift in the Craggie Burn. *Transactions of the Inverness Scientific Society and Field Club,* Vol. 2, 47–53.

MACKIE, W. 1899. The sands and sandstones of eastern Moray. *Transactions of the Inverness Scientific Society and Field Club,* Vol. 5, 34–61.

— 1905. Some notes on the distribution of erratics over eastern Moray. *Transactions of the Edinburgh Geological Society,* Vol. 8, 91–97.

MACKIEWICZ, N E, POWELL, R D, CARLSON, P R, and MOLNIA, B F. 1984. Interlaminated ice-proximal glacimarine sediments in Muir Inlet, Alaska. *Marine Geology,* Vol. 57, 113–147.

MALCOLMSON, J G. 1859. On the relations of the different parts of the Old Red Sandstone in which organic remains have recently been discovered, in the counties of Moray, Nairn, Banff and Inverness. *Quarterly Journal of the Geological Society of London,* Vol. 15, 336–352.

MARTINSON, D G, PISIAS, N G, HAYS, J D, IMBRIE, J, MOORE, T C, and SHACKLETON, N J. 1987. Age dating and orbital theory of the ice ages: development of a high resolution 0 to 300 000-year chronostratigraphy. *Quaternary Research,* Vol. 27, 1–29.

MAWDSLEY, J B. 1936. The wash-board moraines of the Opawica-Chibougamau area, Québec. *Transactions of the Royal Society of Canada,* Vol. 30, Pt. 4, 9–12.

MAY, F, and HIGHTON, A J. In preparation. Geology of the Invermoriston district. *Memoir of the British Geological Survey.* Sheet 73W (Scotland).

MERRITT, J W. 1990a. The lithostratigraphy at Clava and new evidence for the shell-bearing deposits being glacially-transported rafts. 24–40 in *Beauly to Nairn. Field guide.* AUTON, C A, FIRTH, C R, and MERRITT, J W (editors). (Cambridge: Quaternary Research Association.)

— 1990b. The Allt Odhar Interstadial Site, Moy, Inverness-shire: lithostratigraphy. 62–69 in *Beauly to Nairn. Field guide.* AUTON, C A, FIRTH, C R, and MERRITT, J W (editors). (Cambridge: Quaternary Research Association.)

— 1990c. New evidence for glaciomarine deltaic deposits at Alturlie Gravel Pit? 102–105 in *Beauly to Nairn. Field guide.* AUTON, C A, FIRTH, C R, and MERRITT, J W (editors). (Cambridge: Quaternary Research Association.)

— 1992. The high-level, marine shell-bearing deposits of Clava, Inverness-shire, and their origin as glacial rafts. *Quaternary Science Reviews,* Vol. 11, 759–779.

— 1993. Allt Odhar. 159–164 in *Quaternary of Scotland. Geological conservation review series,* 6. GORDON, J E, and SUTHERLAND, D G (editors). (London: Chapman and Hall.)

— and AUTON, C A. 1990. The Dalcharn Interglacial site, near Cawdor, Nairnshire: lithostratigraphy. 41–54 in *Beauly to Nairn. Field guide.* AUTON, C A, FIRTH, C R, and MERRITT, J W (editors). (Cambridge: Quaternary Research Association.)

— 1993. Notes on exposures and natural sections in drift deposits occurring on Geological Sheet 84W (Fortrose), SE of Inverness, Scotland. *British Geological Survey Technical Report,* WA/93/28R.

MILES, R S. 1968. The Old Red Sandstone Antiarchs of Scotland: Family Bothriolepididae. *Palaeontographical Society Monograph,* No. 522, Vol. 122, 1–130.

— and WESTOLL, T S. 1963. Two new genera of coccosteid Arthrodira from the Middle Old Red Sandstone of Scotland and their stratigraphical distribution. *Transactions of the Royal Society of Edinburgh,* Vol. 65, 179–210.

— and WESTOLL, T S. 1968. The placoderm fish *Coccosteus cuspidatus* Miller ex. Agassiz from the Middle Old Red Sandstone of Scotland. Part 1, Descriptive morphology. *Transactions of the Royal Society of Edinburgh,* Vol. 67, 373–476.

MILLER, H. 1841. The Old Red Sandstone or new walks in an old field. (Edinburgh: John Johnstone.)

— 1858. *Cruise of the Betsy and rambles of a geologist.* (Edinburgh: Constable.)

MORAN, S R. 1971. Glaciotectonic structures in drift. 127–148 in *Till: a symposium.* GOLDWAIT, R P (editor). (Columbus: Ohio State University Press.)

MULGREW, J R. 1985. Manganese and associated mineralisation in the Dalroy area. Unpublished MSc thesis, University of Strathclyde.

MURCHISON, Sir R I. 1859. On the sandstones of Morayshire (Elgin & c.) containing reptilian remains; and on their relations to the Old Red Sandstone of that country. *Quarterly Journal of the Geological Society of London,* Vol. 15, 419–439.

MUSSON, R M W, NEILSON, G, and BURTON, P W. 1987. Macroseismic reports on historical British earthquakes, X: The Great Glen. *Report of the Global Seismology Unit, British Geological Survey,* No. 347.

MYKURA, W. 1980. Barytes Quarry at Balfriesh. *Open-file report of the Highlands and Islands Unit, Institute of Geological Sciences.*

— 1983. The Old Red Sandstone east of Loch Ness, Inverness-shire. *Report of the Institute of Geological Sciences,* No. 82/13.

— 1991. Old Red Sandstone. 297–346 in *Geology of Scotland* (3rd edition, revised). CRAIG, G Y (editor). (London: The Geological Society.)

— ROSS, D L, and MAY, F. 1978. Sand and gravel resources of the Highlands Region. *Report of the Institute of Geological Sciences,* No. 78/8.

NEMEC, W, and STEEL, R J. 1984. Alluvial and coastal conglomerates: their significant features and some comments on gravelly mass-flow deposits. 1–31 *in* Sedimentology of gravels and conglomerates. KOSTER, E H, and STEEL, R J (editors). *Memoir of the Canadian Society of Petroleum Geologists,* No. 10.

NICHOLSON, K. 1983.   Manganese mineralisation in Scotland. Unpublished PhD thesis, University of Strathclyde.

— 1987.   Rhodochrosite from Islay, Argyllshire and Dalroy, Inverness-shire, Scotland.   *Mineral Magazine*, Vol. 51, 677–680.

O'BRIEN, C. 1985.   The petrogenesis and geochemistry of the British Caledonian granites with special reference to mineralized intrusions.   Unpublished PhD thesis, University of Leicester.

OGILVIE, A G. 1914.   The physical geography of the entrance to the Inverness Firth.   *Scottish Geographical Magazine*, Vol. 30, 21–35.

— 1923.   The physiography of the Moray Firth coast. *Transactions of the Royal Society of Edinburgh*, Vol. 53, 377–404.

O'SULLIVAN, P E. 1976.   Pollen analysis and radiocarbon dating of a core from Loch Pityoulish, eastern Highlands, Scotland.   *Journal of Biogeography*, Vol. 3, 293–302.

PARNELL, J. 1983.   Ancient duricrusts and related rocks in perspective: a contribution from the Old Red Sandstone. 197–209 in Residual deposits: surface related weathering processes and materials. WILSON, R C L (editor).   *Special Publication of the Geological Society of London*, No. 11.

PAUL, M A, and EYLES, N. 1990.   Constraints on the preservation of diamict facies (melt-out tills) at the margins of stagnant glaciers.   *Quaternary Science Reviews*, Vol. 9, 51–69.

PEACOCK, J D. 1975a.   Depositional environment of glacial deposits at Clava, north-east Scotland.   *Bulletin of the Geological Survey of Great Britain*, No. 49, 31–37.

— 1975b.   Palaeoclimatic significance of ice-movement directions of Loch Lomond readvance glaciers in the Glen Moriston and Glen Affric areas, northern Scotland.   *Bulletin of the Geological Survey of Great Britain*, No. 49, 39–42.

— 1977.   Subsurface deposits of Inverness and the Inner Cromarty Firth.   103–104 in *The Moray Firth area geological studies*. GILL, G (editor). (Inverness: Inverness Field Club.)

— BERRIDGE, N G, HARRIS, A L, and MAY, F. 1968.   The geology of the Elgin district.   *Memoir of the Geological Survey*, Sheet 95 (Scotland).

— GRAHAM, D K, and GREGORY, D M. 1980.   Late and post-glacial marine environments in part of the Inner Cromarty Firth.   *Report of the Institute of Geological Sciences*, No. 80/7.

PEARS, N V. 1968.   Post-glacial tree-lines of the Cairngorm Mountains.   *Transactions of the Botanical Society of Edinburgh*, Vol. 40, 361–394.

— 1970.   Post-glacial tree-lines of the Cairngorm Mountains; some modifications based on radiocarbon dating. *Transactions of the Botanical Society of Edinburgh*, Vol. 40, 436–544.

PIASECKI, M A J. 1980.   New light on the Moine of the Central Highlands of Scotland.   *Journal of the Geological Society, London*, Vol. 137, Pt. 1, 41–59.

— and VAN BREEMAN, O. 1979.   The 'Central Highland Granulites': cover-basement tectonics in the Moine.   139–144 in The Caledonides of the British Isles—reviewed. HARRIS, A L, HOLLAND, C H, and LEAKE, B E (editors). *Special Publication of the Geological Society of London*, No. 8.

— and TEMPERLEY, S. 1988.   The Central Highland Division. 46–53 in *Later Proterozoic stratigraphy of the Northern Atlantic regions*. WINCHESTER, J A (editor). (Glasgow and London: Blackie.)

PLATTEN, I M, and MONEY, M S. 1987.   The formation of late Caledonian subvolcanic pipes at Cruachan Cruinn, Grampian Highlands, Scotland.   *Transactions of the Royal Society of Edinburgh : Earth Sciences*, Vol. 78, 85–103.

POWELL, D W. 1970.   Magnetised rocks within the Lewisian of western Scotland and under the Southern Uplands.   *Scottish Journal of Geology*, Vol. 6, 353–369.

POWELL, R D. 1990.   Glacimarine processes at grounding-line fans and their growth to ice-contact deltas.   53–73 in Glacimarine environments: processes and sediments. DOWDESWELL, J A, and SCOURSE, J D (editors). *Special Publication of the Geological Society of London*, No. 53.

RATHBONE, P A. 1980.   Basement–cover relationships at Lewisian inliers in the Moine Series of Scotland, with particular reference to the Sgurr Beag Slide.   Unpublished PhD thesis, University of Liverpool.

— and HARRIS, A L. 1979.   Basement–cover relationships at Lewisian inliers in the Moine rocks.   101–108 in The Caledonides of the British Isles—reviewed. HARRIS, A L, HOLLAND, C H, and LEAKE, B E (editors). *Special Publication of the Geological Society of London*, No 8.

— and HARRIS, A L. 1980.   Moine and Lewisian near the Great Glen Fault in Easter Ross.   *Scottish Journal of Geology*, Vol. 16, Pt. 1, 51–64.

READ, H H. 1923.   The geology of the country round Banff, Huntly and Turriff.   *Memoir of the Geological Survey of Great Britain*, Sheets 86 and 96 (Scotland).

REID, D J. 1899.   The Inverness section of the Aviemore Line. *Transactions of the Inverness Scientific Society and Field Club*, Vol. 5, 254–263.

— and MACNAIR, P. 1898.   On the genera *Psilophyton, Lycopodites, Zosterophyllum* and *Parka decipiens* of the Old Red Sandstone of Scotland. Their affinities and distribution. *Transactions of the Edinburgh Geological Society*, Vol. 7, Pt. 21, 368–380.

REID, G, and MCMANUS, J. 1987.   Sediment exchanges along the coastal margin of the Moray Firth, Eastern Scotland. *Journal of the Geological Society of London*, Vol. 144, Pt. 1, 179–185.

RENDALL, H, WORSLEY, P, GREEN, F, and PARKS, D. 1991. Thermoluminescence dating of the Chelford Interstadial. *Earth & Planetary Science Letters*, Vol. 103, 182–189.

RICHARDSON, J B. 1965.   Middle Old Red Sandstone spore assemblages from the Orcadian Basin, north-east Scotland. *Palaeontology*, Vol. 7, Pt. 4, 559–605.

— 1967.   Some British Lower Devonian spore assemblages and their stratigraphic significance.   *Review of Palaeobotany and Palynology*, Vol. 1, 111–129.

RINGROSE, P S. 1989.   Palaeoseismic (?) liquefaction event in late Quaternary lake sediment at Glen Roy, Scotland. *Terra Research*, Vol. 1, 57–62.

RITCHIE, W, SMITH, J S, and ROSE, N. 1978.   *Beaches of northeast Scotland*. (Aberdeen: Department of Geography, University of Aberdeen.)

ROBINS, N S. 1987.   Development of groundwater resources in Scotland. *Proceedings of the Institution of Civil Engineers*, Vol. 83, Pt. 2, 747–753.

— 1990.   *Hydrogeology of Scotland*.   (London: HMSO for the British Geological Survey.)

— COOK, J M, and MILES, D L. 1989.   Groundwater near the Moray Firth: the coastal aquifer between Forres and Elgin, Grampian Region.   *Quarterly Journal of Engineering Geology*, Vol. 22, 145–150.

ROCK, N M S, MACDONALD, R, WALKER, B H, MAY, F, PEACOCK, J D, and SCOTT, P. 1985.   Intrusive metabasite belts within the Moine assemblage, west of Loch Ness, Scotland: evidence for metabasite modification by country rock interactions. *Journal of the Geological Society of London*, Vol. 142, 643–66.

ROGERS, D A. 1987. Devonian correlations, environments and tectonics across the Great Glen Fault. Unpublished PhD thesis, University of Cambridge (in two volumes).

ROLLIN, K E. 1992. GM2D: systematic frequency analysis, Slope, Euler and Werner deconvolution of potential field data along profiles. *British Geological Survey Technical Report*, WK/91/11.

— 1993. Grampian Geophysical Interpretation Project: geophysical interpretation on Sheet 84W Fortrose. *British Geological Survey Technical Report*, WK/93/1.

ROSE, J. 1985. The Dimlington Stadial/Dimlington Chronozone: a proposal for naming the main glacial episode of the Late Devensian in Britain. *Boreas*, Vol. 14, 225–230.

RUDDIMAN, W F, and RAYMO, M. 1988. Northern hemisphere climate regimes during the last 3 Ma. *Philosophical Transactions of the Royal Society of London*, Vol. B318, 411–430.

SEDGWICK, A, and MURCHISON, R I. 1835. On the structure and relations of the deposits contained between the Primary Rocks and the Oolite Series in the north of Scotland. *Transactions of the Geological Society*, Second Series Vol. III, 125–160.

SHARP, M. 1984. Annual moraine ridges at Skáelafellsjökull, south-east Iceland. *Journal of Glaciology*, Vol. 30, No. 104, 82–93.

— 1985. "Crevasse-fill" ridges—a landform type characteristic of surging glaciers? *Geografiska Annaler, Series A Physical Geography*, Vol. 67A, 213–220.

SHAW, J. 1979. Genesis of the Sveg tills and Rogen moraines of Central Sweden: a model of basal meltout. *Boreas*, Vol. 8, 409–426.

SIMPSON, I M, and WEST, R G. 1959. On the stratigraphy and palaeobotany of the Late Pleistocene organic deposit at Chelford, Cheshire. *New Phytologist*, Vol. 57, 239–250.

SISSONS, J B. 1966. Relative sea-level changes between 10 300 and 8 300 BP in part of the Carse of Stirling. *Transactions of the Institute of British Geographers*, Vol. 39, 19–29.

— 1967. *The evolution of Scotland's scenery*. (Edinburgh: Oliver and Boyd.)

— 1969. Drift stratigraphy and buried morphological features in the Grangemouth–Falkirk–Airth area, central Scotland. *Transactions of the Institute of British Geographers*, Vol. 55, 145–159.

— 1974. Lateglacial marine erosion in Scotland. *Boreas*, Vol. 3, 41–48.

— 1976. Lateglacial marine erosion in south-east Scotland. *Scottish Geographical Magazine*, Vol. 92, 17–29.

— 1981a. Lateglacial marine erosion and a jokulhlaup deposit in the Beauly Firth. *Scottish Journal of Geology*, Vol. 17, 7–19.

— 1981b. The last Scottish ice sheet: facts and speculative discussion. *Boreas*, Vol. 10, 1–17.

— SMITH, D E, and CULLINGFORD, R A. 1966. Lateglacial and postglacial shorelines in southeast Scotland. *Transactions of the Institute of British Geographers*, Vol. 39, 9–18.

SMALL, A, and SMITH, J S. 1971. The Strathpeffer and Inverness area. *British landscapes through maps, No. 13*. (Sheffield: The Geographical Association.)

SMITH, D I. 1977. The Great Glen Fault. 46–59 in *The Moray Firth area geological studies*. GILL, G (editor). (Inverness: Inverness Field Club.)

— and WATSON, J V. 1983. Scale and timing of movements on the Great Glen Fault, Scotland. *Geology*, Vol. 11, 523–526.

SMITH, J S. 1966. Morainic limits and their relationship to raised shorelines in the East Scottish Highlands. *Transactions of the Institute of British Geographers*, Vol. 39, 61–64.

— 1968. Shoreline evolution in the Moray Firth. Unpublished PhD thesis, University of Aberdeen.

— 1977. The last glacial epoch around the Moray Firth. 72–82 in *The Moray Firth area geological studies*. GILL, G (editor). (Inverness: Inverness Field Club.)

STEPHENSON, D. 1977. Intermontane basin deposits associated with an early Great Glen feature in the Old Red Sandstone of Inverness-shire. 35–45 in *The Moray Firth area geological studies*. GILL, G (editor). (Inverness: Inverness Field Club.)

STRECKEISEN, A L. 1976. To each plutonic rock its proper name. *Earth Science Reviews*, Vol. 12, 1–33.

SUTHERLAND, D G. 1981. The high-level shell beds of Scotland and the build-up of the last Scottish ice sheet. *Boreas*, Vol. 10, 247–254.

SYNGE, F M. 1977a. Land and sea-level changes during the waning of the last regional ice sheet in the vicinity of Inverness. 83–102 in *The Moray Firth area geological studies*. GILL, G (editor). (Inverness: Inverness Field Club.)

— 1977b. Records of sea levels during the Late Devensian. *Philosophical Transactions of the Royal Society*, Vol. 280, No. 972, 211–228.

— and SMITH, J S. 1980. *Inverness field guide*. (Cambridge: Quaternary Research Association.)

TARLO, L B. 1961. Psammosteids from the Middle and Upper Devonian of Scotland. *Quarterly Journal of the Geological Society of London*, Vol. 117, 193–213.

TAYLOR, W. 1910. New localities for Upper Old Red Sandstone in the Moray Firth area. *Transactions of the Inverness Scientific and Field Club*, Vol. VI (1899–1906), 46–48.

TRAQUAIR, R H. 1892. Achanarras revisited. *Proceedings of the Royal Society of Edinburgh*, Vol. 12, 279–286.

— 1894–1914. A monograph of the fishes of the Old Red Sandstone of Britain. Part 2—The Asterolepidae. *Palaeontographical Society Monograph*, Sect. 1 (1884), 63–90; Sect. 2 (1904), 91–118; Sect. 3 (1906), 119–130; Sect. 4 (1914), 131–134.

— 1896. The extinct vertebrate animals of the Moray Firth. 235–285 in *A vertebrate fauna of the Moray Basin*. HARVIE-BROWN, J A, and BUCKLEY, T E (editors). (Edinburgh: David Douglas.)

— 1897. Additional notes on the fossil fishes of the Upper Old Red Sandstone of the Moray Firth area. *Proceedings of the Royal Physical Society of Edinburgh*, Vol. 13, 376–385.

— 1904. On the fauna of the Upper Old Red Sandstone of the Moray Firth area. *Report of the British Association for the Advancement of Science (1904)*, 547.

UNDERHILL, J R. 1991a. Controls on Late Jurassic seismic sequences, Inner Moray Firth, UK North Sea: a critical test of a key segment of Exxon's original cycle chart. *Basin Research*, Vol. 3, 79–98.

— 1991b. Implications of Mesozoic–Recent basin development in the western Inner Moray Firth Basin. *Marine and Petroleum Geology*, Vol. 8, No. 3, 359–369.

— and BRODIE, J A. 1993. Structural geology of Easter Ross, Scotland: implications for movement on the Great Glen fault zone. *Journal of the Geological Society of London*, Vol. 150, Pt. 3, 515–527.

WALKER, M J C. 1975. Lateglacial and early Postglacial environmental history of the central Grampian Highlands, Scotland. *Journal of Biogeography*, Vol. 2, 265–284.

— 1990a. The Dalcharn interglacial site, near Cawdor, Nairnshire: results of pollen analysis on the Dalcharn Biogenic Complex. 54–57 in *Beauly to Nairn. Field guide.* AUTON, C A, FIRTH, C R, and MERRITT, J W (editors). (Cambridge: Quaternary Research Association.)

— 1990b. The Allt Odhar interstadial site, Moy, Inverness-shire: results of pollen analysis. 70–72 in *Beauly to Nairn. Field guide.* AUTON, C A, FIRTH, C R, and MERRITT, J W (editors). (Cambridge: Quaternary Research Association.)

— MERRITT, J W, AUTON, C A, COOPE, G R, FIELD, M H, HEIJNIS, H, and TAYLOR, B J. 1992. Allt Odhar and Dalcharn: two pre-Late Devensian/Late Weichselian sites in northern Scotland. *Journal of Quaternary Science*, Vol. 7, 69–86.

WALLACE, T D. 1883. Shells in glacial clay at Fort George, Inverness-shire. *Transactions of the Edinburgh Geological Society*, Vol. 4, 143–144.

— 1898. Geological notes on Strathdearn and the Aviemore Railway. *Transactions of the Edinburgh Geological Society*, Vol. 7, 416–419.

— 1919. A new locality for manganese. *Transactions of the Geological Society of Edinburgh*, Vol. 11, 135–137.

— 1922. Excursion to Dalroy Burn and manganese mine. *Transactions of the Inverness Scientific Society and Field Club*, Vol. 9, 332–334.

WATSON, D M S. 1908. *Coccosteus minor* Hugh Miller, in the Old Red Sandstone of Dalcross, Inverness-shire. *Geological Magazine*, New Series, Decade 5, Vol. 5, 431.

— 1937. The Acanthodian fishes. *Philosophical Transactions of the Royal Society*, Series B, Vol. 228, 49–147.

WERRITY, A, and McEWAN, L J. 1993. Findhorn Terraces. 187–189 in *Quaternary of Scotland. Geological conservation review series;* 6. GORDON, J E, and SUTHERLAND, D G (editors). (London: Chapman and Hall.)

WESTOLL, T S. 1937. The Old Red Sandstone Fishes of the north of Scotland, particularly of Orkney and Shetland. *Proceedings of the Geologists' Association*, Vol. 48, Pt. 1, 13–45.

— 1951. The vertebrate-bearing strata of Scotland. *Report of the International Geological Congress of 1948*, Pt. 11, 5–11.

— 1979. Devonian fish biostratigraphy. 341–353 *in* The Devonian System—A Palaeontological Symposium. HOUSE, M R, SCRUTTON, C T, and BASSETT, M G (editors). *Special Papers in Palaeontology*, No. 23.

WHITTINGTON, G. 1990. The Dalcharn Interglacial site: results of pollen analysis. 57–59 in *Beauly to Nairn. Field guide.* AUTON, C A, FIRTH, C R, and MERRITT, J W (editors). (Cambridge: Quaternary Research Association.)

WILKINSON, I P. 1989. Quaternary microfossils in a sample from Drummore of Clava. *British Geological Survey Technical Report*, WH/89/335R.

— 1990. Quaternary Foraminiferidae from Clava, near Inverness. *British Geological Survey Technical Report*, WH/90/399R.

WILSON, D, and SHEPHERD, J. 1979. The Carn Chuinneag granite and its aureole. 669–675 *in* The Caledonides of the British Isles — reviewed. HARRIS, A L, HOLLAND, C H, and LEAKE, B E (editors). *Special Publication of the Geological Society of London*, No. 8.

WINKLER, H G F. 1979. *Petrogenesis of metamorphic rocks* (5th edition, revised). (New York, Heidelberg, Berlin: Springer-Verlag.)

WOODWORTH, P L. 1987. Trends in UK mean sea level. *Marine Geodesy*, Vol. 11, 57–87.

YOUNG, J A T. 1977. Glacial geomorphology of the Dulnain valley, Inverness-shire. *Scottish Journal of Geology*, Vol. 13, Pt. 1, 58–74.

— 1978. The landforms of Upper Strathspey. *Scottish Geographical Magazine*, Vol. 94, 76–94.

— 1980. The fluvioglacial landforms of mid-Strathdearn, Inverness-shire. *Scottish Journal of Geology*, Vol. 16, Pt. 2 & 3, 209–220.

ZALESKI, E. 1982. The geology of Speyside and lower Findhorn granitoids. Unpublished MSc thesis, University of St Andrews.

— 1985. Regional and contact metamorphism within the Moy Intrusive Complex, Grampian Highlands, Scotland. *Contributions to Mineralogy and Petrology*, Vol. 89, 296–306.

ZIEGLER, P A. 1988. Laurussia—The Old Red Continent. 15–48 in *Devonian of the World, Volume I: regional syntheses.* McMILLAN, N J, EMBRY, A F, and GLASS, D J (editors). (Calgary: Canadian Society of Petroleum Geologists.)

GENERAL INDEX

BRITISH GEOLOGICAL SURVEY

Keyworth, Nottingham NG12 5GG
0115-936 3100

Murchison House, West Mains Road, Edinburgh
EH9 3LA   0131-667 1000

London Information Office, Natural History Museum
Earth Galleries, Exhibition Road, London SW7 2DE
0171-589 4090

The full range of Survey publications is available through the
Sales Desks at Keyworth and at Murchison House, Edinburgh,
and in the BGS London Information Office in the Natural
History Museum (Earth Galleries). The adjacent bookshop
stocks the more popular books for sale over the counter. Most
BGS books and reports can be bought from HMSO and
through HMSO agents and retailers. Maps are listed in the
BGS Map Catalogue, and can be bought together with books
and reports through BGS-approved stockists and agents as well
as direct from BGS.

*The British Geological Survey carries out the geological survey of Great
Britain and Northern Ireland (the latter as an agency service for the
government of Northern Ireland), and of the surrounding continental
shelf, as well as its basic research projects. It also undertakes
programmes of British technical aid in geology in developing countries
as arranged by the Overseas Development Administration.*

*The British Geological Survey is a component body of the Natural
Environment Research Council.*

HMSO

HMSO publications are available from:

HMSO Publications Centre
(Mail, fax and telephone orders only)
PO Box 276, London SW8 5DT
Telephone orders 0171-873 9090
General enquiries 0171-873 0011
Queuing system in operation for both numbers
Fax orders 0171-873 8200

HMSO Bookshops
49 High Holborn, London WC1V 6HB
(counter service only)
0171-873 0011   Fax 0171-831 1326
68–69 Bull Street, Birmingham B4 6AD
0121-236 9696   Fax 0121-236 9699
33 Wine Street, Bristol BS1 2BQ
0117-9264306   Fax 0117-9294515
9 Princess Street, Manchester M60 8AS
0161-834 7201   Fax 0161-833 0634
16 Arthur Street, Belfast BT1 4GD
01232-238451   Fax 01232-235401
71 Lothian Road, Edinburgh EH3 9AZ
0131-228 4181   Fax 0131-229 2734
HMSO Oriel Bookshop,
The Friary, Cardiff CF1 4AA
01222-395548   Fax 01222-384347

HMSO's Accredited Agents
(see Yellow Pages)

And through good booksellers